INDIAN AGRICULTURE
IN AMERICA

INDIAN AGRICULTURE
IN AMERICA
Prehistory to the Present

R. Douglas Hurt

University Press of Kansas

Published by the University Press of Kansas (Lawrence, Kansas 66045), which
was organized by the Kansas Board of Regents and is operated and funded by
Emporia State University, Fort Hays State University, Kansas State University,
Pittsburg State University, the University of Kansas, and Wichita State University

Library of Congress Cataloging-in-Publication Data

Hurt, R. Douglas.
  Indian agriculture in America.
  Bibliography: p.
  Includes index.
  1. Indians of North America—Agriculture—History.
2. Indians of North America—Land tenure. 3. Indians
of North America—Government relations. 4. Agriculture
and state—United States—History. I. Title.
E98.A3H87 1987   338.1'08997073   87-14764
ISBN 0-7006-0337-9

British Library Cataloguing in Publication data is available.

Printed in the United States of America
10 9 8 7 6 5 4 3 2 1

For
Mary Ellen, Adlai, and Austin

Contents

Illustrations

Preface

The agricultural history of the American Indians is as fascinating as it is complex. Although archaeologists, anthropologists, sociologists, ethnologists, and historians have studied the American Indians for a long time, no one has provided a general survey of their agricultural history. To help fill this gap in the historical literature, I have tried to synthesize the major archaeological, botanical, ethnological, anthropological, and historical scholarship, as well as my own primary research, in order to trace the main developments and major trends of Indian agriculture from prehistory to the present. For the prehistoric era, I have stressed farming methods and the crops raised by the Indians; for later periods I have emphasized the government's agricultural policy. To understand fully the origins of Indian agriculture and the accomplishments of tribal farmers, a host of questions must be asked. For example, what cultigens did the earliest farmers raise? How did they go about the farming process—planting, cultivating, harvesting, and processing? What was the significance of those innovations for their daily lives and for Indian agriculture? Before European contact, for example, Indian farmers were superb plant breeders. Without having a thorough knowledge of genetics, how did they know what to do? Moreover, how did agriculture, rudimentary or extensive, fit into the culture of those particular people?

Such questions are easy to ask but difficult to answer. In the absence of written records, one must rely upon the archaeological record. Yet if precise analysis is impossible, given the nature of archaeological discovery, which often is influenced by accident or chance, broad generalizations are possible. If one accepts the generalization that the story of agriculture essentially involves the history of man's relationship to the environment, then we can focus specifically on particular regions where Indian men

ix

and women made productive use of the plant and animal world by cultivation and domestication. In order for agriculture to be successful, for example, farmers must adopt or develop the plant varieties that will produce abundantly and consistently within the environmental limitations of their region. They also must develop farming techniques that will enable them to use the physical and climatic characteristics of their region to the best advantage. Nowhere among the Indian farmers were these requirements met more successfully than among the prehistoric agriculturists in the Southwest. There, scant precipitation and searing temperatures forced the Anasazi on the plateau, the Mogollon in eastern Arizona and western New Mexico, and the Hohokam in the Gila River Basin to make agricultural adjustments that were unnecessary in other regions of the North American continent. Because of their technical skill in plant selection and irrigation and their ability to use the desert environment to their best advantage, they successfully farmed this harsh region for more than two thousand years before contact with European civilization. Even so, those farmers— like those in other parts of the continental United States before European contact—merely followed the long agricultural tradition in the Western Hemisphere, which dates to at least 7000 B.C., when Mesoamerican farmers were engaged in the rudimentary process of seeding, cultivating, harvesting, and domesticating various plants.

With the development and spread of agriculture, the various Indian peoples who engaged in farming necessarily established systems of land tenure that ensured the wise and orderly use of the land. To understand that process, however, more questions must be asked. How did they develop land-tenure systems? Who controlled the land? Who farmed it? Who owned the produce from the fields? Could the control of specific fields change hands? If so, how did that procedure work? In addition, did the concepts of land tenure differ among the Indian peoples? Other questions must be posed as well. For example, were the needs and rules that governed land tenure for the Cherokee and the Creek in the Southeast the same as those among the Hopi and the Yuma in the Southwest? If not, what influences did environment and culture have upon the creation of different land-tenure systems? Fortunately, ethnologists provide many answers to these questions.

At the same time, we must ask questions about European concepts of land tenure. How did the Europeans treat Indian landholding at the time of contact and thereafter? What differences existed between the Spanish, the French, the Dutch, and the English with respect to the recognition of Indian property rights? How were differences between Europeans and Indians over property rights resolved? Moreover, how did the Europeans, particularly the English, and later the Americans, acquire land? What was the moral, legal, and political basis for their acquisition policies? The answers to these questions will enable us to make a broad sketch of the manner in which Indian lands passed into white control, as well as to explain the long decline of Indian agriculture.

For every question asked, if not answered, more questions quickly come to mind. For example, how did European agricultural practices influence Indian farming? Did cropping patterns and technology change among the tribesmen? Did Indian farmers readily adopt European cultigens and methods? If so, why? Conversely, which European crops and techniques did the Indians reject? Which Indian crops and techniques did the Europeans adopt? What results did European contact with the Indians have for agriculture in both cultures? Many of these questions can be answered from the written record.

The next step of inquiry follows naturally. It involves the relationship of the federal government to Indian agriculture. In general, one must ask: What was the purpose of federal agricultural policy for the Indians? Specifically, why was agriculture important to the policy makers for shaping harmonious social and profitable economic relationships between red cultures and white cultures and between the Indians and the federal government? What place and purpose did agriculture occupy in the general Indian policy of the federal government? If assimilation and acculturation based on agriculture were the goals of the policy makers, for example, why did each fail so miserably? What went wrong? Why was the federal government unable to encourage agriculture among the tribesmen who already had a long tradition of farming and who were among the most skilled agriculturists on the continent? What were the environmental and cultural limitations that prevented Indian farmers from emulating white farmers or, in the case of the nomadic hunting tribes, from becoming agriculturists? In addition, one must ask: How did governmental policy respond to the scientific, technological, and financial needs of the Indians as well as to problems that involved the inheritance, lease, and alienation of Indian lands? Last, how has water law affected Indian agriculture in the semiarid and arid West, and what is the status of Indian agriculture today? The answers to these questions must involve a sweeping stroke of the pen. Nevertheless, these questions, taken together, provide a context for the study of Indian agriculture. The answers, so far as they can be determined, will provide a general familiarity with the major developments in Indian farming as well as a broad understanding of the failure of governmental policies that prevented the expansion of Indian agriculture after contact with European civilization. In particular, this study will show how environmental, political, and cultural restraints prevented the strengthening, and in some cases the creation, of an agricultural economy among the tribesmen. The result was the failure of federal policy makers to achieve the idealistic goals of assimilation and acculturation of the Indians based on agriculture.

I have not tried to write the agricultural history of each tribe; that would be an impossible task. Rather, I have traced the development of Indian agriculture both regionally and topically within a chronological framework. My intent has been to be suggestive rather than definitive. With the exception of an introductory chapter on Mesoamerica, this study concerns only the agricultural activities of the Indians within the forty-eight contiguous states. After the removal period, I emphasize farming in the trans-Mississippi West and the government's agricultural policy for the Indians. I have sought not only to describe those activities and policies, but also to note the differences between various regions and Indian peoples. Moreover, I analyze the significance of agriculture for the Indians in both the prehistoric and the modern worlds. All prehistoric Indian peoples, of course, had their own agricultural problems, which they resolved in their own fashion, depending upon environment and culture. Those problems ranged from concerns about precipitation and soil fertility to trade and peace with their neighbors. After European contact, many of those problems remained as new ones emerged. Until histories have been written about each Indian people, however, specific examples and analysis necessarily must involve those tribal groups that have attracted the greatest scholarly attention. Moreover, much work needs to be completed on a wide variety of subjects, such as the Indian irrigation service, agricultural extension, land purchases, reservation expansion, agricultural labor, tenancy, heirship lands, and the agricultural differences between full bloods and mixed bloods. I hope, however, that this study will provide a broad and useful reference for anyone interested in the agricultural history of the American Indians.

During the course of research for this book, I have received assistance from a number of people at a variety of institutions. I am appreciative of the aid that I obtained at the National Archives, the Library of Congress, and the libraries of the State Historical Society of Missouri, the University of Missouri, the Ohio Historical Society, the Ohio State University, and the State of Ohio. At the Ohio Historical Society, Steve Gutgesell and Joan Jones efficiently located needed sources, and Christopher S. Duckworth prepared two of the illustrations. Laurel Shannon typed the manuscript, and her excellent proofreading prevented a number of errors. I am thankful for their aid and support.

I am grateful as well for the aid of my colleagues in the academic community who provided research materials and critical evaluations of the manuscript during various phases of its preparation. W. David Baird at Oklahoma State University and Peter Iverson at the University of Wyoming permitted me to read manuscripts of their work in progress. Donald L. Parman at Purdue University provided important interview material and, together with Francis Paul Prucha at Marquette University, William E. Unrau at Wichita State University, and Thomas R. Wessel at Montana State University, criticized all or portions of an earlier draft. W. Raymond Wood at the University of Missouri and Gilbert C. Fite at the University of Georgia provided helpful comments for a later version of my work. This study has been improved immeasurably because of their help, and I am indebted to them for their generous and gracious support.

1

Mesoamerican Origins

Simply put, agriculture means raising things on purpose, and the American Indians were the first farmers in the Western Hemisphere. At least seven thousand years before the English colonists founded the settlement of Jamestown in Virginia, the Indians of Central America and Mexico, an area commonly known as Mesoamerica, tilled the soil, planted seeds, and harvested crops. This is not to suggest, however, that those Indians cultivated a multiplicity of crops in vast fields; instead, they depended upon hunting and gathering for nearly all of their subsistence needs. Agriculture met only a small portion of their daily food requirements. Nevertheless, they were beginning the long transition from living primarily by hunting and gathering to residing in settled farming communities. Millenniums passed, however, before they completed that dramatic cultural change.[1]

Over time, the nomadic Indians of Mesoamerica learned the advantages of clearing away unwanted plants in order to enhance the growth of others. This rudimentary weeding process created a new, artificial environment for favored cultigens, which caused genetic changes in these plants and increased their productivity. With these revelations, the basic premise of agriculture had been discovered: planting, cultivating, and harvesting—the agricultural process—provided a much more dependable food supply. As a result, large bands of Indians could settle in a specific area for a longer period of time.[2]

Current archaeological evidence cannot pinpoint a single location where Indian agriculture began or determine precisely why the Indians became agriculturalists. Rather, the archaeological record suggests that the cultivation and domestication of plants originated simultaneously in several Mesoamerican areas. Some of the most complete evidence for early agricultural development comes from the Mexican state of Tamaulipas and

1

from the Tehuacán Valley, located in northeastern and south-central Mexico respectively. Indian farmers may have first raised bottle gourds (*Lagenaria siceraria*) as early as 7000 B.C. in the Tehuacán Valley and in Oaxaca; Tehuacán Valley farmers were perhaps the first to domesticate corn; and Indians may have originally cultivated pumpkins in Tamaulipas. Concurrent agricultural development is important, because it indicates that Indian farmers attempted to meet their own food needs according to the dictates of their environment.[3]

In the late nineteenth and early twentieth centuries, archaeologists believed that agriculture began in the rain forests, where the Indians learned to cultivate root crops. No knowledge of complex food processing was needed to prepare the roots for consumption, because many roots could be eaten raw, roasted, or steamed over a fire. Certainly, the development of agriculture based on root crops would have been a relatively simple process, requiring only a fire-hardened digging stick. Although this scenario for the origin of agriculture may be accurate, an independent location for agricultural development existed in the intermediate altitudes of Mesoamerica. There, in the Tehuacán Valley, archaeologists have found some of the earliest evidence of cultivated plants. As a result of those discoveries, they have traced the evolution of a nomadic people, who depended on hunting and gathering, to a civilization that became dependent on agriculture.[4]

The Tehuacán Valley was well suited to be a center of agricultural development in the Western Hemisphere. Soil and moisture conditions enabled Indian farmers to practice agriculture. The semiarid climate of the valley was an asset, because it had prevented the growth of vegetation. Indian farmers could plant without giving a great deal of attention to clearing the land and to fighting continuous battles with unwanted vegetation that would compete for soil nutrients and moisture. Tehuacán farmers probably planted their crops on the moist alluvial fans at the end of the water courses that spread out from the hillsides. Such sites had several advantages for early farmers. First, floodwaters periodically removed much of the unwanted vegetation. Second, torrents of water from heavy rains washed fertile soil down from the hillsides and replenished the richness of the alluvial fans. Third, when the runoff was minimal, the water irrigated the growing plants. Finally, subsurface moisture watered the crops even in the absence of surface runoff.[5]

No abrupt shift occurred from food gathering to plant raising as the Tehuacán people became more dependent upon a stable agricultural economy. There, as in the regions that are today the Mexican states of Oaxaca, Morelos, and Guerrero and the Basin of Mexico, the Indians, through many millenniums of observing nature, had learned to manipulate their environment by encouraging the growth of certain plants, by removing competing weeds, by clearing the land, and by planting and harvesting. Gradually the cultivation of plants became more important as the Indian farmers throughout the valley adopted more of them. From time to time they added new plants and improved others through the careful selection of the best seeds. Greater agricultural productivity and a wider variety of domesticated plants meant that the Indians needed less time for hunting and gathering and therefore had more time for clearing land, planting, weeding, and harvesting.[6]

The Tehuacán people probably became committed to the agricultural process during the Coxcatlan cultural phase of Mesoamerican civilization (5200 to 3400 B.C.). During that period the Indians became more dependent on plants, seeds, and fruits. The Coxcatlan people grew corn; the tepary, the black and perhaps the common bean; chili peppers; avocados; amaranth; bottle gourds; and two kinds of squash. Each spring, these agriculturists probably came together into large groups, or macrobands, to plant their favorite seeds and to eat the fruits of their labors throughout the summer months until they exhausted the surplus. Then they broke up into small family groups, or microbands, for an autumn and winter of hunting and gathering.[7]

During the Abejas phase (3400 to 2300 B.C.), Tehuacán farmers raised enough food to provide a surplus, which could be stored for use during the fall and winter months, when they traditionally had depended almost entirely upon meat from the hunt. This surplus probably was not enough to enable all of the Indian bands in the valley to remain at the spring and summer campsite year-round, and some no doubt moved away to hunt and gather roots and seeds during the autumn and winter. Still, agricultural production enabled some Tehuacán families to live in permanent villages. Agriculture became so important that those families who moved away for the winter returned to the campsite in the spring to plant their seeds alongside those Indians who had remained. In short, agriculture enabled the Indians of the Tehuacán Valley to adopt a more sedentary life. Indian farmers continued the intensive cultivation of small garden plots of corn, beans, chili peppers, and squash, but they also began to farm large areas or fields once they had become sedentary.[8]

Between 900 B.C. and A.D. 700, Mesoamerican agriculture became more complex. The Tehuacán people cultivated corn and added runner beans (*Phaseolus coccineus*) to their cultivation practices. They also adopted tomatoes, lima beans, peanuts, guava, and domesticated turkeys. In addition to raising a wide variety of plants, the Tehuacán farmers also developed supplementary industries closely related to agriculture, such as salt making and cotton processing. During that period, lowland farmers probably conducted slash-and-burn, also known as milpa or swidden, agriculture, while farmers in the drier areas practiced intensive irrigation farming.[9]

Independent agricultural development also occurred in Tamaulipas, on the northeastern periphery of Mesoamerica. During the Infiernillo cultural phase (7000 to 5500 B.C.), Indian farmers cultivated pumpkins and gourds. Pumpkins provided a very small part of their diet, but the Indians supplemented this food with wild runner beans and chili peppers. Between 2200 and 500 B.C., during the Flacco, Almagre, Guerra, and Mesa de Guaje phases, these Indians became more sedentary as agriculture became increasingly important. They raised corn, sunflowers, squash, and possibly manioc. Between 1500 B.C. and A.D. 900 the Tamaulipas farmers added new domesticates or cultigens—lima beans, squash (*Cucurbita mixta*), tobacco (*Nicotiana rustica*), and cotton.[10]

In contrast to the farmers in the dry Tehuacán Valley and southern Tamaulipas, the Maya people in the region of present-day Guatemala, Belize, and Honduras developed a complex agricultural economy in the rain forests. Indeed, by approximately 1200 B.C. the Mayan people were practicing agriculture, and by 500 B.C. it provided their primary food source. These Maya farmed either extensively or intensively depending upon the amount and the quality of available land. They cleared lands by the slash-and-burn method, and they raised fields from the swampy lowlands. In the open clearings, Maya farmers used a dibble stick to poke a hole in the soil to plant seeds. The topsoil in this region is shallow, varying from a few inches to approximately a foot in depth; and an underlying bed of limestone leaches the nutrients from the soil. This causes soil exhaustion soon after the land has been cleared; nevertheless, Maya farmers cultivated fields intensively for as long as this was profitable. When soil depletion caused decreased yields, more land had to be cleared. By allowing the weeds to cover worn-out land, the Maya conducted a form of soil, rather than crop, rotation. As long as suitable new lands were available for clearing, this farming method met their needs.[11]

The Maya farmers' ability to support a relatively large population from small cultivated plots did not negate the need for large amounts of land. Because the tropical soil became quickly exhausted, the farmers abandoned their croplands every two or three years. They continually cleared new agricultural land and allowed depleted lands to reforest. While four years might be enough to rest the soil after a single crop, if two successive plantings were made, the land might be allowed to lie fallow for six or seven years. Maya farmers often abandoned their plots for fifteen to twenty years—on occasion, for thirty to forty-five years—in order to enable the nutrients to build up in the soil once again. Because the Maya did not have domesticated animals and apparently did not fertilize with human excrement, they could not effectively maintain soil fertility on an extensive basis. Still, this field-rotation system was ideally suited for tropical soils. The return of weed growth reduced soil erosion in a region of heavy rainfall, and the subsequent burning of secondary growth provided the potash needed to help restore soil fertility. While this land-use method certainly depleted the soil, Maya farmers learned that it permitted the consistent production of food necessary to sustain their civilization.[12]

Slash-and-burn agriculture was not the only Maya land-use method. Maya farmers also practiced raised-field, or *chinampas*, agriculture on a large scale. Water covered much of the land during the rainy season, and to cultivate the land effectively and intensively, Maya farmers dug drainage canals and built a series of stone terraces, perhaps as early as 200 B.C. Behind those terraces, they added marl to raise the agricultural plots about one meter above the water level. These raised fields provided several advantages. First, the moist soil of these plots was ideal for seed germination. Second, the soil was sufficiently pulverized and aerated to allow deep root penetration. Third, because topsoil and muck from surrounding drainage ditches formed the raised fields, the soil was extremely fertile. Maya farmers also built hillside terraces, which helped hold organic material on the slopes and thereby retarded soil erosion.[13]

The time, expense, and organizational requirements needed to build an extensive raised-field system dictated the intensive farming of those lands. Furthermore, a large labor force was needed to farm the land and to maintain the terraces, ditches, and stone walkways of the raised fields. For these reasons the raised and terraced fields of the Maya may have been the most productive and intensively farmed lands in Mesoamerica. Moreover, the Maya were skilled farmers who cultivated a wide variety of plants: beans and squash—which they intermixed with the corn—sweet potatoes, manioc, tomatoes, avocados, cacao, chili peppers, cotton, and other crops.[14]

Later, after the Aztec occupation of the Valley of Mexico during the thirteenth century A.D., farmers in that area used slash-and-burn as well as irrigation techniques. In the Xochimilco-Chalco Basin, however, Aztec farmers also practiced *chinampas* agriculture by raising fields from swampy lands. There the Aztecs built raised fields or gardens by carrying sod into the marshes with canoes. Actually, these plots were nothing more than ridges that varied from approximately two to three meters in width. Each farmer planted crops on a number of ridges, then used a canoe to go to the plot. The marsh water irrigated the roots even during the dry season. From surrounding canals, these Aztec farmers obtained muck, rich in organic nutrients, which they applied to their plots. In addition, they fertilized the ridges with a compost of aquatic weeds; they may have applied human excrement as well.[15]

While some *chinampas* farmers lived on the edge of the swamp, most of them resided in island towns such as Mizquic, Cuitláhuac, and Xochimilco. Small villages and scattered houses also stood on foundations among the cultivated ridges. Although the ridge, or *chinampas*, field system probably began in this region before the birth of Christ, its development did not peak until A.D. 1400–1600. Overall, these Aztec farmers possibly reclaimed more than 120 square kilometers of swamp. After excluding the canals from that estimate, *chinampas* farmers had more than 9,000 hectares, composed of winding ridges, available for cultivation. Long before the Spanish Conquest, then, Aztec farmers had developed an intricate land-use system that provided the basis for a highly productive agricultural economy.[16]

Ultimately, these Mesoamerican farmers cultivated a wide variety of small grains, legumes, squashes, and fruits. The squash-bean-corn complex furnished the staple nutritional base, however; and plants in the squash family may have been domesticated first. Archaeological evidence also indicates that these farmers domesticated cucurbits (squash) in several distinct areas. Perhaps as early as 5000 B.C., in Tamaulipas, they raised *Cucurbita pepo*, commonly known as either summer squash or pumpkins. Indian farmers probably domesticated *Cucurbita moschata*, a winter squash or pumpkin, about 4000 B.C. in what is now the Mexican state of Puebla. The area of Oaxaca and Chiapas may have been the center or place of origin of the cultivation and distribution of *Cucurbita mixta* as early as 3000 B.C. The bottle gourd, one of the first plants to be domesticated, however, may have been raised first in the Ocampo cave region of Tamaulipas and at Guilá Naquitz in Oaxaca as early as 7000 B.C. The cultivation of cucurbita indicates that Mesoamerican farmers practiced very conservative crop husbandry, because they tended to raise the same squashes for many generations.[17]

Beans were the second staple in the food complex. Mesoamerican Indians domesticated the common bean (*Phaseolus vulgaris*) about 5000 B.C. in the Tehuacán Valley and about 4000 B.C. in Tamaulipas, though they did not raise it abundantly until agriculture became well established in Mesoamerica. Archaeologists have found the oldest—approximately 200 B.C.— evidence of domesticated runner beans near the Coxcatlan cave in the Tehuacán Valley. Runner beans probably were not domesticated there first because of the hot, arid environment. The beans discovered in the valley may have come into the area through trade with farmers in the humid uplands, possibly from Guatemala. Mesoamerican farmers also started to grow small-seeded lima or sieva beans (*Phaseolus lunatus*) soon after the start of the Christian Era in Tamaulipas and in the Tehuacán Valley as well as on the Yucatán Peninsula. In the Tehuacán Valley they also raised tepary beans (*Phaseolus acutifolius*), which they did not adopt until about two thousand years after they began to cultivate common beans. Tehuacán farmers apparently raised teparies for about three thousand years before this cultigen disappeared from the cropping system for some unknown reason. Perhaps those farmers either reaped greater harvests from the common bean or disease ruined the crop. No one is certain.[18]

Beans were an important agricultural crop for Mesoamerican farmers, one that contributed an important part to the nutrition of the people of that region. Indeed, these four species of beans were closely associated with the cultivation of corn, and Indian farmers may have domesticated some varieties of beans and corn at the same time. Beans have an abundant supply of lysine, an amino acid necessary for good health. Corn lacks both lysine and the amino acid tryptophan. Beans complement corn by adding these two amino acids to zein, the amino acid in corn, thus creating a protein of high nutritional value. This union of amino acids occurs with any combination of corn and beans. While Mesoamericans lacked a scientific explanation, about seven thousand years ago they recognized that a combination of corn and beans provided a healthy diet, and thereafter, both crops remained basic to their agricultural process. Because of insufficient archaeological evidence, no one can yet state precisely when Mesoamerican farmers first cultivated beans or when they began making careful selections for the development of specific varieties. Nor can anyone identify the origins of these cultigens. Beans, however, were important to Mesoamerican agriculture, and they spread northward to become just as significant to the Indian farmers within the continental United States.[19]

Corn, the most important Mesoamerican cultigen, also remains a "mystery" crop, although archaeological work has provided some insights into its origin. Archaeologists have not yet absolutely determined the ancestor of today's corn, but some have contended that it was a now-extinct wild corn. Others insist that corn developed from a related wild grass known as teosinte. Multiple origins for the domestication of corn may exist, and the Tehuacán Valley probably was one center. Warm temperatures and a rainy season from April through October provided ideal circumstances for seed germination after the dry winter months. Archaeologists have recovered prehistoric corn from five caves—Coxcatlan, Purron, San Marcos, Tecorral, and El Riego—in the Tehuacán Valley, which show the evolutionary development of the plant over a period of sixty-five hundred years.[20]

The earliest corn yet discovered has been dated at approximately 5000 B.C. from the Tehuacán Valley, but no clear evidence exists to indicate that the Indians practiced agriculture in this region at that time. In fact, no one is certain when or where Mesoamerican farmers began cultivating this plant. Between 3400 and 2300 B.C., during the Abejas cultural phase of Mesoamerican civilization, corn became part of the agriculture of the region, and Indian farmers were raising it along with bottle gourds, squashes, tepary, common and jack beans, chili peppers, and avocados. By 900 B.C., corn provided the basis for irrigation projects, large villages, and more extensive agriculture than ever before. It was a dietary staple for Mesoamericans by that time.[21]

Overall, then, Mesoamerican farmers were superb plant breeders. Even though they did not understand the precise principles of heredity or hybridization, they used their skills at seed selection to improve their corn crops. These farmers changed corn from a multistock to a single-stock plant. This enabled the plant to produce greater yields and also made harvesting easier. By selecting seed from plants that branched the least and had the fewest ears, Indian farmers could reproduce corn plants with strong stalks and long ears. They also planted their corn in hills and continued to heap the soil around the plants during cultivation. This process discouraged the growth of sucker branches that would weaken the plant and prevent it from reaching maximum productivity. Strong, healthy plants produced the best seeds, which the Indians used during the next planting season. While the Mesoamerican Indians did not control pollination and probably did not understand how it worked, they discovered that when they crowded corn plants into a relatively small area, the plants tended to produce not only a single stem but also an excellent condition for cross-pollination. The result was a plant that had a high tolerance for inbreeding. Although the archaeological record proves that Mesoamerican farmers were expert corn breeders, no one can yet say with certainty just how they knew what they knew, because they did not leave a written record for study; but they obviously knew that their method of seed selection for corn worked. The corn grew, and they achieved better crops with stronger plants that produced larger ears and bigger harvests. Further improvements in corn production would not come for more than five thousand years.[22]

Not only did Mesoamerican farmers become highly skilled at clearing forested lands, at raising fields from marshlands, at domesticating a wide variety of plants, and at breeding a number of varieties of corn, they also developed a sophisticated irrigation system. For example, by the Santa Maria cultural phase of Mesoamerican civilization (900 to 200 B.C.), Tehuacán farmers were busy with the painstaking process of selecting and raising improved avocado trees, which did not bear fruit for seven years but did require irrigation during that time. At first, these farmers probably planted avocado trees along the banks of streams and dug irrigation ditches to water the trees during the dry season. Small dams, dating from 800 to 600 B.C., also impounded runoff and stream flow for irrigation; they indicate an increasing technological sophistication. In time, canals were developed to transport water from dependable sources—a technique that these farmers evidently preferred, since they did not emphasize the construction of dams to impound large amounts of water. By approximately 200 B.C., large irrigation projects dotted the Tehuacán Valley; and from then until the Spanish Conquest, Tehuacán farmers irrigated most of their crops.[23]

In the Valley of Oaxaca, in the southern Mexican highlands, canal irrigation was impractical, because the springs and streams could not provide enough water. Here, as early as 700 B.C., Indian farmers were using "pot-irrigation," which involved digging shallow wells that the farmers drew water from with jugs. The water table was only about three meters below the surface, so wells could be dug with little difficulty. They provided enough water so that Oaxaca farmers could plant three corn crops annually. While this system was not practical in the Tehuacán Valley, where the water table was too deep, it was ideal in the Valley of Oaxaca. The individual farmer could dig his own wells, so a highly organized labor system was not required for the construction of extensive irrigation canals.[24]

On the eve of the Spanish Conquest, if not before, Indian farmers in northwestern Mexico's valley of Sonora also practiced irrigation farming. They diverted the runoff from heavy rains across small agricultural plots in the arroyos of the valley. Stone and dirt terraces slowed the runoff sufficiently to irrigate the crops while also allowing the silt in the water to settle on the fields, thereby increasing soil fertility. Those prehistoric Indian farmers also diverted streams on the flood plain of the valley floor by constructing weirs out of branches and soil. The weirs diverted part of the stream into irrigation ditches, which channeled the water among the plants in the fields.[25]

Mesoamerican agriculturists emphasized the cultivation of plants, rather than the raising of animals. With the possible exception of the domestication of turkeys and dogs in North America and the llama in South America, the prehistoric Indian farmers of the New World paid almost no attention to raising animals or fowl. Mesoamerican farmers, nevertheless, were highly skilled. They made specific adaptations to local environmental problems by developing irrigation systems in the dry Tehuacán Valley and elsewhere, by building raised fields in areas too wet for cultivation, and by growing the kinds of plants that produced the most abundant harvests. As their food production increased, the farmers met the growing demands for food by expanding the size of their fields if similar lands were readily available; or by intensively farming the lands already under cultivation if they could not acquire more suitable land.[26]

By 1000 B.C. the Mesoamerican people were full-time farmers who were producing enough food so that large numbers of people could form perma-nent villages and develop distinctive cultures and civilizations based on agri-culture. Although cultivated fields generally did not extend farther beyond the villages than the farmers could walk both ways in a day and still work, the variety of plants that the Mesoamerican farmers domesticated and culti-vated was very large: corn; common, tepary, jack, and lima beans; pump-kins; squashes; chili peppers; amaranth; cacao; tomatoes; and avocados pro-vided both staple foods and seasonings. The farmers understood and prac-ticed the principles of selection, irrigation, and terracing—all with the aid of stone tools for clearing, planting, and cultivating. They also raised surplus crops and stored grain in underground pits or granaries.[27]

Without a doubt, no single "hearth" or center of agricultural origin exists in Mesoamerica, nor did this region experience an agricultural "revolution" during the prehistoric period. The development of agriculture evolved as Indian farmers acquired, synthesized, and applied their knowledge of cultivation, domestication, and plant breeding. The evolutionary development of agriculture in that region gave the Indians increasing control of their environment and altered their lifestyles dramatically. Agriculture made food supplies more certain, decreased starvation and want, improved nutrition, and made possible the development of villages and complex civilizations. Given the absence of modern science and technology, the achievements of the Mesoamerican agriculturists are nothing less than astonishing. Far to the north, in what is now the continental United States, Indian farmers later built on those achievements.[28]

2

Prehistoric Agriculture

No one can yet say with certainty when agriculture began in the region of the continental United States, but Indian farmers in the Midwest and the Northeast cultivated crops at least five millenniums before Christ. At least by 5000 B.C. some Indians in the present state of Illinois were raising squash. Several millenniums later they were probably cultivating sunflowers, goosefoot, and sumpweed as well, and about A.D. 1000 they had developed a complex agriculture based on three major crops—corn, beans, and squash—while a host of other plants provided supplemental crops.[1]

Most archaeologists are now certain that agricultural practices diffused into eastern North America from Mesoamerica. In time the Indians added local plants to the cultivation process. The rich soil of the midwestern flood plains fostered the growth of many plants whose fruits could be gathered annually. In these areas, Indian peoples learned to cultivate locally important plants, such as goosefoot, sumpweed, and sunflowers, by removing competing weeds in a primitive cultivation process and later by actually planting seeds and tending the modest crops until harvest time.[2]

The first Indians who learned to cultivate Mesoamerican and local plants were probably women. Ethnological studies indicate that in the tribes of the historic period, women were the food gatherers while the men devoted their time to hunting and fishing. If this theory is true for the prehistoric period, it suggests that women, as food gatherers, had a better understanding of and more interest in the plant world than did men. Furthermore, because their domestic duties kept them closer to home on a daily basis, women had a better opportunity to learn to raise and cultivate certain plants. Because women probably assumed the responsibility for planting seeds and tilling the soil, they may also have developed the first agricultural tools. In time, women progressed from using their fingers and

11

hands to using sharp-pointed wooden digging sticks and later to fashioning hoes from the shoulder blades of deer or from antlers or clam shells for planting and cultivating. If these suppositions are true, then as the Indians increased their dependence on food production, agriculture became one of the most important occupations of Indian women.[3]

The development of agriculture came quite slowly, however; some archaeological evidence suggests that it began during the Late Archaic Age (ca. 4000 to 1000 B.C.). Plants such as the sunflower and sumpweed may have been planted and cultivated on a local basis. Certainly, the Late Archaic peoples extensively used food plants, and as they increased their reliance on seed foods, they became more sedentary. A dependence on food plants and a sedentary life, in turn, mutually supported the further development and growth of agriculture. In all probability, though, no one will ever know when the first cultigens were adopted or by which people. At the Phillips Spring site on the western Ozark Highland in Missouri, archaeologists have recovered the remains of cultivated squash that date about 2280 B.C. The Late Archaic people at the Carlston Annis and Bowles sites in western Kentucky also raised squash at least as early as 2230 B.C. Indian farmers were also cultivating squash and bottle gourds in the Lower Tennessee River Valley by 2000 B.C., which they were using extensively for food and utensils by 1500 B.C. Although no one can state precisely how much squash the Late Archaic farmers raised or estimate the percentage of food that it contributed to their diet, Indian farmers were cultivating it for more than one thousand years before the earliest known native plants, such as sunflowers, were domesticated in North America. By the first millennium B.C., however, Indian farmers were probably cultivating sumpweed after first tolerating its growth on the nitrogen-rich middens. In time, it became an important crop, because its seed contained a high oil content which contributed fat to their diet. Eventually, they cultivated sumpweed from northeastern Arkansas to North Carolina. Sometime between A.D. 500 and 1000, however, corn superseded sumpweed as a major cultigen.[4]

In the Midwest, the Adena, who lived in the Early Woodland period (1000 to 100 B.C.) of the prehistoric age, also practiced agriculture. They cultivated squash, pumpkin, sunflower, maygrass, sumpweed, and probably goosefoot. Evidence that they cultivated plant food has come from the Florence Mound site in Pickaway County, Ohio, where fragments of pumpkin and squash rind date to approximately A.D. 530 ± 250 years. A second excavation of a cultivated vegetable plant has come from the Cowan Creek Mound in Clinton County, Ohio, where archaeologists found fragments of a squash rind, gourd rind, and goosefoot seeds. Analysis of the squash rind indicates that it probably was pumpkin, and it dates at A.D. 440 ± 250 years. Archaeologists, however, now believe these dates are from five hundred to one thousand years too late, because squash found at the Adena site of Newt Kash Hollow cave shelter in Kentucky dated at 650 B.C. ± 300 years. These dates are among the earliest yet determined for the cultivation of nonlocal plants in the Midwest, that is, for the cultivation of plants of Mesoamerican origin.[5]

By 100 B.C. the Hopewell culture had largely replaced the Adena. Like the Adena, the Hopewell lived along the streams that drained into the Ohio River Valley. The principal evidence of this culture can be found along the Scioto, the Little Miami, the Great Miami, and the Muskingum rivers in the southern part of Ohio. The Hopewells' economy was based on agriculture; this enabled them to live in permanent villages but to supplement their farming activities with hunting, fishing, and gathering. Like the Adena of the Early Woodland period, the Hopewell apparently raised plants that were native to the region, such as maygrass, sumpweed, and sunflowers; they may also have cultivated some corn.[6]

About A.D. 800, Indian agriculture changed from botanically diverse garden plots to an extensive field system. By A.D. 900, a time known as the Mississippian period (A.D. 900 to 1600), the Indian tribes of the Ohio country, the Upper Great Lakes, and the Mississippi River Valley began placing even greater reliance upon farming. Archaeological evidence indicates that the Mississippian culture probably began in the St. Louis, Missouri, area, known as the American Bottoms, and spread up the Mississippi River into the Ohio country. As the population grew, corn production increased, and women maintained their dominance in agricultural practices.[7]

Mississippian farmers probably introduced beans to the eastern agricultural complex. They may have raised this legume as early as A.D. 800 ± 100 years, but the earliest archaeological evidence for beans in eastern North America dates at A.D. 1070 ± 80 years. In any event, beans appear to have been the last cultigen adopted by these prehistoric farmers. The late inclusion of beans from the Southwest may mean that beans were less easily adapted to different environmental conditions, or it may mean that Indian farmers simply were slow to use this plant east of the Mississippi River.[8]

Corn, however, was the primary crop of Mississippian farmers, and they tended to reside where the annual frost-free period ranged from 120 to 215 days. They raised corn varieties that would often mature in 90 days, enough time from planting to harvesting to provide an acceptable safety margin against short growing seasons. Certainly, these farmers knew from experience when to plant their seeds. If they made a miscalculation, however, and planted too early, a late spring frost might kill their crops and force them to replant. Or if they planted too late, an early autumn frost might ruin the harvest. Consequently, a margin of approximately twenty-five days existed at the beginning and end of the growing period. Although

The Hopewell people in the prehistoric Midwest are known as the Mound Builders because of their extensive earthworks. A horticultural people, they raised native seed plants, corn, and squash. Library of Congress.

this is an adequate safety margin for an average year, unseasonably late or early frosts could wreak havoc on a crop.[9]

Some Mississippian tribes that resided in areas with more than 190 frost-free days annually seem to have planted two corn crops with a staggered seeding time. This farming method assured the harvest of at least one crop and provided the possibility of reaping two harvests to guarantee adequate reserves during the winter months. The Mississippian farmers grew twelve-row northern flint corn, which eventually became the most popular corn among these farmers, because it required a shorter growing season. Much later during the historic period, the Indians taught the European settlers to grow northern flint, the variety from which geneticists have developed many of the modern hybrids.[10]

During the Mississippian period, Indian agriculture in the South resembled that in the Midwest and the Northeast. The skill and knowledge required to grow corn spread in the South where Indian farmers supplemented their food production by hunting, fishing, and gathering. The inhabitants of the Warren Wilson site on the Swannanoa River in North Carolina probably were typical Mississippian farmers. They raised flint corn, which they harvested in the late summer or early autumn and stored in pits; consequently, they probably lived on that site all year. About A.D. 1400 ± 200 years the Indians who occupied the Fort Walton Mound site on the bank of the Chattahoochee River in Alabama also engaged in agriculture. They raised one variety of flint corn, while the farmers at the Seaborn Mound site, fifty-six miles up river, raised a different variety. Although the people who inhabited these sites belonged to the same culture, no one knows why they raised different varieties of corn.[11]

Perhaps the best evidence for prehistoric Indian agriculture in the South comes from several Mississippian sites along the Georgia coast—Couper Field on St. Simon Island, Kenan Field on Sapelo Island, and Pine Harbor—where they raised corn and beans on the middens. We cannot be certain how long these farmers worked their lands, but the archaeological evidence indicates that they rotated their crops from field to field in order to maintain productivity. When they let their lands lie fallow, they usually had to move their villages relatively great distances, because good coastal soils were widely scattered.[12]

Archaeological evidence does not suggest that these Mississippian farmers cultivated vast fields of highly specialized crops; rather, they farmed the flood plains of the rivers and streams. There, the naturally fertile sandy loam soil was easy to till. Indeed, the technological development of the southern agriculturists dictated in part that crops would be grown along the river valleys. For example, the flint hoes and spades that have been recovered in what is now Kentucky were well suited for cultivating river-bottom lands. Indian farmers used stone and wooden tools to loosen or to stir the soil, rather than to turn it over; but these tools were not suitable for working the heavy clay soils of the uplands. The Indians also farmed the flood plains because they knew these lands were the most productive. The southern Indians lacked extensive lands that were good for agriculture, so one archaeologist has speculated that the attempts of the various Indian people to capture the farmlands of others may have been the cause of a great deal of Indian warfare.[13]

Beyond the Mississippi River in the prehistoric Southwest, agriculture had begun as early as 1300 B.C. By the time of Christ, the Indian farmers of the Southwest had made the seed selections and developed the plant varieties that were best suited for the climatic conditions in that region— from the cool, moist mountains to the hot, dry desert. The corn, for example, that the people from northern Mexico had first carried through the central Sierra Madres into the Southwest by 1000 B.C. grew best in the wetter, high elevations. This corn was a Chapalote-like small-cob corn. Soon thereafter, the Mogollon farmers in the Bat Cave region of present-day New Mexico developed a new type, or local "race," of corn. That variety was more variable and drought-resistant. It could be grown in the wetter high elevations as well as in the dry lower regions. Not only did this new variety produce larger ears with more rows of kernels, it also sprouted from deep planting. Dry, sandy soil necessitated this development. Planting depths of a foot or more prevented germination before late June or early July, so that the summer rains would water the new plants. If germination occurred in May as a result of occasional spring rains, the ensuing high temperatures and dry conditions would ruin the crop. This corn became the staple of the Hohokam-Basketmaker (Anasazi) complex, and it provided the subsistence basis for southwestern Indian civilization.[14]

These farmers also cultivated several varieties of squash and beans. The squash *Cucurbita pepo* was being raised along with the earliest corn at Bat Cave and as far north as Flagstaff, Arizona, by A.D. 1000. It was an important cultigen for the Hohokam. About A.D. 900 the southwestern farmers also began cultivating two other varieties of squash, *Cucurbita moschata* and *Cucurbita mixta*, using the seeds and flesh for food. Seeds provided twenty times the calories and twice the calcium of the flesh, as well as more vitamins. Because both parts could be dried for winter use, squash was an important food year-round. Until the Indians began cultivating beans, squash supplemented corn by providing the amino acid tryptophan, which corn lacked, for their dietary needs. Squash did not become an important crop until about 300 B.C., however, and was not cultivated on a widespread basis until about A.D. 900.[15]

Perhaps as early as 500 B.C. the people of Bat Cave were raising beans. The lack of a suitable technology for preparing beans for consumption may have influenced the relatively late adoption of this cultigen to their agricultural practices. Much later, sometime between 100 B.C. and A.D. 100, the Hohokam began to cultivate the common, or kidney, bean. Because beans usually have to be soaked and cooked for a relatively long time before the seeds are edible, they may not have been practical until the invention of pottery made such soaking and cooking possible. This is only a guess, however, because the Indians could have eaten beans in the prepottery age after parching and grinding them. The southwestern farmers primarily raised two varieties of beans: the common bean, in the plateau area, and the tepary bean, which the Hohokam favored.[16]

The tepary bean was particularly well suited for southwestern agriculture, because in the absence of irrigation, it had superior drought resistance, whereas high temperatures and evaporation rates tended to wilt the other varieties of beans. Indian farmers therefore preferred the tepary, since it tolerated hot, arid conditions and because it yielded more than the common bean by a ratio of four to one. Still, the adoption of teparies took time. Prehistoric southwestern farmers had relied on the common bean for as long as three thousand years before the drought-resistant tepary was developed, perhaps by the farmers in the Sonoran Desert or in southern Mexico. By A.D. 1000, however, the Hohokam farmers at Snaketown cultivated teparies, which had spread to the Mogollon in New Mexico by A.D. 1100.[17]

The southwestern farmers also raised two other bean varieties—lima and jack—on a small scale, adding both varieties relatively late in the precontact period. About A.D. 1000 the lima, or sieva, bean probably entered the Southwest from along the western slopes of the Sierra Madre by passing from farmer to farmer down the San Pedro Valley into the Gila River area. From there, it spread northward to the Anasazi. Sometime between A.D. 700 and 1200, the Hohokam also began to cultivate jack beans (*Canavalia ensiformis*) for food or for ceremonial purposes.[18]

In contrast to the eastern farmers, the southwestern agriculturists did not cultivate beans among their corn plants. Instead, they developed bush varieties that were self-supporting rather than vining. The development of bush beans was important, because in the Southwest, closely planted cultigens could not compete successfully for the limited supply of soil moisture in the absence of irrigation. Consequently, Indian farmers seeded the bean and corn crops in separate locations where the plants did not compete with one another.[19]

Besides raising corn, squash, and beans, the southwestern farmers also cultivated cotton, which had been domesticated in the Tehuacán Valley sometime between 3400 and 2300 B.C. It spread rapidly as people learned how to use it. Cotton probably reached the Hohokam settlement of Snaketown about A.D. 500. Because southern Arizona is well suited for cotton cultivation and because the Hohokam were skilled at weaving it into cloth, they probably were instrumental in disseminating cotton among farmers throughout the Southwest.[20]

The earliest woven cotton yet discovered in this area came from the Tularosa Cave in New Mexico. It dates about A.D. 700, but cotton may have been cultivated there earlier. After A.D. 1000, the Anasazi, Mogollon, and Hohokam, among others, cultivated cotton. The southwestern Indians valued cotton seeds for eating and for vegetable oil. The dual value of cotton—for food and for fiber—in addition to its light weight, made it important for domestic use and for trade. No one is certain how much acreage the Indians devoted to cotton or how much seed and fiber was produced from the annual crop, nor has anyone determined the precise range of cultivation.[21]

The cotton that the southwestern farmers raised—*Gossypium hopi*—has the shortest growing season of any variety. The bolls ripen within eighty-four to one hundred days, and it can tolerate relatively high altitudes and arid conditions. Southwestern farmers planted this cotton variety along the flood plains and at the mouths of the arroyos, where runoff from seasonal rains spread across the alluvial fans and provided enough moisture for the plants. Because the arroyos contained water for several days after a heavy rain, the cotton fields stored enough moisture to last until the next precipitation. If the Hohokam used methods of cotton raising similar to those of the Pima and the Papago, who now occupy a portion of that area, they probably planted the seeds with the aid of a digging stick, in a manner similar to seeding corn. They may have seeded one crop in March, which they harvested in July, then planted another crop in April, which they picked in September. The Hohokam also probably harvested the green bolls, which they placed on the roofs of their houses to ripen in the sun. When the cotton dried, they removed the fiber and seeds from the pods and stored it in their houses until needed. If frost threatened the crop, the Hohokam probably cut the entire plant for drying before they removed the bolls. No one is certain, though, precisely how the prehistoric Indians stored the fiber.[22]

Corn, beans, and squash contributed to the development of a sedentary life in the Southwest, and cotton gave the Indian farmers there a reliable source of fiber for weaving. In addition to these crops, as early as A.D. 630, western farmers may have sown the seeds of the tobacco plant *Nicotiana attenuata* in the ashes of burned-over areas. The tobacco variety *Nicotiana rustica*, however, which the eastern farmers raised, was not introduced into the Southwest until after the time of Spanish contact.[23]

The searing temperatures of the lowlands in the Southwest, where no irrigation system existed, meant that farmers had to contend with a short growing season. Corn had to mature within one hundred days, or else the crop might be lost to heat and drought. Indeed, without irrigation, the combination of high temperatures and scant precipitation gave the Gila River area of the Southwest the shortest growing season in North

America. The higher elevations of the Southwest receive between ten and twenty inches of precipitation annually, which falls in July and August, during the middle of the growing season. Consequently, southwestern farmers planted their crops so that the plants could make the best possible use of that moisture when it did arrive. Along the flood plains of the Colorado River, however, planting began in June or early July, after the rivers had receded from overflows in May and early June. Although this procedure meant that the planting season came during the hottest time, the moist soil enabled the crops to make a quick start, promoted rapid growth, and usually ensured a harvest, even if no other moisture fell until the crop had matured. Without irrigation, though, these farmers could raise only one crop annually. When they learned the principles of irrigation, they were able to use the long frost-free or growing season of the lower elevations (e.g., 357 days at Yuma, Arizona) to produce two or three crops annually. Until they began to irrigate, however, the areas that were nearly frost free were little more suitable for agriculture than were regions where frost restricted the growing season.[24]

In the Southwest the Indian farmers stored their crops in a variety of ways. The Hohokam, for example, stored corn in baskets and in pottery vessels. Pottery storage jars were particularly convenient, because the openings could be sealed to prevent damage from rodents, insects, and moisture. The Anasazi, in contrast, used storage pits and fashioned store rooms or granaries among the cracks, niches, and overhangs in the canyons. If the soil of these protected areas was too sandy or if it crumbled easily, the Anasazi scooped out the holes and lined them with flat rocks or with clay, in a manner similar to the eastern Indians, who lined their storage pits with bark. The Anasazi used flat stones, clay, or mud to cover the openings. These granaries usually were about two to three feet in diameter and three feet deep; some granaries had walled fronts. Indian farmers probably used granaries, such as those found in Monument Valley in Utah and northern Arizona, as temporary storage places until the corn could be carried to their homes.[25]

While the farmers in the prehistoric Southwest adopted the plant varieties that would survive and produce abundantly in the hot, dry environment and while they stored their crops for time of need, they also enhanced the success of their agricultural activities by using irrigation. These farmers developed several methods to apply supplementary water to their crops. One method was the very basic use of floodwater. This procedure simply involved planting crops in areas that were inundated with floodwater during periods of rainfall. Farmers such as the Pueblos planted their fields at mouths of arroyos; then as the floodwater emptied from these small gullies or canyons, it spread across the alluvial fan and watered the crops. Some farmers also planted their crops on the broad

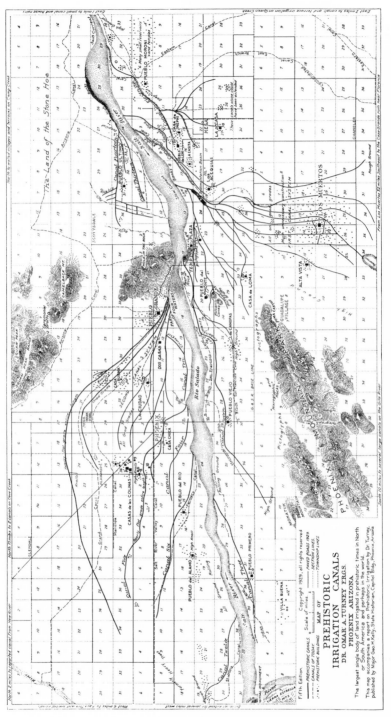

The prehistoric Indian farmers in the Southwest built extensive irrigation canals along the Salt River. The Hohokam, the most skilled of these canal builders, did not construct reservoirs to store water; instead, they relied on the overflow of the streams. *Arizona Historical Society.*

flood plains of the river valleys. When a river overflowed its banks, it gave the soil a moisture supply which enhanced the planting process, or the water irrigated the growing plants. In either case, the overflow had to saturate the soil enough but not wash away the crops. This water-use method required the southwestern farmers to have an excellent understanding of the physiographic features of their environment and to master the principles of runoff and stream flow.[26]

Among the southwestern farmers, the Hohokam exercised the greatest irrigation skills. In the valleys of the Salt and Gila rivers, the "land of the canal builders," they constructed major systems of irrigation canals that stretched for 150 miles or more. These canals rival ones that Old World farmers made along the Euphrates and Tigris rivers. The longest single canal in the Salt River Valley extended approximately 14 miles, measured thirty feet from crown to crown of each bank, and reached depths of ten feet or more. The Hohokam did not, however, build reservoirs to store water. Therefore, without a large water reserve, they had to channel overflow or stream flow from the rivers into the canals. Although the Hohokam were able to control water flow over a vast area, they did not build reservoirs to impound water because they knew that evaporation would waste more water than they could store. Nevertheless, their technical feat in constructing such extensive canals is remarkable, and no other Indians surpassed their engineering techniques until the nineteenth century.[27]

Although the Hohokam did not have modern surveying equipment, they apparently had an understanding of the principles of irrigation. We cannot be certain about the details of the construction of their irrigation projects, but it undoubtedly involved simple gravity flow, as they used their engineering abilities to make the water flow where they needed it. The construction of these canals, though difficult, might not have taken as much effort or as many workers as one might expect. The Hohokam probably solved the problem of determining the grade for the main canals by admitting water into the excavating area from the rivers upstream. The water flow would have indicated the direction for further excavation, and it would have softened the soil to make digging easier. Hohokam men and women probably dug the irrigation ditches one to two feet wide at the beginning, gradually increasing the depth and width of the ditches. In other words, the Hohokam probably did not set out initially to construct a canal twelve miles long, ten feet deep, and thirty feet wide. Rather than undertake such an incredible task, it is more likely that a small group of men and women undertook canal building on a more modest scale. Fifty men and women, for example, removing one lineal meter of earth per day, could have constructed a canal two meters wide, a meter deep, and two to three miles long, such as the Pioneer Canal near Snaketown, in about 100 days. This is not to say that canal building was easy for the

Hohokam. Indeed, it must have been excruciatingly hard work, because their implements were handleless stone hoes, ironwood digging sticks, cottonwood spades, and baskets.[28]

Although the Hohokam probably constructed these canals over a long period of time, they apparently built them without the aid of a highly centralized hydraulic society, even though the repair of leaks and the dredging of silt probably required some sort of organized communal effort. An intervillage decision-making process perhaps determined water allocations. Archaeological evidence suggests that small family groups may have been responsible for the repair of specific portions of the canals. Certainly, some system of control and maintenance was necessary for a canal network that spread in veinlike fashion over several hundred miles. Still, the precise methods of control and organization remain a mystery.[29]

The archaeological evidence clearly indicates that the Hohokam built brush-and-sand dikes or weirs to channel the water from the rivers into the main canals. If the canals began to leak, they used adobe to patch the seepage area, then they burned brush on top of it to harden the clay. The Hohokam lined part of one canal near Pueblo Grande, Arizona, with clay to minimize the loss from percolation as water flowed to the fields. Although the lining of canals with clay was probably not a common practice because of the amount of labor required to complete the task, it indicates the technical ability of the Hohokam to solve problems associated with irrigation agriculture. Today, this ancient principle can be seen where modern irrigation canals are lined with concrete.[30]

No one is certain how many acres the Hohokam irrigated, because only the basic engineering principles of this sytem are known. No one knows how they applied the water to their fields. Did they use furrows to channel the water from the lateral ditches into crop rows? Did they flood the entire field with a sheet of water? Or did they use some other method of water application? No one knows whether they used water to flush away the accumulated salts from the surface of the fields; but they probably did, because otherwise the fertility of the soil would have been depleted long before their civilization became extinct in the fifteenth century. Still, no one knows how frequently the Hohokam irrigated their lands or how much water they applied to their crops at any one time. Moreover, dry and silted-in canals would have affected the amount of acreage under irrigation. Thus, without knowing the method of application and the amount of water used, no one can accurately estimate the acreage under irrigation. Archaeologists conservatively estimate that the Hohokam irrigated their fields for at least 500 to 600 years, approximately A.D. 800 to 1400. This method of farming in the Southwest was well established long before it peaked sometime between A.D. 1200 and 1400.[31]

After the thirteenth century the Hohokam irrigation system declined for some reason. Hohokam farmers probably experienced decreased productivity from their croplands because of alkalinity, which occurs over time when salt from irrigation water builds up in the soil. Some areas may have become unusable because of waterlogging from overirrigation. Drought may have been a cause; or erosion may have lowered the river channels and left the canal intakes above the water source. Cultural and social developments that are still not known, as well as hostile neighbors, may have played a role in the decline of their civilization. Irrigation agriculture, however, did not end with the waning of Hohokam civilization. Rather, the Indian farmers who came after them continued to practice the irrigation techniques of the Hohokam, only on a smaller scale.[32]

In addition to floodwater and irrigation farming, the people of the Southwest also developed important conservation techniques that aided agricultural productivity and stability. In contrast to the Hohokam, most farmers, such as the Anasazi and Mogollon, lived where water was scarce. To farm successfully they had to develop water-conservation techniques that permitted the maximum use of the available moisture, so they developed several conservation methods that capitalized on floodwater runoff. Water was so important that they devoted considerable time and effort to developing the conservation practices that helped ensure a bountiful harvest, even during times of drought. They built walled terraces in a stair-step fashion by placing boulders along hillsides. These walls trapped the runoff and held the water until as much of it as possible had sunk into the soil. The plots that these terrace walls formed were small; some were no more than eight by ten feet. The Anasazi also constructed similar terraces, called check dams, in the upper parts of water courses where the grade was not particularly steep or the runoff swift. They carefully anchored the ends of these dams to hold the structures against the water when it came rushing down the stream or arroyo bed. The Indians built these rock dikes high enough so that the water would deposit silt behind them to a depth of three to four feet, thereby creating a well-irrigated garden plot with rich soil. Excess water passed down the hillside, where it was trapped behind other terraces until as much water as possible had been used. These terraced and check-dam plots were suitable for planting supplementary crops of corn, beans, and squash or for a specialty crop of some sort.[33]

At Point of Pines, Arizona, and at the Beaver Creek community in Utah, Indian farmers also placed rocks in lines to form linear borders on gently sloping land to help hold the soil and to slow the runoff so that more water would sink into the fields, thereby enhancing crop productivity. Grid borders, made from stones, ran parallel to the slope of the land and probably served as boundaries for individual and communal fields and reduced soil erosion, retained runoff, and increased moisture penetration. Farmers at the Beaver Creek community also set flat and oblong stones in rows on high ground to deflect the wind upward away from the crops. This conservation technique prevented blowing sand from lacerating the plants, and it retarded the loss of soil moisture from the hot winds. These farmers also used stone-lined ditches to conduct runoff to adjacent croplands. In addition to conserving soil and water, these techniques encouraged the removal of rocks from the cultivated area.[34]

Indian agriculturists used these methods of conserving soil and water in other parts of the Southwest as well. Terraced fields frequently were located several miles from the villages. Certainly, these conservation measures enabled the southwestern farmers to cultivate more land more efficiently in a region where water is scarce than they could have done without using such skills. Farmers at Point of Pines, for example, cultivated as many as one thousand acres comprising small terraced or stone-bordered plots. Like the irrigation canals of the Hohokam, the terraces and linear borders took time to build as each farmer used skill and ingenuity in the absence of sophisticated technology to solve his own problems of soil and water conservation.[35]

In retrospect, then, millenniums before the arrival of white settlers east of the Mississippi River, Indian farmers had learned to cultivate plants of Mesoamerican and local origins. They discovered how to select the seeds that would yield maximum harvests according to local soil and climatic conditions, thus making great strides toward farming in harmony with nature. White farmers did not make similar progress in plant breeding until the development of the agricultural experiment stations in the late nineteenth century. Indian farmers also knew how to clear land in the best manner their technology allowed, and they learned how to cure corn and squash and to store those crops for future use. They discovered the narcotic effect of tobacco, which they raised primarily for ceremonial purposes.

Although the agricultural Indians emphasized corn, beans, and squash, they also extended the range of other cultivated plants such as the sunflower and the Jerusalem artichoke. Corn was by far their most important crop, and they raised a number of varieties—flint, dent, soft, and popcorn—all of which they selected according to the requirements of the local growing season. No one can say for sure how much corn the Indians raised annually before the first contacts with the Europeans. If fifteen bushels of shelled corn would support one individual for an entire year, the number of acres under cultivation in eastern North America might have equaled the number of persons inhabiting that region, or roughly 365,000.[36]

Such estimates are guesswork at best. One can be more accurate by stating that agriculture was important to the food supply of the Indians prior to the arrival of white settlers. Even though the Indians had not developed a farming system based on raising a combination of plants and animals similar to that which developed in the Old World, their achievements are nevertheless astounding. The Indian farmers showed the European settlers which plants to cultivate, particularly corn, beans, pumpkins, and tobacco. When one considers that Indian women in the East transmitted the New World's agricultural tradition to white men, the significance of this cultural transfer is particularly important.[37]

Similarly, the Hohokam, the Anasazi, and the Mogollon were the major agricultural people in the prehistoric Southwest. They adopted plant varieties that were well suited to the severe growing conditions of that region. Over time, they improved their crops by using great skill in the selection of seeds. In the case of corn, for example, they chose the ears with the largest kernels, usually taken from the center of the ear, where the seeds were the most uniform. Dry-land farmers, such as the Anasazi, selected seeds from the ears that matured first. Centuries later, the Zuñi maintained this practice, which the Mormons also adopted. Moreover, the Indian farmers learned to plant their crops in locations that had different exposures to the sun and moisture to ensure at least a partial harvest during times of abnormally hot or cold temperatures and moist or dry years. For example, if a flood destroyed their crops along a stream or arroyo, they would still harvest a crop from the drier hillsides. Or in drought years, perhaps only the stream-side fields would produce a crop. Thus, by using highly specialized crops, by carefully selecting field sites, and by depending upon soil fertility, elevation, and runoff, the Anasazi and the Mogollon successfully raised crops for centuries in a region that farmers still consider unsuitable for agriculture without irrigation.[38]

The Hohokam particularly left a distinguished mark on the history of Indian agriculture. These "canal builders" engineered an irrigation system on a twentieth-century scale. They not only established the only true irrigation culture in prehistoric North America, but they also made agriculture practical where farming would have been tenuous at best without irrigation. Certainly, irrigation requires farmers to have greater knowledge, such as an understanding of the behavior of water and the technical skill needed to control it, than dry-land farming requires. Indeed, the ability of the Hohokam to solve their farming problems, which the environment created, makes them an important agricultural people. During the 1870s the Mormon settlers in the Salt River Valley cleared the weeds, silt, and wind-deposited soil from several Hohokam canals and used them for irrigating their own fields. In 1887, immigrant white farmers discovered so many ancient irrigation canals in that area that land values declined because of the expense of leveling the ditch embankments.[39]

The prehistoric farmers in the Southwest gained increasing control over their environment as they adopted and modified cultigens so that early maturing and drought-resistant crops would furnish a dependable source of food and fiber. By A.D. 500, agriculture was providing an important subsistence base for the Anasazi, the Hohokam, and the Mogollon peoples. During the following nine centuries, the southwestern farmers continued to refine their agricultural techniques. By A.D. 1100, agriculture probably had superseded hunting and gathering in importance, and it provided approximately half of the Indians' dietary needs, thereby giving them greater security.[40]

After contact with the Spanish in the mid sixteenth century, Indian agriculture became more diverse with the adoption of European crops and livestock. Although Indian agriculture underwent fundamental change, these farmers maintained many traditional techniques, and they continued to pass that knowledge on from generation to generation.

3

Eastern Farming
at the Time of European Contact

By the time of Spanish exploration the southern Indians had engaged in farming for a long time, and they possessed a well-developed agricultural economy. When De Soto's expedition landed in 1539, the Spaniards found the Indians in the Apalachee region (modern Tallahassee) raising large quantities of corn, and one chronicler reported that the province of Mocozo was "cultivated with fields of Indian corn, beans, pumpkins and other vegetables, sufficient for the supply of a large army." Indian agriculture was so extensive near present-day Ocala, Florida, that the Spaniards took a three-months' supply of corn from the fields. Indeed, Indian agriculture and the Spanish appropriation of corn and other produce made the four-year De Soto expedition possible.[1]

When the Spanish arrived, the Apalachee and the Timucua Indians of northern Florida were already well-established agriculturists, living in permanent settlements, planting two crops annually, and rotating their fields. The Indians cleared their land by burning the brush, and the men used hoes fashioned from sticks, shells, and bones to prepare the soil for planting. The women planted the seeds with the aid of dibble sticks. These farmers, like those in the Northeast, erected watchtowers from which sentries guarded the crops against marauding birds and animals. The Spaniards were amazed that the northern Florida Indians could cultivate such vast fields. Near Ibitachuco and Ucahile, De Soto's army passed through cultivated fields, presumably of corn, which were said to extend as many as twelve miles.[2]

Besides raising corn, beans, and squash, the Florida Indians cultivated a native root (*Zamia*), which could be made into bread. They also may have cultivated tobacco, because at about this time the English adventurer John Hawkins reported that they smoked something that he called a dried herb. Although these farmers did not fertilize their fields, they took care

In the mid sixteenth century, Indian agriculturists in Florida raised corn and beans, which they planted with the aid of shell and wooden hoes and digging sticks. This 1564 engraving by Theodore DeBry depicts Indian farmers preparing the land and planting seeds. Library of Congress.

to preserve the corn crops from spoilage and insects by drying and storing them in stone warehouses with palm-leaf roofs.[3]

De Soto's discovery of an Indian population with a rich agricultural base probably seemed heaven sent, because the expedition's food supplies were nearly exhausted when the Spanish landed. On one occasion, the Spaniards were so hungry that they ate the ears of corn from the stalks, including the cobs, when they came upon a field. If the southern Indians had not been good farmers, De Soto's expedition might have perished or been forced to turn back, because the Spaniards became dependent upon Indian crops, particularly corn, for a major portion of their food supply.[4]

The Indian farmers whom De Soto discovered tending fields, harvesting crops, and storing grain had picked their lands well, for the Indian fields of the mid sixteenth century have become part of the fertile farmlands in the vicinity of Dade City, Ocala, Gainesville, Lake City, Madison, and Tallahassee today, as have the Indian lands along the Savannah, Coosa, Alabama, and Tombigbee rivers. These lands were not only fertile and easy to till, but the long growing season of six months or more also made possible two crops per year, should floodwaters destroy the first planting.

By the time of European contact, Indian farmers in Florida had well-built storehouses for their crops. Between 1539 and 1542, De Soto's expedition relied upon Indian farmers for survival. Library of Congress.

An acre of this land, with a mixed crop of corn, beans, and squash, might have provided enough food to feed one person for the entire year.[5]

Farther to the North, in 1584, a visitor in the ill-fated settlement of Roanoke reported that the Indians were growing white corn that was "fair and well tasted" and beans of "diverse colors." They planted corn in May, June, and July and harvested it in July, August, and September. Several years later, Thomas Harriot reported that the Algonquin tribes of the North Carolina sound and the Virginia coast also farmed. There the men used long-handled wooden mattocks to prepare the soil for planting. The women shared this work, using short-handled paring tools, which they pushed from a sitting position to break the soil, kill the weeds, and loosen the old cornstalks from the ground. After completing this task, they allowed the weeds to dry in the sun for several days before gathering them into piles for burning. These farmers did not use ashes or dung for fertilizer. With the soil tilled and the weeds burned, they planted corn in hills with the aid of a digging stick. Between the hills, which were spaced about a yard apart, they also seeded beans, pumpkins, orache, and sunflowers.[6]

Toward the interior, farming became more important for the southern Indians, and agriculture probably contributed from 25 to 50 percent of

During the late 1580s the Indians in the village of Secoton, near Sir Walter Raleigh's colony of Roanoke in present-day North Carolina, raised abundant corn crops as well as some sunflowers and squash. This engraving by Theodore DeBry is after a water color by John White, the original leader of the colony. Library of Congress.

their food needs. Away from the coastal plain, the Indians farmed the sandy loam soils of the alluvial valleys in the lower elevations of the Piedmont, the Interior Low Plateau, and the Cumberland Plateau. There, floodwaters deposited silt that enriched the soil and made continuous cultivation possible. These river-bottom soils also held enough moisture, even during dry summers, to sustain agriculture without the use of irrigation. These soils were intensively cultivated and highly productive—all of which enabled Indian farmers to support large sedentary populations.[7]

The Algonquin farmers in this region planted three varieties of corn, two of which ripened in about twelve weeks, and the third, two weeks later. The Algonquin also raised two varieties of pumpkins, as well as sunflowers and chenopods. Their yields from an "English acre" (an area approximately 660 feet by 66 feet) averaged two hundred bushels for the combined harvest of the corn, bean, pumpkin, sunflower, and chenopod crops. Although the Algonquin planted pumpkins and squash between the corn hills, they also raised these crops in separate fields. Besides tending large fields, the Algonquin also cultivated small family plots (about 100 by 200 feet), which an English observer reported they kept "as neat and cleane as we doe our gardein beds." These gardens provided fresh vegetables until the larger fields could be harvested.[8]

The Algonquin also raised tobacco (*Nicotiana rustica*) in separate plots. It grew about three feet high and produced a short, round leaf. They prepared the tobacco for smoking by drying the entire plant over a fire or in the sun and by grinding it into small pieces for their pipes. By the seventeenth century the Algonquin of Virginia had adopted a milder variety of tobacco (*Nicotiana tobacum*), which John Rolfe introduced from the West Indies.[9]

Captain John Smith observed that in Virginia the Indians girdled the trees and grubbed the roots to clear new agricultural lands. They planted corn and beans in the same hole, so that the cornstalks would support the climbing beans. During the course of the growing season, the women and children weeded the crop three or four times by hoeing earthen hills around the stalks. These hills gave the stalks support in windy and wet weather and encouraged the growth of bracer roots at the base of the plant. The Indian farmers in Virginia did not till the soil between the hills. Smith also noted that these farmers planted corn from May until mid June and harvested it from August to October. Each stalk bore two or occasionally three ears, some of which were picked for roasting before they were ripe. Indians harvested the ripe corn by picking the ears, which they then dried in the sun. Each night they raked the corn into piles and covered it with mats to protect it from the dew. When the corn had dried thoroughly, they shelled it by twisting the cobs between their hands and stored it in baskets under their roofs. Women performed all of these tasks.[10]

By the eighteenth century the Virginia Indians were raising four varieties of corn. Two of these grew only three or four feet high, produced small ears, and ripened in midsummer; the other two grew ten feet high, produced ears eight inches long, and matured in the autumn. Although they raised flint and dent corn, these farmers preferred the latter, because it produced the largest yields, sometimes a thousand kernels for each one planted. One of the late-ripening varieties may have been popcorn. These Indians also planted tobacco, whose leaves they cured in the sun and stored for later use; and they raised pumpkins, gourds, cushaw squash, macocks, and watermelons. One observer found the pumpkins finer than any that he had seen in England. They also raised peaches, which they probably obtained from the Spaniards in Florida.[11]

By the mid eighteenth century, agriculture also was important for the Cherokee, Choctaw, Chickasaw, and Creek, who cultivated small private gardens and large communal fields. The gardens, located close to their houses, were usually fenced in order to keep their horses from trampling the crops. These farmers seeded corn in early spring, about the time the wild fruit ripened, so that the newly planted seeds would not unduly tempt the birds. They planted cornfields communally, one after the other, until all of the seeding had been completed. They also seeded pumpkins, beans, and occasionally sunflowers, as well as tobacco, cabbage, potatoes, peaches, peas, leeks, and garlic.[12]

The women of these tribes cultivated the corn crop several times during the summer, and when the crop ripened, each family harvested its allotted area. Before these farmers deposited any of the crop in their private granaries, however, each family contributed a certain amount to the crib of the village chief. This donation was entirely voluntary, but the crib provided a surplus from which any family could draw when its own supply became exhausted. They also used this reserve to help neighboring towns whose crops had failed and to provide food for travelers or for the men who were away from the village on hunting or war expeditions.[13]

The Cherokee and Choctaw men assumed a greater role in agriculture than did many Indian males in the South, because they commonly helped clear the land and plant and harvest the crops; nevertheless, the women did most of the farming. In the mid eighteenth century, for example, one observer noted of the Chickasaw that "the girls and women do more work than the men and boys." The women and children probably weeded the crops, as they did among the other southern tribes at that time, and the Cherokee women owned the crops and other agricultural produce. During the early colonial period, English traders and soldiers purchased corn, poultry, and pork from the women, not from the Cherokee men.[14]

Although the Cherokee and the Creek adopted European agricultural methods, such as raising cattle for beef, hogs for salted meat, and poultry for eggs, they did not use the plow because they believed it would lead to what twentieth-century Americans call technological unemployment. By the last quarter of the eighteenth century, the agricultural economy of the Cherokee and the Creek was closely linked to white civilization. Many elderly tribesmen planted crops in order to sell the surplus to white settlers. Rather than viewing the plow as an important agricultural tool that would make them more efficient and ease their labors, they believed the plow would enable one or two farmers to produce larger crops with yields far greater than they could possibly sell. Those farmers who used the hoe and the dibble stick would not be able to compete, and the result would be unemployment and starvation for the elderly. Indian farmers in the South would adopt the plow eventually, but it did not become a common agricultural tool until the nineteenth century.[15]

Farther to the north, in 1535, the French explorer Jacques Cartier reported that the Indians living in the vicinity of Montreal had large corn-fields and that they stored their crops in garrets under the roofs of their houses. Nearly a century later, in 1610, Samuel de Champlain noted that most of the Indians in the vicinity of Lakes Erie and Ontario grew corn for their chief food as well as for trade with the hunting tribes. The Huron actually became the main corn suppliers for the Ottawa Valley Algonquin in Quebec, as well as for the nonagricultural tribes in the region of the Upper Great Lakes.[16]

In the latter region the Huron and the Tobacco Huron particularly depended on agriculture for about 75 percent of their nutritional needs. Among the Huron the men participated to a greater extent in the farming process than was the case among the other northern tribes. Huron men cleared the trees and brush from the fields by cutting and burning. The women still had the task of clearing the land between the tree stumps and grubbing the roots from the ground. When a stump rotted, they removed it. Women also cultivated and harvested the traditional crops of corn, beans, squash, sunflowers, tobacco, and, by 1687, watermelons, which the Europeans introduced. Corn was still the primary crop, and the women gave careful attention to it. Before planting time in the spring, they selected the best seeds, which they soaked in water for several days to speed germination. The Huron planted the kernels in holes spaced about a step apart, and as the corn grew, they hoed up the soil around the stalks to give the plants support and to prevent sheet erosion. After the corn had sprouted a few inches, the Huron planted beans in the same hills, so that the cornstalks would provide support for the vines. The Huron pri-marily raised flint corn, which matured in one hundred days, and flour corn, which ripened within 130 days. At harvest time the Indians picked the ears, stripped back the husks, and tied the corn to a rack in bundles, which they suspended from the ceilings of their lodges. When the corn dried, the women shelled the ears and stored the kernels in large baskets in a corner of their houses or on the porch.[17]

It is impossible to state just how much corn the Huron raised, but they tried to grow enough each year to provide a two-to-four-year surplus to guard against crop failure and to furnish an adequate commodity for trade. A village of about one thousand inhabitants needed at least 360 acres of corn. One estimate placed the annual harvest at 390,000 bushels on approximately 23,300 acres, for an average yield of almost 17 bushels per acre. Another anthropologist estimated that each Huron woman produced about 1.3 pounds of corn for each family member's daily allotment, an amount that met about 65 percent of their diet. This evidence is based, in part, on Roger Williams's observations that Indian women in New England raised as much as 45 bushels of corn annually for each family member. The Huron cornfields apparently were so large that at least one Frenchman got lost as he walked through them from one village to another. Squash was also an important crop, and the Huron women took care to sprout the seeds in bark containers filled with damp powdered wood, which they then warmed over a fire. After the seeds sprouted, the Indians planted them in hills. They also raised sunflowers, the seeds and oil of which they added to their corn soups; and by the late seventeenth century, because of French influence, some Huron families were milking one or two cows. Still, corn was their most important crop, and when famine struck, as in the spring of 1639, they sold their children to procure it.[18]

The Huron apparently did not use fertilizer, instead, they prolonged soil fertility by burning, cultivating, and planting beans among the corn hills. By burning the brush from the fields before planting time, they unwittingly were adding magnesium, calcium, potash, and phosphorus to the soil. Burning also reduced soil acidity and thereby promoted bacterial activity and the formation of nitrogen. The bean crop also provided nitrogen, and frequent weeding helped to prevent the growth of grasses that could not be chopped away with stone, bone, or shell hoes. At the same time, the prolonged practice of burning decreased the organic soil material and consequently forced the Huron to relocate their fields and villages every six to twelve years, depending on soil conditions. Since the Huron, like the other agricultural tribes, did not plant corn in straight rows, the irregularly spaced hills helped to retard soil erosion. Huron cornfields reportedly yielded between twenty-five and thirty bushels per acre, but when the yield had declined to seven or ten bushels per acre, they abandoned their fields.[19]

Corn also was an important crop for the Iroquois-speaking tribes in what is now New York State. There, the Mohawk cleared fields by killing the trees and underbrush with fire. About two years later, they planted their first corn crop among the dead trees. The Mohawk women selected seed from the best ears of the previous harvest and, under the supervision of a matron, assembled in the fields, where they planted the hills about five or six feet apart. After having seeded a field, the group moved on to the next field, thereby efficiently using their numbers and organization at planting time. The Mohawk farmers planted corn deep enough to prevent the crows from pulling up the young plants but shallow enough to permit it to sprout.[20]

After the Mohawk had seeded their corn, they erected watchtowers or platforms in the fields, from which the women and the older children guarded the crop against marauding birds, especially crows. The Mohawk also attempted to prevent crop damage from these pests by soaking their seed corn in a potion of hellebore, which poisoned the crows. As an added precaution, the Mohawk snared the crows and hung the live birds by their feet to frighten away other pests. In addition, the Mohawk used snares and deadfalls to trap raccoons, woodchucks, and deer, which raided the corn crop. The Mohawk harvested the corn by pulling the ears from the stalks or by uprooting the stalk for removal of the ears later; they also stripped the husks with the aid of a wooden peg. Like the Huron, the Mohawk left several husks attached to the ears for braiding, so that the ears could be hung from the insides of their houses. These Indian farmers stored shelled corn in bark containers or in underground storage pits. The Iroquois used the husks, which they removed from the ears, to make mats, baskets, and moccasins and to wrap around bread for baking and boiling. The Mohawk also used hollowed-out cornstalks for the storage of maple sugar, salt, and medicines.[21]

Farther to the west, during the late seventeenth century, reports indicate that corn was the primary crop of the Illinois tribe. When their corn ripened in late August, the women gathered the crop and husked the ears, which they spread on mats to dry out of doors. Each night, they gathered the ears into piles, which they covered with blankets to keep the evening dampness off the harvested crop. After a week's time, the Illinois threshed the ears by beating them with sticks. These Indians also grew watermelons and pumpkins. They hollowed out the latter, cut them into slices, and dried the pieces in the sun for a day. Then, they tied the slices together and hung them from a rack to dry for several more days. Pumpkin slices preserved in this manner could be kept for several months, during which time the Indians commonly cut chunks for boiling in pots of meat and corn.[22]

About this time, in 1699, one explorer said that the Quapaw tribe, located along the Illinois River, were "living on nothing scarcely but Indian corn," which grew from twelve to twenty feet high. Corn cultivation also was important to the Sac Indians living along the north side of the Rock River where it empties into the Mississippi. There, Chief Black Hawk re-

The Indian agriculturists in North America customarily erected watch-towers from which the women scared away marauding crows and other pests from the cornfields. Ohio Historical Society.

ported that the cornfields of his village extended parallel along the Mississippi for two miles. He recalled: "We had about eight hundred acres in cultivation, including what we had on the islands of the Rock River. The land around our village, uncultivated, was covered with bluegrass, which made excellent pasture for horses. . . the land being good, never failed to produce good crops of corn, beans, pumpkins, and squashes. We always had plenty—our children never cried with hunger, nor our people were never in want." Here too, the women cleared the fields, planted corn, and built and repaired fences around the fields.[23]

By the time of the French explorations, then, agriculture was an important endeavor for most of the tribes. Nicholas Perrot, a French commander in the Old Northwest, clearly stated the critical importance of agricultural products in the daily diet of the Indian population. Sometime between 1680 and 1718, he wrote that the Indians were "naturally industrious, and devote themselves to the cultivation of the soil." Corn, beans, and squash were the chief crops, and he noted that "if they are without these, they think they are fasting, no matter what abundance of meat and fish they may have in their stores, and Indian corn being to them what bread is to Frenchmen." By the late eighteenth century, corn was so important to the Miami and their allies living along the Maumee River that, according to General Anthony Wayne, the Indians were planting it continuously from the present location of Fort Wayne, Indiana, to Lake Erie.[24]

The Indian tribes of the Midwest, like those of the Northeast, traditionally stored their corn in underground pits or caches to prevent other tribal members from stealing it. Those storage pits were about eight feet in diameter and from five to six feet deep. Each pit had an opening two to three feet wide, the top of which the Indians covered with grass and earth when the cache was full. Indian women also had the task of placing the corn in storage. One white traveler observed that the women received "neither sympathy nor assistance from the brawny fashionable bedaubed youths who [were] sunning themselves in the plain below, whilst these poor creatures [were] toiling."[25]

Corn, squash, and beans were not the only crops important to the Indian farmers of the Midwest. In the region of the Upper Great Lakes, many tribes gathered wild rice. While most tribes simply collected this grain on an annual basis, the Ojibwa, or Chippewa, and the Assiniboin sowed wild rice in the marshes surrounding the Great Lakes to guarantee a harvest the following year. The Chippewa typically sowed as much as one-third of the rice crop in waters averaging about two feet deep. The Assiniboin also weeded the rice by pulling out the water grasses that grew among the rice stalks. Because the rice stalks might rise more than eight feet above the water and because the muddy bottom was impossible to walk upon, this grain had to be harvested in canoes. About the time the rice was ready for harvest, the Indians poled their canoes into the rice fields and tied the stalks into bunches with basswood fiber strings. The Chippewa women gathered a bunch of rice stalks with a curved stick called a rice hoop, then tied a bark string around it. By tying the wild rice into sheaves or bundles before it was ripe, the heads did not shatter into the water as the stems moved back and forth in the breeze. Three or four weeks after the women bound the rice into bundles, about the end of September, each family returned to harvest the rice from its allotted or recognized area. Chippewa women marked the boundaries of their rice fields with stakes about ten days before the rice was ready to harvest. A man usually poled the canoe up to the bunches of wild rice, while a woman bent the heads over the center of the canoe and beat the grain out with a stick. A skilled rice gatherer would leave the unripe grain in the heads for harvest later. By so doing, these Indian farmers gleaned a rice field several times until they had harvested as much grain as possible.[26]

Wild-rice farming also was largely the domain of the Indian women, who harvested the grain and spread it out to dry on skins in the sun or who dried the grain by parching it in a kettle or by putting it in a bag that they hung over the smoke or the heat of a fire. Sometimes the women spread the rice on a mat attached to a three-foot-high scaffold. Then, they

In the Great Lakes region, the Chippewa sometimes sowed, weeded, and harvested wild rice, which was high in carbohydrates and well suited to the dietary needs of the Indians in the cool North. Ohio Historical Society.

built a slow fire underneath the scaffold and periodically shook or turned the rice. By the end of the day, the rice was dry and ready to thresh. In order to thresh the rice, the Indians dug a hole about a foot and a half deep and two feet in diameter, which they lined with a deer skin or with wooden staves. Next, they poured the rice into the pit, and the men threshed the grain either by treading upon it or by pounding it with a pestle. Once the rice had been threshed, the women winnowed it with a basket and with the prevailing winds or a birch-bark fan. When the chaff had been removed, they placed the rice in birch-bark boxes or in skin bags and stored it in underground pits. If the rice had been dried properly, it would keep for many years. Although it is difficult to estimate the average family harvest with any certainty, approximately five bushels were harvested per family. Some women, however, harvested as much as twenty-five bushels annually and used any surplus for trade.[27]

While only a few Indian farmers raised wild rice, none fertilized their fields. Contrary to popular belief, the Indians east of the Mississippi River did not fertilize their corn with fish or teach the Europeans to do so. Squanto apparently showed the Pilgrims how to use fish for fertilizer in 1621, but the tribesmen did not follow that practice. Instead, when their lands became exhausted, they abandoned them and cleared new areas for cultivation. Squanto knew about fertilization, but he probably gained that knowledge while he was a captive in Europe. In 1614 an English captain kidnapped Squanto and sold him to a Spaniard. Under circumstances that are not entirely clear, Squanto eventually made his way from Spain to England and then to Newfoundland. On 16 March 1621 he strolled into the Pilgrim settlement at Plymouth. Therefore, Squanto had lived among the Europeans for seven years, during which time he may have observed farmers fertilizing their fields with fish—an agricultural technique that dates back at least to the Roman period. Squanto, then, probably did nothing more than inform one group of Europeans about the agricultural practices of another group with a similar heritage. Had the Indians along the Eastern Seaboard and in the Midwest and the South fertilized with fish or another substance, the colonists would have commented on it, as they did about other aspects of Indian agriculture. With the exception of the Pilgrims, however, the colonists did not. Apparently, the Indians preferred to rotate their fields instead of fertilizing to maintain crop productivity. After the colonists had expanded their farmlands, the Indians did not adopt European methods of fertilization, which perhaps reflects a cultural resistance to that practice.[28]

The Indian agriculturists east of the Mississippi River seem to have been similar. They cleared land by girding trees and by removing brush with the aid of fire, axes, and mattocks. No one can state with certainty how much acreage these farmers cultivated. The communal fields no doubt varied in size, depending on the number of people who needed food, the work force available for agricultural labor, and the fertility and openness of the land. In early-seventeenth-century Virginia, for example, the communal fields often varied from twenty to two hundred acres in size. Yet the Kechoughton Indians cultivated as many as three thousand acres of corn. Productivity varied, depending on soil fertility, rainfall, plant disease, and insect attacks. Without a doubt, though, the Indians were skilled agriculturists who not only raised a substantial portion of their annual food but also provided a surplus for times of crop failure or for trade among themselves and white settlers. In 1614, for example, the Chickahominy Indians in Virginia agreed by treaty to supply the nearby colonists with one thousand bushels of corn annually. The Indians required payment in iron hatchets. The Europeans recognized that the Indians were knowledgeable farmers, because one observer noted that in the tidewater region of Virginia, "wherever we meet an Indian old field or place where they have lived, we are sure of the best ground." The colonists appropriated Indian fields, because such lands only needed minimal clearing. Or if those fields had been abandoned, the colonists knew that the soil was good, because the Indian farmers did not waste their time cultivating poor lands.[29]

Soils were exceptionally rich in some areas east of the Mississippi River, but the southern Indians probably could not cultivate fields in the uplands for more than a decade before crop yields would begin to decline because of soil exhaustion. When that happened, the Indians had little choice except to let their fields lie fallow or to clear new lands and start over. Most of the southern farmers confined their agricultural endeavors to the "riverine" areas of the region in order to avoid yellow and red clay soils that were difficult to till with flint and wooden tools and that lost fertility quickly when cultivated.[30]

On those riverine soils, Indian farmers used two agricultural methods—intercropping and multiple cropping—to achieve maximum yields from their small fields along the river valleys. By intercropping, they planted several cultigens, such as corn, beans, and squash, together in the same fields. Thus, they used all of the available space and enabled the crops to complement each other. The beans, for example, replaced the nitrogen that the corn withdrew from the soil. The cornstalks provided a place for the beans to climb on, and the squash vines covered the ground and stunted the growth of weeds. By using several plantings, these farmers could seed two crops in the same fields, if the growing season permitted that practice. They used this method primarily for planting early and late varieties of corn.[31]

For the eastern Indians, corn, squash, and beans superseded all other crops in importance. The men prepared the soil, but the women usually had the responsibility for planting, weeding, and harvesting the crops. The women also were responsible for cultivating the small fields or garden plots. The technology of these farmers was also similar. Mattocks or hoes, made from sticks, bones, and flint, and wooden dibble sticks remained their primary tools, even after white traders had provided iron tools. For example, the French and Spanish in Natchitoches, Louisiana, provided iron hoes to the Pascagoula Indians, but the latter preferred their lighter wooden mattocks instead. These various tribes also harvested their corn and prepared it for eating in a similar manner. They distinguished several strains of corn based on color, shape, and size, and they recognized two broad categories—early and late. They picked the early corn while it was still green and roasted it on the cob, and they used the late-ripening corn for bread.[32]

During the late seventeenth century, some Indian farmers also began to raise livestock, which they acquired from the English colonists. They did not take particularly good care of their beef cattle, because they did not put up hay or maintain pastures. An abundance of game may have hindered their livestock-raising endeavors, but one should not be too critical of the Indian farmers in this respect, because many white farmers also neglected their livestock. In any event, the integration of livestock raising into Indian agriculture took time, just as did the adoption of the plow.[33]

Without a doubt the Indians made important contributions to the agricultural development of North America. They showed the Europeans which seeds to plant on which lands. The Indians also taught the white farmers the importance of protecting the corn crop from rodents and from the weather by storing it in cribs. Today the agriculture of the eastern Indians continues to have a profound influence upon cultivation and diet. White farmers still raise tobacco, although it is a different variety from the one that the Indians taught the Virginia colonists to cultivate and use. Corn, beans, and squash are commonly intercropped in gardens. Moreover, gardeners often hill their corn during the cultivation process, just as the Indian farmers were doing long before European contact. Hominy, cornbread, dumplings, succotash, and beans remain common foods, while corn on the cob is a household term.[34]

Although no one is certain how much agriculture contributed to nutrition, farming, together with hunting and gathering, helped meet the food needs of the Indians east of the Mississippi River. After the arrival of the Europeans, Indian farmers adopted additional crops, such as peaches, figs, and watermelons. In time they also accepted the European practice of raising poultry, swine, cattle, and horses. Domesticated animals and fowl did not, however, supplant wild meat in their daily diet until the nineteenth century. In the meantime, Indian farmers cultivated their crops in a centuries-old manner, while they resisted increasing pressure for their lands by white farmers. In the trans-Mississippi West, Indian farmers also continued their traditional agricultural practices at the time of the first contact with European civilization.[35]

4

The Trans-Mississippi West

When the Spanish explorers reached the American Southwest, they were particularly impressed with the extent and quality of Pima, Yuma, and Pueblo agriculture. The southwestern farmers did not cultivate fruit trees, but staples such as corn, beans, and squash met their needs and provided sustenance for the Spaniards as well. Francisco Vásques de Coronado, for example, like Hernando De Soto in the East, relied upon Indian agricultural products while his party spent two years (1540–42) searching for Quivira. As the Spanish explored the Southwest, they observed the Pueblos, who farmed without the aid of irrigation. In contrast, they watched the Pima irrigate their fields and perhaps saw them raising turkeys. Along the lower Colorado River, they noted that the Yuman-speaking tribes planted quick-maturing crops that tolerated high temperatures on the alluvial flood plain.[1]

The Spanish also observed that cotton was a major crop in the Southwest, and Coronado noted that the Indian farmers cultivated it from the Rio Grande in the south to the Hopi country in the north. As in prehistoric times, these farmers cultivated the most rapidly maturing variety, because scant precipitation in the lower elevations and cold temperatures in the high country gave the region the shortest growing season on the North American continent. The Zuñi Pueblo, for example, planted cotton seeds about an inch and a half deep in irrigated gardens. Then they scraped dirt borders around the cotton plants to help retain water from hand irrigation. They watered the cotton plants for three days, then allowed the plants to dry for three days, at which time they began the irrigation cycle again. This procedure continued until they harvested the crop in September, when the Zuñi and Pima farmers preferred to snap the bolls from the plant, rather than to pick the fiber. Then they dried the bolls in the sun before they separated the fiber from the pod by flailing the bolls between two blankets.

Next, they removed the seeds, sand, and trash from the cotton before it was ready for spinning. The Hopi particularly were known for producing a large amount of cotton, much of which they traded to the Zuñi. The nearby Pima also grew a considerable amount of high-quality cotton. These tribes considered the cotton harvest to be women's work. After the arrival of the Spanish, however, the cultivation of cotton declined rapidly, because the Spanish introduced sheep, which provided a warmer and more durable fiber. By the late nineteenth century, the southwestern farmers had nearly eliminated cotton from their agriculture. The Hopi continued to raise cotton, but they only used it for ceremonial purposes.[2]

After contact with the Spanish explorers and missionaries, Indian agriculture in the Southwest changed dramatically. During the seventeenth century, Spanish missionaries introduced wheat, oats, barley, chili peppers, onions, peas, watermelon, muskmelon, peaches, apricots, apples, cattle, sheep, goats, donkeys, horses, and mules—all of which the Indian farmers adopted over time. Indeed, the introduction of new crops and livestock did not immediately alter the nature of southwestern Indian agriculture. Rather, a long time passed before the various Indian farmers adopted new types of crops and animal husbandry. The Anasazi, for example, did not learn to raise livestock until after the Spanish colonized New Mexico in 1598; and as late as 1702 the Spaniards had difficulty in establishing cattle raising among the Indians in southern Arizona. The Spanish did quickly succeed in introducing cattle among the Indians in what is now the Mexican province of Sonora, but they were never successful in convincing the Indians there, or in the American Southwest, that swine raising was beneficial to their agricultural endeavors. Ignaz Pfefferkorn, who observed Spanish influences on the Indians during the mid seventeenth century, wrote: "The industry cannot be expected to develop in Sonora . . . because no one wants to be a swineherd. To expect a Spaniard to become one would be a sovereign offense. And no Indian can be induced to do it, not because his pride stands in the way, but because of his inherent, implacable hatred of swine. The animal is so abhorrent to him that he would suffer the severest hunger rather than eat a piece of domestic pork." Eventually, livestock raising would become important to the Indian farmers of the Southwest, but the lowly pig would not be part of it. The desert climate, of course, was unsuitable for raising hogs, but culture as well as the environment prevented the Indians from engaging in this agricultural endeavor.[3]

Although European crops became important, no one is certain about exactly which varieties the Spanish introduced. In the case of wheat, the Spanish left no record of the varieties that they raised or the Indian farmers adopted. The archaeological evidence, in the form of the straw used for making adobe bricks, indicates that the Spanish introduced two kinds of wheat early in the contact period. One variety was a bearded wheat called Propo (*Triticum vulgare graecum*); the other was Little Club (*Triticum compactum Humboldtii*). The Spanish missionaries carried these varieties into the Southwest from Sonora and Baja, California, and Indian farmers cultivated Propo and Little Club in the region throughout the Spanish and Mexican periods. Late in the Spanish period, some of the Indian farmers in California began to raise a bearded red-chaff, club wheat, now known as California Club (*Triticum compactum erinaceum*), which they grew until the early American period.[4]

During the late seventeenth century the Spanish began a major effort to introduce European civilization to the southwestern Indians by establishing missions in the province of Nueva Vizcays, which they renamed Pimeria Alta, or land of the Upper Pima. This included all of what is now northern Sonora and southern Arizona, from the Altar River in the south to the Gila River in the north and from the San Pedro River on the east to the Gulf of California and the Colorado River on the west. Within this vast region the Spanish organized the amenable Pima around the missions, to which the Pima were to donate three days of agricultural labor weekly. During the remaining three work days, they were to cultivate their own fields. At the missions, the priests distributed seeds, iron hoes, and axes, taught the Pima to raise cattle, and maintained oxen and plows for community use. The Pima sold a portion of the crops that they produced on misson lands to the silver miners in the area.[5]

At the time of Spanish contact, however, the digging stick was still the most important farm tool for the Pima, who fashioned it from branches of ironwood or mesquite. They also used an ironwood hoe with a saber-shaped blade, which they used in a kneeling position to cut weeds, to loosen the soil around the plants, and to dig irrigation ditches. After the arrival of the Spanish, the Pima began to fashion shovels from cottonwood trees and to craft fire-hardened three-pronged rakes from mesquite branches. The Pima did not use bone, horn, or stone for their agricultural implements.[6]

For the Pima farmers who had access to running streams or springs, planting time began after the danger from late-spring frosts had passed. The other Pima, who depended upon floodwater to provide enough moisture for their crops, had to delay planting until the summer rains ended, usually in late June or early July. Neither group planted later than mid August, however, because early autumn frosts might ruin the crops. First the Pima planted corn, teparies, pumpkins, bottle gourds, and cotton; then they seeded a second crop after the first one had been harvested. They also made single plantings of kidney and lima beans and chili peppers. The Pima did not plant any crop in a straight row, nor did they plant in exactly the same spot year after year.[7]

Usually, the Pima cultivated their crops three times. Wheat needed cultivation only once, because it grew during the cool months of winter and early spring before the weeds could become well established. The Pima preferred to hoe the weeds before they became six inches high. If the weeds got out of control, however, the Pima pulled the larger ones by hand. They also gave the soil between the corn, bean, and pumpkin hills a shallow hoeing to discourage weed growth and to keep the ground from cracking and losing soil moisture through evaporation. The Pima left the cut weeds on the grounds, instead of raking and burning them. These agriculturists were careful not to overplant; consequently they did not need to thin their crops during the growing season to ensure a bountiful harvest. The Pima considered cultivation, like planting, to be men's work, although the women and children usually helped.[8]

Like the eastern Indian farmers, the Pima sent their children into the fields to frighten the birds away from the young plants, but they did not build watchtowers or platforms. They tried to prevent rabbits from eating their chickpeas and lentils by planting those vegetables in the middle of their wheat fields, and they attempted to keep the squash bugs off the vines by sprinkling the plants with ashes. After the Pima began to raise wheat, they sometimes burned the stubble to rid their fields of grasshoppers—a desperate technique that their white counterparts also utilized. Wheat rust also periodically troubled the Pima, but they had no remedy for this disease other than to pull up the plants in the hope that the smut or rust could be kept from spreading.[9]

In contrast to planting and cultivating, harvesting was considered to be women's work, although the men helped by pulling the cornstalks out of the ground. With the introduction of wheat, however, the division of labor changed, and the men primarily became responsible for harvesting that crop. Families within a neighborhood helped each other with the wheat harvest, or specially favored relatives were invited to take part in return for a share of the crop. The Pima considered the corn ready for picking when most of the stalks and husks had dried. After the arrival of the Spanish, the Pima began to cut the cornstalks at ground level with a knife, or they pastured their cattle in the fields of stalks rather than pull up the plants by hand. Sometimes the Pima picked and husked the ears directly from the stalks, but usually they roasted the unhusked ears, then sun dried the crop on their housetops for about ten days. Both the Pima and the Papago shelled their corn before placing it in storage. The most common method of shelling was to place the ears on a mat and then to beat or thresh them with a club. The women had the responsibility for storing the corn in basketlike granaries. With the introduction of wheat, the Pima used straw to make storage baskets that could hold from ten to fifteen bushels of grain. These large baskets, the walls or coils of which were three to five inches thick, were stored in separate buildings.[10]

In October the Pima also harvested pumpkins, which the women placed in storehouses where the squash kept without spoiling until February or even as late as April of the next year. These storehouses were actually pits about eight feet wide and two feet deep. An arrowwood roof, supported by mesquite poles, covered each pit. The Pima only stored the mature pumpkins. They cut the immature ones into narrow strips for drying, or they cooked and ate them immediately. Both the Pima and the Papago considered pumpkin seeds a delicacy, which they roasted and ate much as contemporary Americans eat peanuts.[11]

The Pima and Papago women harvested the tepary-bean crop in October by pulling the vines from the soil. Next they threshed the pods from the vines with a pole and allowed the pods to dry for several days; then they flailed the beans from the pods with a short stick and winnowed the chaff by using a basket. They allowed the beans to dry in the sun for several more days before storing the crop, according to color, in covered baskets or ollas. The tepary beans were an important crop for the Pima and the Papago, because the yield was high, and the beans tolerated desert conditions very well. If the tepary crop was not irrigated, it yielded from 450 to 700 pounds per acre; under irrigation, however, the crop averaged as much as 800 to 1,500 pounds per acre.[12]

Sometime between 1697 and 1744 the Pima began to raise wheat intensively; and by 1770 it rivaled corn as the most important crop. In 1774, one Spaniard wrote that the wheat fields along the Gila River were "so large, that, standing in the middle of them, one cannot see the ends, because of their great length. They are very wide, too, embracing the whole width of the valley on both sides." By the nineteenth century, the Pima planted two principal varieties, Sonora and Australian, in November or December. They did not begin to broadcast, or sow, their seed until they could prepare a uniformly pulverized seedbed with a plow. At first, when harvest time came in May or June, they pulled the entire plant from the soil. Soon, however, they adopted the Spanish technique of cutting grain with a sickle, which was used by both the men and the women. One bushel of seed returned from twenty-five to fifty bushels of grain, and the Pima were able to produce satisfactory crops by irrigating their fields, just as the Hohokam had done in that same area centuries earlier. After the Pima cut their wheat, the women carried it in baskets to the threshing floor, where it dried for several days before the women threshed the grain from the heads with their feet or with poles. Later, during the nineteenth century, when the Pima increased their wheat crop because of the American market, they used horses to tread the grain from the heads, because they raised too much to handle easily with a flail or by treading with their own feet. No matter how the Pima threshed their wheat, though, they winnowed it with baskets and sun dried it before storing it in baskets.[13]

The Pima and the Papago also planted tobacco between late March and early April, but only the old men cultivated it. The Papago planted tobacco in a basin, where the soil was silty, rather than among other field crops. The basin served to collect rainwater, so they did not irrigate the crop by hand. The Pima, on the other hand, planted tobacco on their irrigated lands by loosening the soil with a digging stick, then blowing the seed from the palms of their hands to scatter it. After the seed had been broadcast in this fashion, they harrowed it into the soil with a branch. The Pima weeded their tobacco once or twice, but they did not fence this crop, as the Papago did with cholla cactus. The Papago used the same plots each year for tobacco until the soil became depleted; then they cultivated another site until it, too, became exhausted. The Pima, however, rotated their tobacco crop. When the tobacco leaves turned yellow or brown, the Papago harvested the crop and dried it in the sun, whereas the Pima preferred to dry the leaves in their houses. When the tobacco dried, both tribes rolled and stored the leaves in ollas or gourds to protect the crop from moisture and insects. The Pima selected their tobacco seeds from the tallest plants that had the largest pods, but the Papago usually planted all the seeds they could obtain from the previous crop. The Pima did not plant tobacco annually, since one crop generally provided a sufficient supply for several years. Only the old men and occasionally the women smoked tobacco, because both the Pima and the Papago believed that tobacco not only made young men cough, but also made them lethargic and unable to withstand cold. Neither the Pima nor the Papago chewed tobacco.[14]

Cattle raising did not become important among the Pima until about 1820—long after they had begun to raise horses. Their slowness to tend cattle was partly because of poor range conditions in the Gila River area, which, in the absence of irrigation, did not provide enough grazing for large numbers of cattle, and partly because the Pima developed a tradition that upon the death of the head of a household, any livestock that had not previously been given away were to be slaughtered and eaten. This custom, of course, kept the number of cattle low. During the gold-rush years, the Pima gave greater attention to cattle raising, because they quickly saw a new market developing for meat and other agricultural products among the miners. Nevertheless, the Pima never developed large holdings of oxen for draft purposes, even though they increased their herds of cattle for beef.[15]

Instead of livestock raising, the Gila River Pima, who occupied the only land that had a dependable source of water, chose to raise crops that provided the greatest return for the amount of labor required. Consequently, they did not adopt many of the cultigens that the Spanish brought into the Southwest, especially fruit trees and grapes. The care of orchards and vineyards required knowledge and skill that they did not possess, and they apparently lacked the desire to learn those new techniques. Moreover, while the Gila River Pima adopted the Spanish practice of horse raising, they did not adopt the horse-powered threshing mill; in fact, the agricultural technology of the Pima did not change substantially after contact with the Spanish. The Pima adopted the iron hoe but changed its form to meet their own ideas about proper shape and use.[16]

Except for tobacco, the Pima did not practice crop rotation, because the silt from the river during flood stage periodically enriched the fields, and the problem of soil exhaustion never plagued them. Although the Gila River provided the needed water for irrigation, it also created the problem of salt accumulation, because the river water contained high amounts of alkali. The Pima apparently were skilled at solving this difficulty, because they periodically flooded their fields with a swiftly moving sheet of water which washed the salt away from the surface. Evidently, this procedure was effective, because the Pima reportedly claimed that they never had to abandon a field because of the build-up of salt.[17]

Like the Hohokam centuries before them, the Pima also used irrigation canals to transport water to their crops. Where the river water entered, the canals were about ten feet deep and four to six feet wide. The canals did not have head gates, although brush dams might have been used to stop or slow the water flow. Consequently, as long as the river was high, water flowed in the canals. Low earth borders surrounded each field and kept the water on the crops as it flowed from the lateral ditches. The Pima men were responsible for the repair work on the communally owned main canals, which they cleaned and repaired every spring; but each farmer was responsible for maintaining the lateral ditches that irrigated his own fields. To facilitate the irrigation process and to ensure a uniform distribution of water, the Pima leveled their fields by carrying baskets of soil to low spots or by flooding their lands when the Gila River contained large amounts of silt. The Papago also used ditch irrigation, if they lived by a running stream. Frequently they had to rely on floodwater farming for crop moisture, and occasionally they used ditches and dikes to convey water from arroyos to their croplands.[18]

The Pima irrigated their crops when necessary, usually from three to five times during the growing season. If the river was low, the land might receive only one irrigation prior to planting, and the moisture that was stored in the soil had to support the crops. Generally, the Pima irrigated most when the crops were growing and less after the plants had reached maturity. The corn crop needed more water than teparies, pumpkins, and melons. Usually, the Pima irrigated the wheat crop about four times. The only plants that the Pima and the Papago watered by hand were the bottle gourds.[19]

During the nineteenth century, each Pima family irrigated from one to five acres, although some who had acquired land through inheritance farmed as many as twenty acres. The fields were laid out in rectangular fashion along irrigation canals. Whenever the Pima prepared a new tract of land for irrigation, the men used digging sticks and shovels to clear the land and to excavate the irrigation ditches. Then, a committee of half a dozen men made land allotments to those individuals who had helped to dig the ditches. Among the Pima, the men also had the responsibility of irrigating the crops. During the late nineteenth century, Pima canals were capable of irrigating from thirteen thousand to sixteen thousand acres.[20]

By the mid nineteenth century, Pima farmers had increased contacts with white traders, who offered various wares and textiles at excessive prices. The Pima countered that practice by selling damp wheat to those traders. Because the traders lacked moisture-testing equipment, which is commonly used at grain elevators today, they frequently did not know that the grain was damp. The result was that the wheat both weighed and measured more than it should have, and the Pima therefore received extra money for their crop. Apparently, they were not much different from their white counterparts in other sections of the country, because some of the Pima also tried to get more money for less wheat by adding sand to the middle of their wheat bags or by concealing poor wheat under a layer of better grain. In any event, because of the traders' demand for wheat, the Pima were raising considerable quantities of it by 1850. Estimates of production and sales, though inaccurate, give some insight into the extent of Pima wheat production. In 1858, for example, the Pima reportedly sold 100,000 pounds of wheat to the Overland Mail Line, which had established a route through their region, as well as large quantities of pumpkins, beans, squash, and melons. The next year, they sold as much as 250,000 pounds of wheat to the firm, in addition to a large supply of other vegetables. In 1860 the mail line purchased an estimated 400,000 pounds of wheat from the Pima. The Pima still had more wheat than they needed, so they sold it to government and private teamsters operating between Yuma and Tucson. In 1861 the Pima sold some six hundred chickens as well.[21]

The Pima expanded wheat production rapidly to meet this market, because they adopted a new, although still primitive, technology. About 1850 they began using an ardlike plow to prepare the seedbed, instead of the wooden digging stick, hoe, and spade. This wooden plow was a European innovation, which the Spanish introduced into the Southwest soon after their arrival. Until the mid nineteenth century, the Pima had had no use for it, because the wooden plow would have enabled them to expand their acreage to such an extent that excess production would have been a major problem. Nothing would have been gained if they had produced surpluses that would only have spoiled in time. With the arrival of white miners and teamsters, however, those surpluses could be traded to the advantage of both parties. Like their other tools, the plow was fashioned from mesquite or ironwood. The tongue was made from cottonwood, and rawhide straps were attached to a yoke of oxen. The Pima covered the point of the plow with an iron strap to prevent the wood from splitting and to improve its cutting ability. A single handle, set at an angle of about fifty to seventy degrees, made at least some control possible as the point gouged a shallow furrow across the field. Wooden pins and wedges held the parts together. About 1880 the Pima acquired steel or iron plows, which enabled them to prepare a better seedbed with less effort.[22]

Thus, the effect of Spanish and, later, American cultural and technical influences on Pima agriculture was significant. Wheat provided a staple that soon superseded corn. Wheat, in turn, necessitated a new form of technology—the plow—which enabled the Pima to produce large crops to meet the demands of the newly developing market that the incoming white settlers created. New crops gave the Pima a more diverse and nutritious diet. The Papago, however, who were close relatives of the Pima, remained less advanced in agricultural ability largely because they lacked access to water. Generally, the Papago worked for the Pima farmers at harvest time and traded desert products, crafted items, and Spanish goods for additional agricultural produce.[23]

The riverine tribes of the lower Colorado River, such as the Yuma, the Cocopa, the Maricopa, and the Mohave, also were agricultural at the time of the Spanish contact, and the Mohave were the most skilled farmers of all the Yuman-speaking people. The riverine farmers depended upon the Colorado and the lower Gila rivers to overflow onto the flood plain, thus depositing silt and moistening the soil. They did not practice canal irrigation, but the Cocopa did control the water flow by constructing earthen walls along their fields to protect their lands from floodwater. They used an opening, or gate, in the walls to admit water to their fields. When the fields were sufficiently irrigated, the Cocopa closed the gates to prevent damage to the land or crops. After the water had receded and the ground had begun to dry, the riverine people planted their corn, bean, squash, and cotton crops. When the mud no longer clung to their feet, they stopped planting, because that was a sign the soil was too dry to permit germination. Because of the annual or semiannual floods, these farmers could not maintain permanent fields, because the water destroyed boundary markers or even the fields themselves.[24]

In 1775 the Spanish observed that the Yuma annually raised two crops of corn, as well as cowpeas, tepary beans, cantaloupes, watermelons, calabashes, peaches, onions, cauliflower, anise, and manzanilla. In the spring the Yuma planted tepary beans first, because the crop needed at least minimal moisture for a long period to foster germination. They did not thin their crop stand after it sprouted. The Yuman-speaking farmers also planted, cultivated, and harvested their corn in a manner virtually identical with that of the Pima. The men pulled the stalks from the ground; or if the ears were to be picked from the stalks, the women did the harvesting. The ears were always husked in the field and then carried home and stored on their housetops according to color. These farmers shelled only the seed corn immediately for storage in large jars or baskets. Later, they shelled the remainder of the crop as it was needed. To protect the ears from moisture and rodents, they covered the containers with grass or mud lids and placed them on a covered platform, which stood about six feet off the ground. These storage facilities were not as extensive as those among the Pima, because flooding discouraged the building of numerous platforms. During flood time, the riverine tribes frequently abandoned the river valley and moved to higher ground.[25]

The Yuma also pulled up the tepary vines, allowing them to dry for several days before trampling the pods and winnowing the seeds for storage in baskets. Both the men and the women harvested pumpkins and melons, but families reaped the wheat crop by reciprocal agreement. They threshed the grain from the heads by trampling, and they winnowed the wheat for storage in pottery jars or arrowwood baskets. The Yuma, in contrast to the Pima and the Papago, never threshed with a flail or club or by trampling the crop under horses' hooves. Generally, these people considered the harvesting and threshing of the wheat crop to be men's work. Among the Mohave, however, the women conducted all phases of the wheat harvest, from reaping to storing the grain. Usually, the Yuma women selected the seed for the next crop, although the men might aid in this process if they so desired. The Yuma attempted to keep the corn and tepary crops pure by sorting the seed by color and planting each in separate fields. They were more successful, however, in keeping their wheat varieties pure. All of this work enabled the Yuman-speaking farmers to raise from 30 to 50 percent of their food.[26]

Of all the farmers in the Southwest, the Hopi, the westernmost Pueblo Indians, were the best dry-land farmers. Corn was their most important crop. Although they did not know about cross-pollination, the women were skilled at seed selection, and they did not choose kernels from an ear that they believed to be a mixture of two varieties. Consequently, Hopi corn varieties remained fairly pure. Like the other farmers in the Southwest, the men planted, cultivated, and harvested the crop, but the women performed the tasks of shelling, storing, and trading it. The Hopi made their first planting of corn in mid April for harvest in July. The second crop was sown in mid May for harvest in late September and early October. Moisture from the winter rains was critical for sprouting the seeds and sustaining the crop until the summer rains arrived.[27]

The Hopi planted their corn crops on the flat land below the mesas, where the sand insulated the soil and prevented the moisture from evaporating. They also seeded corn where natural flooding occurred, such as where rain water poured over the cliffs and spread across the fields. Because water was vital, the Hopi planted their fields wherever it was available, even if the fields might be many miles from their villages. The amount of moisture in a given area determined field size, but the largest fields probably were no more than several acres. When planting time came, as many as a dozen seeds were placed in holes from twelve to eighteen inches deep. By using a large number of seeds for each hill, the Hopi ensured against total loss from destruction by field mice, cutworms, and sandstorms. Deep planting permitted the formation of a root system in moist soil, prevented early germination, and protected the sprouts from being washed away by heavy rains. Hopi farmers spaced the hills about three feet apart, and as the plants grew, they thinned the stalks to five or six per hill.[28]

The Hopi, like the other agricultural Indians, harvested a portion of the corn crop when it was green, but they allowed most of it to mature on the stalks. They harvested the corn by snapping the ears from the stalks, which grew only about five feet tall. Then, they carried the ears to the village, where they removed the husks for use as tamale wrappers. Next, the ears were spread in the sun to dry. When these tasks had been completed, the Hopi cut or pulled the cornstalks and tied them into bundles for livestock feed. When the ears dried, they were stacked by color, like cordwood, in the owner's house or storehouse. Customarily, the Hopi tried to store one year's supply of seed corn. This practice ensured at least a year's food supply, if for some reason a crop was destroyed. Like the other southwestern farmers, the Hopi did not practice crop rotation. Instead, soil fertility was maintained by spacing the plants far apart and by not planting in the same spot every year.[29]

The Hopi were also skilled at raising sunflowers, which they probably got from American traders or Mormon settlers in the nineteenth century and which they used in making a purple dye for basketry materials and textiles. To a lesser extent, the Hopi also used sunflower stems for various items that required light wood, such as ventilation hoods for bake ovens. Sunflower seeds were planted in early May, along with beans and watermelons. Because one or two rows were sufficient, entire fields of sunflowers were never cultivated. The Hopi kept the soil well tilled so that the plants could sprout easily, and they placed brush or stones on the windward side of the new plants to protect them from blowing sand. After the sunflowers had become well established, the Hopi gave them no other attention until the heads were full of seeds and ready to fall. Then they cut the heads and knocked them together to remove the seeds.[30]

Hopi tradition required that a married man live with his wife's family, and he took the seeds of his mother's house with him. He planted those seeds to raise a crop for his wife and her family, if they so demanded. After he had demonstrated his success as a farmer, he was allowed to plant the seeds that belonged to his wife's household. As a result of this practice, crop varieties were distributed throughout the community. Corn and beans were primarily the seeds from the man's household, while other vegetables and fruits were the contributions of his wife's family. Fruit trees were inherited through the women in this matriarchal society, where land also descended from mother to daughter. As a result, the Hopi developed a complex division of labor between men and women, one that is reflected by the emphasis on specialized crops in the various families.[31]

The Hopi revolted against the Spanish in 1680 and drove the padres from the missions, but they did not reject the cultigens that the Spanish had introduced to the Hopi agricultural system. Peach orchards, for example, appeared in great numbers, and fruit became an important part of their diet. Chili peppers, watermelons, and onions gave the Hopi a greater variety of foods, and safflower and coxcomb produced dyes that gave color to their bread, which heretofore had been white or grey-blue. Red and yellow bread gave visual testimony to Spanish influence on Hopi foods and agriculture. In contrast with the other Pueblo farmers, however, the Hopi never adopted wheat as an important crop.[32]

The agriculture of the Pueblos, particularly that of the Jemez and perhaps that of the Zia and Tewa, had a major influence on the farming practices of the Navaho. Indeed, the Navaho may have learned to farm from the Pueblos about A.D. 1500. Like the Pueblos, Navaho farmers located their fields where they could take advantage of streams or floodwaters to irrigate their crops. They, too, preferred sandy soil, which insulated the moisture in the ground below. And like the Pima, the Navaho tasted the soil before planting, to determine whether it was salty and unsuitable for cultivation. They tested potential field areas by planting clusters of corn. The areas that supported the best clusters then were chosen for the corn crop the following year. In the highlands the Navaho planted wheat, oats, and potatoes; and in the valleys they sowed vegetables, such as squash, melons, and corn.[33]

In April the Navaho cleared new lands for planting by burning the brush and trees. The men were responsible for this task, but they did not remove the stones from the fields. They also prepared the land for irrigation by building small dikes around the fields, so that rainwater would be held there. Sometimes they built dikes across arroyos to divert runoff onto the fields. Where perennial streams existed, the Navaho dug ditches to irrigate the land. In times of drought, they carried water in pots and in goatskins from wells and springs in order to save their crops.[34]

Planting time lasted from the arrival of the full moon in May until the appearance of the new moon in June. The leafing of the aspen trees and the blooming of the yucca also signified that seedlings would be safe from frost. Like the Hopi and the Zuñi, some Navaho relied upon the position of the sun to determine seedtime. Corn, their most important crop, was the first one planted, and they seeded several varieties based on color—white, yellow, blue, black, red, and variegated. Navaho farmers planted watermelon seeds at intervals of about one month after the corn crop was in the ground. They recognized two varieties of watermelon, the seeds of which they obtained from the Hopi—dark green melons with black seeds, and striped melons with pink seeds. Muskmelon and squash were planted next in sequence. Beans were the last seeds to be planted, preferably in the valleys.[35]

Planting was a communal event. In large fields, Navaho farmers began in the center and placed the seeds in a spiral row which progressed clockwise toward the edge of the fields. Or the corn might be set out in straight rows, with the planting beginning either at the edge of the field or at its center. Each variety of corn was planted separately after the Navaho men had used a digging stick, from either a kneeling or a standing position, to open the soil. The women followed, dropped the seeds into the holes, and covered them with loose moist soil.[36]

The Navaho planted melons and squash together, but they did not plant these crops among the corn rows, because they believed that such a practice would prevent the melons and squash from bearing fruit. They used digging sticks to loosen the soil, and they planted the melon seeds at a depth of about three inches. A basin was then hoed around the seeds to catch the rain water. The basins were spaced about twelve feet apart to ensure enough room for vining. A second crop was sowed later to provide a staggered harvest throughout the growing season. Before being planted, squash seeds were soaked in a sack containing wet manure. The sack was buried near a fire, and when the seeds sprouted, the Navaho transplanted them. An alternative method was to place only a few seeds in a hole to avoid crowding and a poor crop stand. Like the Hopi, the Navaho placed wind guards of brush and stone, about a foot and a half high, around their melons and squash to keep blowing sand out of the basins. They also planted beans separately by placing a half dozen seeds in hills spaced two to four feet apart.[37]

The Navaho weeded their crops two or three times before harvest with implements similar to those used by other farmers in the Southwest. They considered weeding men's work, but the women used the hoe if the men were absent. In contrast to many Indian farmers, the Navaho did not hill their corn during the weeding process, but they did thin the clustered plantings. Like other southwestern farmers, the Navaho did not fertilize their soil or rotate their crops, but they apparently did not experience problems of soil exhaustion. A major hazard for the corn crop was the cutworm, which the Navaho commonly picked from the roots. Grasshoppers also presented problems, and the Navaho eliminated them from their fields as best they could by driving their sheep through the rows of corn. The sticky lanolin of the wool trapped the grasshoppers that alighted on the animals. Once the sheep were out of the fields, the Navaho picked off and destroyed the insects. Squash plants were sprinkled with a mixture of urine and goats milk to prevent damage from chinch bugs and cockroaches, and scarecrows and watchers were used to keep pests out of the fields.[38]

In Navaho country, farmers harvested the bean crop first. In September, the plants were pulled from the soil and threshed with a wooden club or by trampling with feet. The Navaho women winnowed the beans by using a basket. Next, the Navaho men harvested the corn crop by pulling off the ears. Like the Hopi, the Navaho cut the stalks and bundled them as winter feed for the livestock. The ears were husked, either in the field or back at the family hogan, spread in the sun to dry, then placed in a pit, around which the men sat with clubs about four feet long. As the men beat the ears, the women raked away the kernels with their hands, placed the shelled corn in baskets, and winnowed it in the wind. When the first frost came, the melons and squash were harvested, most of which were cut into thin slices and dried.[39]

After the corn had been threshed and the melons and squash dried, the Navaho stored the crops in fire-hardened storage pits, which were lined with cedar bark, covered with branches or flat rocks, and sealed with about a foot of dirt. These storage pits averaged five or six feet deep and were carefully concealed in the fields or near the hogans. The shelled corn was either poured into the pit or placed in elk-hide or goatskin bags. If the pits were dry and the corn was properly cured, it could be stored for two years without danger of spoilage. The Navaho also stored uncut squash and watermelon in pits or in unused hogans; the squash was wrapped in cedar bark or cornhusks to prevent bruising and freezing. Commonly, beans were sacked and stored with the corn.[40]

After the arrival of Spanish settlers, the Navaho began cultivating peaches and wheat, which gave them a more varied diet and a more diverse agricultural system. The cultivation of peaches was emphasized in the Canyons del Muerto and de Chelly and in the Nazlini region of Arizona. Peach seeds were planted in the early spring or late autumn by cracking the pits that had been saved and dried from the previous year. Seeds were planted by the handful in holes dug about a food deep near a cliff or overhang for protection. The Navaho fashioned basins around the seeds, to collect rain water and to hold the water that they carried to the plantings by hand. When properly cultivated, irrigated, and protected from grazing cattle, peach trees bore fruit in three or four years. The Navaho apparently never learned to prune their trees, though they did trim off the dead limbs.[41]

The Navaho raised two varieties of peaches—clingstones and freestones—but preferred the former, because they were heartier and could withstand rough handling. When the peaches were about the size of marbles, watchers guarded the orchards against pests until the crop was harvested. When the crop ripened, about the end of September, the Navaho picked the fruit, removed the pits, and placed the flesh on the rocks to dry in the sun. Then, they stored it in buckskin or goatskin sacks. Peaches were in great demand by Indians who lived where orchards could not be grown, so the Navaho were able to trade this item for needed goods such as mutton, coffee, and sugar. Navaho peaches from Canyon de Chelly also provided food for the United States Army during the Navaho wars of the 1840s and 1850s. During the 1860s the army destroyed several thousand Navaho peach trees in a scorched-earth policy designed to force recalcitrant tribe members to submit to reservation policy. Soon after the Navaho had returned to their homeland in 1868, they quickly reestablished their peach orchards.[42]

Wheat was more important, because it provided food in summer, at a time when the Navaho were usually hungry. Wheat could also be grown at relatively high elevations, so the Navaho were able to cultivate more land than ever before. According to tradition, soon after 1860 the Navaho began to raise wheat, a smooth-headed or nonbearded variety of spring wheat that they planted in March or April. It was seeded in rows, with the hills a foot or two apart. The Navaho preferred to plant their wheat where mountain streams provided water for irrigation, but they did not plant it where floodwaters would inundate the fields. The wheat, which was planted in hills, could be weeded with a hoe, but the weeds had to be pulled out by hand if the seeds had been broadcast. When the wheat matured in July or August, the Navaho cut the heads with a knife and carried the harvest in a blanket to a cleared area, where they threshed it with clubs or under horses' hooves. Then it was winnowed by hand, to remove most of the coarse and heavy chaff, and winnowed again with a basket. The Navaho stored their wheat in pits and ground it into flour as it was needed.[43]

Soon after the arrival of Juan de Oñate's expedition into the Southwest in July 1598, the Navaho began raising sheep—a skill that they probably learned from the Pueblos, who, in turn, had learned how to raise livestock from the Spanish. These sheep were descended from hearty Spanish stock which quickly adapted to the desert environment. The long, straight wool of the Navaho sheep did not have much natural yolk, or grease, and it could be woven without washing. The Navaho grazed their sheep by day and corralled them at night to guard them from predators. The sheep belonged to the women, perhaps because the women were the chief herders or because they had the responsibility for weaving the wool into cloth. The women also sheared the sheep by hacking off clumps of wool with stone- or iron-bladed knives. Sheep raising substantially improved the diet of the Navaho, because the Southwest had little game to furnish an adequate meat supply; and mutton soon became a staple. The Navaho also captured horses that had escaped from Spanish ranches and added them to their livestock-raising practices. The men owned the horses, which they kept primarily as a symbol of wealth. By the nineteenth century, livestock raising, particularly of sheep, had superseded all other agricultural endeavors of the Navaho.[44]

To the North, Indian agriculture on the Great Plains resembled that east of the Mississippi River. Planting time came when the thickets of wild plum blossomed. Corn, beans, squash, sunflowers, tobacco, and watermelon were the major crops. In the Missouri River region of Nebraska, the farming tribes, such as the Pawnee, raised a dozen or more varieties of corn. Dent, flint, flour, and popcorn were common, and these farmers retained the purity of each variety as best they could by carefully selecting the seed and by planting the varieties far apart to avoid cross-pollination. The agriculturists of the central Great Plains also raised fifteen or more varieties of beans and at least eight varieties of squash.[45]

Among the Pawnee the men and women both cleared the land, but the women planted the crops in fields of one to three acres. Tough prairie sod forced them to plant on the soft creek bottoms and flood plains, where their scapula hoes could break and stir the soil. Like the other Indian farmers in North America, the Pawnee did not fertilize their fields; but unlike most Indian farmers, the Pawnee hoed their crops only once or twice. If they hoed twice, the second cultivation usually came in June, so that they could leave their villages in time for the summer hunt. When they returned in the late summer or early autumn, they harvested their crops, which were then dried and stored in caches.[46]

The Pawnee produced a corn crop estimated at 20,000 to 40,000 bushels annually. Production such as this certainly indicates a major dependence upon agriculture for their livelihood; but after they had acquired horses, at about the turn of the eighteenth century, their reliance upon agriculture decreased rapidly. Horses soon became a symbol of wealth as well as a means by which the buffalo hunt could be made more productive. As the horse became more important to Pawnee culture, agriculture received less emphasis. As the Pawnee became more mobile, they had less use for croplands, which tied them to a specific location. The Pawnee refused to adopt livestock into their subsistence patterns, because those animals could not keep pace with their mounts in time of need. Because the Pawnee did not cut hay, they had to move frequently during the winter to provide pasture for their horses, an activity that further hindered agricultural development.[47]

The Indian farmers in the central Great Plains also raised tobacco. They did so by first burning dry grass on the plot where the seeds were to be sown. This procedure killed the weeds and retarded further unwanted plant growth. Then, they scattered the tobacco seeds relatively heavily to produce a thick crop stand. Apparently, the Pawnee did not cultivate this crop once it had sprouted. About two months later, when the leaves appeared brown and ripe, the entire plant was pulled up and the unripe seed capsules were picked and dried separately. The Pawnee considered those seeds particularly flavorful for smoking. Next, the leaves were dried in the sun and crumbled into a fine powder, which resembled tea in texture. This tobacco, known as *Nicotiana quadrivalvis*, apparently had a pleasant flavor, because the Indian farmers to the east, who raised *Nicotiana rustica*, preferred the Plains' variety; and although the eastern farmers tried to trade with the Plains tribes for the seeds, the Plains farmers jealously kept the tobacco variety to themselves.[48]

The hoe was the most important farm tool among these agriculturists, but it differed from the implements that the eastern and western farmers used, because it was fashioned from the shoulder blade of a bison. Because bone hoes were not strong enough to break the tough, fibrous prairie sod, these farmers confined their agricultural activities to the alluvial bottoms. Because only the flood plains were farmed, the fields were small, averaging from one to four acres, and the women frequently traveled as far as ten miles to tend the fields spread along the creeks and rivers.[49]

The Mandan and the Hidatsa, who used these techniques and implements, were the most agricultural people in this region. They lived on the eastern fringe of the northern Great Plains, in what is now North Dakota, occupying permanent villages along the tributaries of the Upper Missouri, where they farmed the rich alluvial soil. In this portion of the Great Plains, agricultural activities began as soon as the ground thawed in the spring. When the ice in the Missouri River had broken up and the geese had appeared, the Indians cleared the ground with antler or willow rakes. The women, and sometimes the old men, cut the brush from the field site, spread it evenly across the ground, and burned it. This procedure softened the soil and made it easier to dig.[50]

Like other Indian farmers, the Mandan and the Hidatsa used digging sticks to loosen the soil and to poke holes for seeding the crops. Then they seeded corn in hills. The Mandan used the same cornfields every year, but they moved the location of the hills. This procedure helped to prevent soil exhaustion, because they too did not fertilize or practice crop rotation. In contrast, the Hidatsa planted their hills in exactly the same spots where they had seeded the year before.[51]

Like the Pawnee farmers in the central Great Plains, these northern agriculturists cultivated their crops twice. During the first hoeing, they broke the soil between the hills. The second hoeing, called "weeding the corn," also involved hilling the crop. Hoeing was an early-morning and a late-evening chore for the women, but the men frequently provided an armed guard to prevent the women from being killed by an enemy— particularly the Sioux. Other watchers guarded the crop from towers and chased the birds from the fields. By the early nineteenth century the Mandan had replaced their bone hoes with iron ones, which they acquired from traders and which were in common use in the late 1830s. Iron hoes gave the Mandan greater flexibility in choosing field sites, because areas could be cleared that had a heavier cover than could be handled with a bone hoe.[52]

Harvest time came in early August, when the women picked the squash, then dried and stored it in caches. About two weeks later, they harvested green corn, which they boiled or roasted for immediate use or which they dried, shelled, and stored in rawhide bags. In September or October the Mandan and Hidatsa women harvested the remainder of the corn crop after the ears had dried on the stalks as much as possible. The ears were picked and tossed into piles, then carried in baskets to the drying platform in front of the lodges. The outer husks were removed, and the ears were braided into a string, which customarily reached from the knee to the foot and back up to the knee—some fifty or more ears. This braid was the standard measure for ear corn. The Mandan and the Hidatsa used baskets to measure shelled corn. While the women husked the ears, they watched for exceptionally long, straight-rowed, good-colored ears, which they braided separately for seed. After the ears had dried on the platforms, those of lesser quality were threshed on the ground or on skins with a stick. Then, the cobs were removed, and the chaff was winnowed by dropping the kernels from an elevated place, such as a lodge roof or a drying platform. The Mandan and the Hidatsa stored the best ears in braids, but they placed the threshed corn in hide sacks, which they kept in the lodges for immediate use or poured into caches for storage.[53]

The Mandan and Hidatsa caches were jug-shaped, with narrow openings just wide enough to permit one person to enter. These pits were lined with grass, and the floors were covered with willow branches and grass. Dry willow ribbing and wooden pegs held the grass against the walls. In these pits the Hidatsa stacked braided ears against the walls; then they poured the shelled corn into the center and piled squash slices on top of it; finally they dumped more shelled corn onto the squash to help protect it from moisture. The openings were sealed with a hide cover, grass or another hide cover, and a layer of earth. The openings were concealed by raking ashes or refuse over the spots to hide the caches from any enemy. These pits lasted for many years and were used over and over again. Customarily the Hidatsa kept a two years' supply of seed corn in reserve with the aid of their caches.[54]

Both the Mandan and the Hidatsa planted squash in late May or early June after they had sprouted the seeds, which they did by wetting the seeds and mixing them with the leaves of sagebrush or buckbrush. Next, the seeds and leaves were spread over a grass mat which had been placed on a buffalo hide. Then, the hide was gathered up and suspended from a pole near the fire. After about three days, the seeds sprouted, and the women planted two pairs of squash sprouts per hill on a sloping site, so that the rain would not pack the soil and prevent the sprouts from growing.[55]

The squash harvest began just before the green corn was ready for picking. Like the Indian farmers elsewhere, the Hidatsa and Mandan sliced the squash, strung it on spits, and dried it on platforms. After three days of drying, the women strung the slices on dry grass cords and hung them in their lodges. On sunny days, the squash was taken outside for more drying. When the slices were thoroughly dry, they were placed in parfleche bags and stored in caches. For seed, the Hidatsa and the Mandan saved the largest squash, which they picked at the time of the first frost. The women removed the seeds, which they dried by a fire, sacked, and stored in caches. They also boiled the seeds and flesh of nonseed squash for eating.[56]

The bean crop on the northern Great Plains was planted after the squash had been set out; and both were planted on hillsides. When the crop ripened in autumn, the women pulled up the vines and dried them for about three days, at which time the beans were threshed from the pods by treading on them and then by flailing the remaining unopened pods with sticks. The beans were winnowed and dried for another day before being poured into sacks for storage. The Hidatsa recognized five varieties of beans—black, red, spotted, shield-figure, and white—which they planted, threshed, and stored separately to maintain purity. These women farmers were careful to save the largest and best-colored beans for seed.[57]

Each Mandan and Hidatsa family usually planted one variety of sunflower, such as black or white, in the spring. The crop was harvested when the petals fell away from the seeds and when the backs of the heads turned yellow. The women cut the heads before the seed dried and placed the heads face down on their housetops. As the seeds dried and loosened, the women threshed them from the heads with sticks and then stored the seeds in skin sacks.[58]

Like the Pawnee in the central plains, the old men of the Mandan, Hidatsa, and Arikara tribes also planted tobacco; the old men were the only ones permitted to smoke it. The young men were taught that tobacco smoke would make them "short winded" and "poor runners" in an environment where speed and endurance were matters of life and death. Tobacco was never raised in the same field with corn, because it had a strong smell that was believed to affect the corn crop and to turn the stalks yellow prematurely. The men planted tobacco in the spring about the time that the women planted sunflower seeds. A hoe was used to loosen the soil where the seeds were to be planted, then the area was smoothed with a rake. Next, a stick was used to mark rows, which were spaced about eighteen inches apart and into which the seed was sifted rather thickly. Then the men covered the tobacco seeds with a thin layer of soil. When the crop sprouted, they thinned it and scraped earthen hills around each plant with a buffalo rib.[59]

When the tobacco plants blossomed in mid June, the men picked the new flowers every fourth day, right before the blossoms began to seed, and dried them on the lodging floor. Just before the first frost, the men harvested the crop by pulling up the plants, which they hung and dried in their lodges. The men preferred to smoke the dried blossoms first, after which they smoked the leaves and the stalks. The farmers in the northern plains did not use any special procedure for selecting tobacco seed; they merely saved enough for the next year's crop.[60]

Of all their agricultural endeavors, corn raising was the most important, and the farmers on the northern plains raised several varieties—flint, flour, and dent. The distribution of these varieties depended upon the environmental conditions that governed the agricultural practices of the specific tribes. Flint corn, which matures earliest, is the heartiest type; the most northern tribes cultivated a variety that matured in about ten weeks. Most of these farmers, though, preferred to raise one of the five or six varieties of eight-rowed flour corn. This corn type was easier to grind, and it tasted better when eaten green than did flint corn. In the southern Great Plains, these farmers primarily raised flour and dent corns. No family planted more than two or three varieties, and a different member of the family planted each variety in a separate plot, which was spaced as many as one hundred yards from another to prevent cross-pollination as much as possible.[61]

Corn was important to the Great Plains farmers not only for sustenance but for trade as well, because the hunting tribes of the high plains
craved vegetables, particularly corn. Therefore, the agricultural villages
in the northern plains became important trading centers. By trading, the
agricultural tribes acquired buffalo meat, antelope hides, and flour made
from the prairie turnip. The Mandan, the Hidatsa, and the Arikara often
traveled into the high-plains region to meet and trade with the hunting
tribes, such as the Cheyenne and the Arapaho. A white observer noted
that on one occasion the Indians of a trading caravan were so loaded down
that they looked like "farmers going to the mill." Once the trading had
begun, it had the excitement and bustle of a "country fair." Only the northern farming tribes, however, traded substantial amounts of agricultural
products with the hunting tribes. In the central plains, the Pawnee, the
Omaha, the Ponca, and the Oto conducted little or no agricultural trade
with the tribes of the high plains or with their northern farming neighbors.
Among the northern agriculturists, this trade was beneficial but hazardous.
The Sioux, for example, who were good customers of the farming tribes,
were temperamentally unpredictable: the Sioux have been called the "Picts
of the Upper Missouri," an indication that any trade with them had to
be conducted with the utmost caution. With the arrival of the white fur
traders, the Indian farmers gained a new market for several hundred bushels of corn annually. In 1833, for example, the white fur traders at Fort
Union at the mouth of the Yellowstone River converted a portion of the
Mandan corn crop into a "fine sweet liquor." Thus the Great Plains farmers
were efficient, and they were able to raise at least as much as half of their
food.[62]

The harsh environment of the Great Plains did not prevent at least
some tribes from developing important agricultural economies. On the
northern plains, the Indians transformed corn from a warm-weather plant,
which required high daytime and nighttime temperatures during a growing
season of 150 days or more, to a tough, compact plant that matured in
60 days and resisted drought, wind, cool temperatures, and even frost.
No one can be certain how long it took these farmers to make this development, but it surely ranks as one of the greatest achievements in plant breeding. They also developed a similar heartiness with squash and beans.[63]

When the white settlers moved onto the plains, they were quick to
appreciate the achievements of the Indian farmers. The corn that the settlers brought with them did not produce as well as that which the Plains
tribes raised, and the whites soon adopted the new varieties. The Indians
also gave the settlers the beans that the white plant breeders developed
into the Great Northern variety, which became one of the major commercial field beans during the early twentieth century.[64]

With the coming of the white settlers, the Indian farmers began to raise new crops, such as potatoes, onions, and turnips. In the case of the Mandan, potatoes became a part of their agricultural endeavors about 1832, and this crop became a favorite food among all the farming tribes in the Upper Missouri region. After 1860 the Omaha, as well as some other tribes, raised more potatoes than corn. Of course, these Indian farmers encountered problems. The Plains agriculturists, like the white settlers who followed them into that region, battled two major problems—drought and grasshoppers—although early and late frosts also presented difficulties. Because they did not practice irrigation or have effective controls against insects, the best adjustment these farmers could make in face of natural disasters was to harvest their crops early to save as much as possible before the sun cracked the soil and withered the plants or before the grasshoppers descended from the sky and devoured the crops. One must also remember that the Plains tribes beyond the ninety-ninth meridian did not practice agriculture, because the lack of precipitation made cultivation too uncertain. Moreover, the mounted tribes west of the meridian were among the most nomadic on the continent; they preferred to hunt buffalo and to trade with the agricultural tribes on the eastern edge of the Great Plains for vegetable products, rather than to live in fixed villages and to farm for themselves.[65]

Agriculture and cultural preference enabled the tribes on the eastern edge of the Great Plains to develop fixed villages, because the crops provided a dependable food supply and ended the necessity of nomadic wandering in search of game, vegetables, seeds, and fruits. In time, surplus production made possible the development of an important trading system, not only among the nomadic hunting tribes of the high plains but also among the traders and settlers who migrated into the region from the east. Agriculture also provided a stable economic base, which gave Indian farmers time to develop complex political and ceremonial organizations and practices.[66]

When the Spanish arrived, then, Indian agriculture in the trans-Mississippi West was complex and efficient. The Spanish did, however, change the traditional characteristics of Indian agriculture. In the Southwest, Indian farmers began to plant wheat and barley in the winter, a time of customary idleness. Both crops required different planting, cultivating, and harvesting techniques, which the tribesmen learned from the Spanish or from other Indian farmers whom the missionaries had taught to cultivate in the manner of the Europeans. The Acoma, the Laguna, the Zuñi and the Hopi Pueblos, as well as the Navaho, adopted the Spanish practice of grazing sheep, whose wool they used for blankets. The Pueblos, however, did not give much attention to fencing, and the livestock continually damaged their fields. The introduction of cattle, horses, and sheep required greater crop productivity for feed and forage. Pack animals, such as donkeys and horses, stimulated agricultural trade and increased the value of the corn crop. When the Indian farmers in the Southwest learned that they could sell or trade their crops to the Spanish, they gave increased attention to farming for profit. With the arrival of the United States Army, traders, and settlers during the nineteenth century, commercial agriculture became even more important for many of the tribes.[67]

Certainly, European influences brought fundamental change to Indian agriculture in the West, as well as to other regions of North America, when Indian farmers adopted new crops, new livestock, and different methods of cultivation. The most fundamental influence that the Europeans had on Indian agriculture, however, did not concern crops, livestock, or cultivation techniques. Rather, it involved a concept of landownership that was completely alien to the Indian farmers, and this altered their agricultural system for all time.

5

Land Tenure

In contrast to our knowledge about the agricultural methods of North American Indians, our understanding of their land-tenure system is limited. At best, we can make several generalizations which apply to the farmers of the various regions. Part of the difficulty in studying land-tenure patterns is that historians and other scholars still use tribal conceptualizations when studying Indians. More correctly, one should think, not of tribes, but of small groups that are similar in language and custom and that possess political and social unity. These small groups are the individual villages that functioned autonomously within a cultural whole, much as the city states did in ancient Greece. These villages were self-governing, and each owned or controlled a particular territory.[1]

The Europeans did not understand this organizational structure; instead, they referred to the Indians who shared a similar culture as tribes, thereby giving the Indians, at least in the minds of the Europeans, far more political unity and much greater control over the land than the various villages or groups had in reality. The Europeans, and later the Americans, particularly the emissaries of the federal government, mentally preferred to collect the related nations or villages into larger groups, which they called tribes, because they preferred the convenience of dealing with a few representatives of a "tribe," rather than with representatives of numerous villages. As a result, white civilization treated the tribes as sovereign nations or states, thereby endowing them with a de facto system of land tenure which the Indians never claimed or even understood.[2]

Before this development occurred, the various tribes, to use the European concept of the Indians' cultural relationships, developed different solutions to problems of land tenure. Those solutions varied from region to region. In the northeastern United States, for example, the Indians adopted two forms of land tenure. First, villages claimed sovereignty or exclusive ownership over an area that other bands recognized. When the soil, firewood, or game became depleted near the village, the Indians moved to another location within their territory. They might not return to that village site for generations, if ever. Second, in contrast to the communal ownership of a large amount of land, another concept of land tenure involved the individual control of the gardens and fields within the general territorial boundary. Among the Iroquois and the Huron, for example, the family's lineage controlled and cultivated the land. Uncultivated land separated the fields of each lineage. The eldest woman of each lineage exerted overall control of the land, although the other women had usage rights to particular plots. Each lineage retained the right to use those fields as long as the village remained on the site and as long as the women cultivated the fields. When the village moved, the cultivation rights to a specific parcel of land expired, and the village chiefs, or headmen, allotted new fields to each lineage, based on the size of the lineage and its food needs.[3]

If an Iroquois family felt safe from attack, its members might not live in the village but instead would reside outside its confines on the land that the family farmed. In this case, each family also controlled the land that it cultivated, and the village probably claimed adjacent land which the family intended to farm at a later date. Although this system of land tenure—that is, the private control of land—could present problems of conflicting claims, land was plentiful, and disputes over the individual control of land probably were rare.[4]

Thus, ultimate land tenure depended upon village sovereignty over a particular area, while the immediate individual control of a field depended upon the actual occupation and use of the land. If a plot of land was cleared of trees and brush and planted with crops, it was automatically removed from the communal domain as long as the family continued to use it. One might call this principle "squatter's rights." If a field was abandoned, however, it reverted to communal or village control, after which it could be freely allotted to another lineage or could be claimed and farmed by someone else. Communal lands that were not under cultivation could be used by all village members for hunting, fishing, berry picking, or wood gathering. Generally, land could not be sold or inherited. This concept of land tenure can be termed "use ownership," because land belonged to an individual farmer for as long as he or she cultivated it. The effect of this system of land tenure was to ensure an equitable distribution of property, so that every family would have enough field space to meet its needs; at the same time, it prevented the development of a concept of land hoarding for wealth and status.[5]

Overall the Indians in the eastern United States believed the Great Spirit had given the land to them; and because the land was a gift from him, only he could take it away. Chief Black Hawk expressed this philosophy of land use best when he said: "My reason teaches me that land could not be bought or sold. The Great Spirit gave it to his children to live upon, and cultivate as far as necessary for their subsistence; and so long as they occupy and cultivate it, they have the right to the soil—but if they voluntarily leave it, then any other people have the right to settle upon it. Nothing can be sold, but such things as can be carried away." Black Hawk, of course, was referring only to the transfer of lands among the Indians; he did not mean that whites could claim unused tribal lands for their own use. According to Black Hawk's view of land tenure, land could be ceded temporarily to white men as a gift; the Indians considered this to be a reciprocal transaction that did not involve compensation for the use of land. In this sense, there could be no land transfers from Indians to whites in the same sense that we now understand land sales. Indeed, Indian land could not be sold, because it did not belong to the present generation, which was acting only as trustee of the land for the generations yet unborn. Consequently, the land belonged only temporarily to the generation presently inhabiting it, but it could be loaned to other Indians who needed it, subject to their good behavior. In this respect, then, the Indians could neither accept nor understand the white man's concept of land sales and absolute ownership.[6]

Among the Iroquois and the Wyandot, the women who farmed the land controlled it. Agricultural plots were sometimes marked to forewarn others not to intrude. Iroquois women, for example, set up poles, which they had painted with the clan's totem or with their name sign, around the boundary of their melon fields. This marking indicated ownership, and although any member of the clan might pick the melons, the individual whose name was indicated had prior right to the fruit. During the late eighteenth century, the Shawnee, who lived in what is now southern Ohio, occasionally combined their fields with those of close relatives. The women, and sometimes the men of the clan, joined together to plant, weed, and harvest the corn crop. According to this system of land use, the crop was divided among all clan members by some leader who assigned individual family plots before the corn had matured.[7]

Land tenure among the southern Indians resembled the land-use pattern in the Northeast and the Midwest. No fixed rules existed concerning the size of the plot or field that an individual member or family might cultivate. Each village member could clear and farm as much land as he or she desired. As long as the land was cultivated, village custom protected it from encroachment by others. Indeed, no one would think of seizing the land of another; but if a field was abandoned, anyone could claim it and begin to farm. In this respect, no one could claim absolute right or title to a portion of land, because it ultimately belonged to the village that owned it communally.[8]

In the South as well as in the North, the women had immediate control of the fields that they cultivated. After the arrival of the Europeans and the subsequent loss of lands to them, the southern Indians began to strengthen their principles of communal ownership. Just before the American Revolution, for example, the Cherokee prohibited individuals from selling land to a colonist without the approval of the headmen, who spoke for the entire Cherokee nation. The intent was to prevent the sale of land, and it gave sanction to the concept of ultimate communal ownership. In reality, however, communal property rights meant very little to the Cherokee—except for their hunting grounds—and the concept of private land control predominated among these agriculturists. Indeed, even after contact with white civilization, when a Cherokee woman married a white man, she continued to control the land that she cultivated—in sharp contrast to the legal constraints that denied white women the right to own property.[9]

Thus, in tribal terms, which knowledgeable Englishmen and Americans should have understood, the Cherokee occupied the land and farmed individual fields, but the right of use theoretically was subordinate to the nation's fee-simple title to the land. Moreover, no one could claim more land than he or she could cultivate. This principle prevented the accumulation of land for its own sake and forestalled the development of a large landed monopoly by any individual or group. Before the arrival of white civilization, then, the principles of land tenure among the Cherokee were both communal and private, but the concept of private control, based on possession and use, was, in reality, the more important.[10]

On the Great Plains the garden plots along the rivers and streams belonged to individual households. Among the Pawnee, each plot belonged to a woman for as long as she cultivated it; but if she stopped cultivating it or died, the land could be claimed and used by someone else. Strips of unused land divided the fields to prevent disputes over the boundaries. Among the Hidatsa, those strips were about four feet wide. Each woman had the responsibility for hoeing the weeds that sprouted on her half of the dividing line. Sometimes the owner of one plot would ask a neighbor for permission to cultivate the median, or boundary strip. If that permission was granted, the woman had the responsibility of keeping the weeds off the entire earthen divider. In contrast to the land-tenure system of the Northeast and the South, each household retained the right to its land even though the fields might lie fallow.[11]

Among the Plains tribes in general, land was inherited from mother to daughter. In the case of the Hidatsa, if relatives did not cultivate the land after the death of the owner, anyone was free to claim it. Protocol required, however, that the woman request permission to do so from the deceased's relatives. The dead woman's son, mother, daughter, husband, or sister could grant that permission, although the entire family probably consulted over the matter. If another woman took the land, the transfer was not one of sale. Rather, it was granted as a gift.[12]

In the desert Southwest, land tenure differed slightly from that in the northern and southern portions of the contiguous United States. There, individual males could acquire land in four ways. First, the headman of a village might give land to an individual as compensation for work on an irrigation canal that brought new lands under cultivation. Second, a man could apply to the headman of the village for a plot of land. Third, a man might develop a piece of unclaimed land for himself without a specific grant or allotment having been made to him. Last, he might acquire land through inheritance. Inheritance and individual effort were the usual methods for obtaining land. The allotment method may have been developed or stimulated by the reservation system.[13]

In any event, the size of the fields or holdings depended upon need and the amount of land that an individual could farm effectively. And this depended upon access to water. Prior to contact with the Spanish, Pima and Papago farmers cultivated from one to five acres of land. After the arrival of the Spanish and the introduction of new technology, those individual holdings sometimes increased to fifteen acres. Similar to the lands of Indian farmers in the East, Pima and Papago lands belonged to the tribe communally, but the fields were individually controlled, and the patriarchal head of the family determined how the fields would be utilized. In contrast to the eastern Indians, the women in the Southwest usually did not own the land. Their labor in the fields, however, gave them a right to a portion of the crop and the freedom to dispose of it without permission from their husbands.[14]

Each individual who controlled land that he did not need could loan it to someone else. Like the Indian farmers east of the Mississippi River, however, the Pima and the Papago did not recognize the right of sale. In the case of land disputes, two men who knew both parties settled the disagreement. In the event that the farmer of a plot of land moved to another area, he gave a close relative the right to cultivate his fields. In return, the relative usually gave the farmer a portion of the crop. When the farmer returned, he resumed cultivation of the property. If the move was permanent, the relative traditionally gave a gift, such as a horse or cow, to the farmer, which formalized the transfer of the right-of-use to the land. These transfers only occurred among close relatives, and the family always tried to maintain control of its traditional lands.[15]

Besides control, inheritance was an important aspect of Pima and Papago land tenure. Since title to the land was vested in the men, it was inherited through the male line of the family. Theoretically, the entire family inherited it, and the land remained undivided as long as the widow lived.

During that time, the eldest son was in charge of the farming operation. When the widow died, the male heirs divided the land, although an unmarried daughter had the right to claim a portion of it. A daughter who married a poor farmer also might be given a portion of the land to help her out. If a couple was childless, the nearest relatives of both divided the land among them; but if a woman returned to her family after the death of her husband or a divorce, she forfeited all rights to his land. Usually, unmarried daughters only had the right to a portion of the produce from the land. When they married, they lost that benefit, because this Indian culture assumed that husbands would provide for their wives.[16]

Among the Yuman-speaking tribes, land was controlled individually rather than communally. Although lands were inherited patrilineally, anyone could claim and clear unused fields. The chief or headman did not grant allotments, but an individual usually asked permission to expand a plot if it was in an area where his neighbors also might want more land. Because no communal restrictions were placed upon the land, it could be freely traded howsoever the owners pleased. Among the Mohave, earthen ridges with arrowweed fences marked the field boundaries.[17]

The Yuman women—that is, the daughters of a farmer—theoretically had the right to inherit land or at least to make some claim to it. If a woman did not marry, someone would plant a portion of her father's land for her. The widow also had a "presumptive right" to use a portion of the land. Upon the death of the farmer, though, the land usually was allowed to remain fallow for a season or two, although this custom seemed to vary, because the Mohave felt free to cultivate after a death as soon as the new planting season arrived. Among the Cocopa, whose lands were particularly susceptible to flooding, boundary markers were of little value, so they did not formalize any concept of landownership, nor did they develop strict rules of inheritance.[18]

In contrast to the Indian farmers east of the Mississippi River or even to the other ones in the Southwest, the Yuman-speaking agriculturists used brute force to resolve land disputes. When the tribes of the lower Colorado River area were unable to settle disagreements over land peacefully, they resorted to a ritualized form of combat. If one farmer claimed a portion of his neighbor's land, relatives and friends of each party would gather at the disputed site. Then each rival would grasp the other at the shoulders or waist and, with the aid of his supporters, try to push the other back across the point of dispute. Whenever one of the rivals was thrown to the ground and the knee of the other was placed in his stomach, the dispute was settled and the new boundary line between the fields was drawn there. Both sides might try to push their claimant across the disputed area, while the other side resisted and tried to shove or drag its claimant in the opposite direction. The object of the struggle was to be the first one to reach the farthest end of the contested land. In either case, once the roughhousing was over, if the defeated party agreed to the outcome, everyone considered the dispute settled. If he did not, however, then both groups met at the site the next day to settle the matter with clubs. Formality limited the size of the clubs to between two and three inches in diameter. Women did not participate in the fighting, which was too rough: broken ribs and arms were common results of the fighting. The individual who lost the club fight had to move away or give up the land, because the winning party arbitrarily fixed the new boundary as he saw fit.[19]

Among the Hopi and the First Mesa Tewa, the headmen divided the arable land within the territory that the village claimed and apportioned it among the clans. These allotments might vary in size from several hundred square yards to more than a square mile. The lands of each clan were specifically marked with boundary stones, placed at each corner or at the junction point where the fields touched the land of another clan. These stones or rock slabs might have the clan symbol carved on their faces. No markers were used to define family fields within clan lands. Disputes were rare, because each clan claimed several fields that were not contiguous with those of the other clans. This landholding system also helped reduce the risk of crop failure, because if one field failed for some reason, another located elsewhere still might provide a harvest.[20]

In contrast to the Yuman-speaking farmers, the Hopi women controlled the land, which they inherited through a matrilineal system. In addition, whenever a woman's daughter married, the mother gave her a portion of land. Still, absolute control always resided with the clan, and the individual women merely possessed the right to use it. While a woman's daughter inherited the land, the men, who took the primary responsibility for cultivating the fields, owned the livestock and passed that property to their sisters, brothers, or other clansmen at the time of their death. Land disputes were rare, and a man would not demand the recognition of his rights to cultivate a field if another infringed upon it. The reluctance to do so, much in contrast to the Yuman-speaking farmers, was based upon a superstition that if the Hopi men did press their rights to the land, they would shorten the lives of their children or themselves. If a dispute did occur, the senior woman of the clan, who had control over the lineage and its lands, settled it. Large families were given the right to cultivate the surplus lands or unneeded fields of smaller families. No rent or interest was required for that use.[21]

Hopi men sometimes owned land in their own right. They could do so if they cleared wasteland and planted fruit trees or crops on it. Because wasteland was considered part of the village's land or territory, it was not under the control of a particular clan. Upon the death of the owner, this land either reverted back to the village or passed to a son or to the dead man's relatives. If this land was not cultivated continuously, another person might ask permission to farm it; but the owner could renew his right to cultivate that plot at any time. If the owner never reclaimed the land, however, everyone considered it to belong to the man who was now cultivating it. If a clan became extinct, a related clan took control of the land and merged it with its own. Or the fields might be passed to the woman who cared for the last survivor of the clan, as payment for her services. Thus, while Hopi land tenure was based on rigid principles of matrilineal inheritancé, in practice it was flexible to meet the needs of a particular family or clan.[22]

Among the Zuñi Pueblo, the concept of land tenure based on female ownership and inheritance also prevailed; but as among the Hopi, this principle was not an absolute rule. A man worked the fields of his wife's parents, but he might also cultivate land that he had obtained from his relatives. If a man did own land, his daughter's and sister's children inherited it. Usually, a man who owned land gave a field to his sister's son when the young man was old enough to cultivate it; he also gave a field to his sister's daughter at the time of her marriage. Zuñi women, however, always had first claim to the land, because they were expected to remain with the clan all of their lives. Matrilineal inheritance kept the family lands together under one household, thus preventing fragmentation. Occasionally, the Zuñi family might loan land to another family that needed more crop production. These loans did not require payment or reimbursement. In contrast to most Indian farmers in the East or the West, the Zuñi recognized the sale of land among their own people, and they transferred it from one family to another on a permanent basis for payment in sheep, horses, blankets, or turquoise beads.[23]

Land tenure among the Havasupai also differed slightly from the traditional concepts of ownership among the Indian farmers in North America. Havasupai families and individuals controlled the fields by right of use, but in the case of serious disagreement regarding patrilineal inheritance, the headmen of the tribe intervened to settle the dispute. Usually the hardest working and neediest sons were given preference in inheritance disputes. In addition, any land that a Havasupai controlled but did not cultivate could be occupied and farmed by any land-poor family. After a period of time the community recognized the new farmer or family as the legitimate user of the land. The individual farmer or user had no absolute voice in the inheritance deliberations of the family. Instead, the principles of relationship, personal merit, and need—which were public or communal conceptualizations of land tenure—were preeminent considerations for the inheritance of land. Relationship tended to keep land within the family, but the concepts of merit and need ensured the right of the community to transfer land to others. As a result, a strong feeling of community rights among the Havasupai kept the concept of patrilineal land inheritance from being an absolute rule of their land-tenure system. It also ensured an easy transfer of land to a family that was in need of more field space. Grazing lands were the communal property of the village.[24]

Among the Navaho also, the control of land was based on use. Whenever a man or woman cleared a tract of land, he or she was assured of the right to use it thereafter, and that property passed to the clan. If a man died, his sister's son inherited the land; but if the man did not have a nephew, the land went to a brother or sister. If a woman held the right of use, it passed on to her mother, sisters, or brothers at the time of her death. The first user of the land, or anyone who inherited it, had the right to farm it as he or she pleased. If a farmer did not plant the land or even moved away for a period of years, he had the continuing right to cultivate it upon his return, no matter whether someone else was farming it at the time. Generally, a farmer would ask a relative to cultivate the land if he was going to be away for a considerable time.[25]

The Navaho were careful to mark boundaries between their fields. They often planted a double row of corn—that is, two rows spaced eight to twelve inches apart—to mark the boundary of an adjacent field. Sometimes a row of yellow corn marked the boundary for a field of white corn. Usually, the Navaho tied the leaves of the two outside rows together to show control, or they picked the ears from the outside row first to help indicate the property line. After the introduction of livestock, the Navaho used fences to mark field boundaries. This development was intended to prevent grazing sheep, cattle, and horses from eating the crops, rather than being designed to indicate property lines. Like the Zuñi, Hopi, and Yuma, the Navaho used earthen ridges and rock piles to mark field boundaries. Those boundaries were laid off, not only to distinguish family lands from those of strangers, but also to keep the lands of a father, his sons, his sons-in-law, and his relatives and friends clearly distinguishable. No one could cultivate land beside a field that was already planted without the permission of the farmer who was using the land. If that permission was granted, the earthen boundaries enabled everyone to know exactly where their rights of use or occupancy began and ended. Not until the twentieth century did the Navaho practice the concept of buying and renting land.[26]

In contrast to the Navaho's concept of land-use occupancy for cultivated lands, they did not distinguish control of any kind for grazing land. Indeed, they regarded grazing land as belonging to everyone, and an individual was free to pasture his cattle or sheep wherever he wished. No one, however, would graze his stock on range land that was being used to full capacity by someone else. The Navaho also used water communally for stock watering as well as for domestic purposes. Only with the coming of white civilization did the Navaho farmers learn the concept of private ownership of springs.[27]

Overall, then, several generalizations can be made concerning the nature of Indian land tenure. Title to a general territory was a group right, not an individual right. North American Indians usually did not think of private property as an absolute individual right or consider land as a commodity that could be bought, sold, or permanently alienated in some fashion. Simply put, the community owned the uncultivated land, while the individual created a control or use claim by cultivating a specific field or plot of land. If arable land was plentiful, the individual's claim lapsed whenever the land became exhausted or abandoned. This characteristic of land tenure was common in what is now the eastern United States. In the Southwest, however, where the climate limited the arable land, an individual's claim to the fields remained valid, even during periods when the land lay fallow.[28]

Land tenure or control was not vested in an individual; it was vested with the lineage or household. The control unit could be either patrilineal of matrilineal, depending upon the particular culture. While a family or individual could claim additional land by clearing wasteland for cultivation, the lineage or clan had a paramount right to it. In this sense, tribal property was much like corporate property: every village member was a landowner, but he or she did not have absolute control of the land. An individual could not sell it, and among certain people, an individual could lose his right to the land if he left it untended. If an individual cultivated the land, however, he or she could do so indefinitely without interference from any other person.[29]

After contact with white civilization, various individuals or groups in a village occasionally gave whites permission to occupy their territory and to use certain lands. Those transactions usually involved the payment of a fee in some form. While whites almost always considered such transactions to be sales by which they obtained the right of exclusive ownership, the Indians invariably regarded such proceedings as being nothing more than temporary permits to use the land, pending compliance with the terms of the agreement. Because the Indians did not recognize individual rights to the land other than the right of use or occupancy, thus making absolute ownership by individuals an impossibility, the individual could not sell or alienate the land in any fashion. Moreover, because the village did not have sole ownership of the land any more than past generations had had absolute ownership of it, the tribal group could not alienate it either.[30]

While the Indians did not recognize permanent land transfers to whites, they seldom acknowledged land transfers from one Indian to another through sale. Usually, land was transferred through inheritance or as gifts. Furthermore, Indian farmers neither developed a system of land tenancy nor recognized the right of easements or the concept of mortgage. Consequently, they did not need to develop rules concerning the relationships between mortgagors and mortgagees or between landlords and renters. Still, the Indian farmers recognized such contemporary aspects of land tenure as trespass, squatter's rights, waste, ejectment, or reversion and escheat. The concept of trespass sometimes caused war among the tribes. Squatter's rights, while a white American term, was practiced in principle by Indians who cleared land and claimed it for their own use and occupancy. Among the Navaho, the concept of ejectment or reversion was recognized, because a person could reclaim his land after a lengthy absence, even though someone else was cultivating it at the time. The Indians also recognized the principle of escheat, because in the absence of heirs, the land in question reverted to the village for redistribution and use by others. Indian farmers, however, with the possible exception of the Havasupai, apparently did not recognize the concept of eminent domain, and the village did not interfere with individual or family land tenure.[31]

The Indian concept of land tenure enabled the various villages to make the best possible use of the land in order to meet their own specific needs. Each people or culture also developed a rational system for transferring land after the death of the owner. The major problem with the American Indians' concept of land tenure was that it differed from that of white civilization. Specifically, the problem of Indian land tenure was one of degree or scale. It was not that the Indians did not have a rational system for land use; rather, it was that they did not use their lands on a large-enough scale. Put another way, the problem of Indian landownership was not that they did not use the land for farming, but rather that they did not use enough of it. European and, later, American technology in the form of plows, cultivators, seeders, and planters, together with the aid of draft power from oxen and horses, meant that white farmers could cultivate more acreage than could the Indian farmer. The white farmers were also able to sell their produce in a market system that artisans, merchants, and others created at home and abroad. Market demand meant that commercial agriculture was possible, and white farmers wanted more land in order to earn profits that would improve their standard of living. The subsistence farming and limited agricultural trade of the Indians resulted from an inadequate or improper use of the land—at least in the minds of white farmers. Because Indian lands were undercultivated, whites were convinced that Indian lands should give way to a new tenure system that would make the land more productive. Soon after contact with white civilization, the Indian farmers along the eastern seaboard learned that landownership among the newcomers depended, not upon use, but rather upon "pen and ink." Moreover, for the new farmers, the possession of land was exclusive, and all transfers were understood as irrevocable sales. The various methods by which this new civilization acquired Indian lands ultimately put an end to agriculture as a way of life for the Indians between the Atlantic shore and the Mississippi River.[32]

6

The Alienation
of Indian Lands

With the arrival of European settlers, the Indian system of land tenure changed for all time. As the Spanish, French, Dutch, and English claimed territorial rights in North America, they imposed new concepts of property ownership on the Indians who had already claimed the land and were cultivating the soil, because the Europeans immediately decided that the newly found lands belonged to the nation that discovered them. This rationale did not give full recognition to Indian title to the land, even though the colonizing powers did not arbitrarily ignore Indian land claims or remove the Indians from the desired lands at will.[1]

At first, the European nations attempted to deal with Indian title to land in what they believed to be an equitable manner. Theoretically, Spanish policy required that payment be made for the acquisition of Indian lands and that a deed be issued. Although the Spanish did not uniformly enforce this policy and while they treated Christian Indian landholders more equitably than non-Christian ones, the Spanish government at least recognized Indian title to land by right of occupancy. By so doing, the Spanish did not officially sanction the acquisition of Indian lands by force of arms.[2]

In 1532, only forty years after Columbus had discovered the New World, Francisco de Vitoria, professor of theology at the University of Salamanca in Spain, declared that the Indians "were true owners, both from the public and private standpoint," of the lands that they occupied and used. Five years later, Pope Paul III legitimized Vitoria's philosophy in a papal bull when he wrote that the Indians "are by no means to be deprived of their liberty or the possession of their property, even though they be outside the faith of Jesus Christ." As a result, the Spanish soon recognized the right of Indian land title in the legal code that applied to the New World. In essence, that code required Spanish officials and citizens to respect Indian land title and to make restitution whenever Indian property rights had been violated.[3]

In Louisiana, for example, the Indians could not sell their lands without the approval of the Spanish governor. His ratification of Indian land sales made it possible for title to pass to the purchaser, but it also symbolized the crown's absolute title to Indian lands. In the Southwest, the Spaniards treated the Indians as though they were subjects of the crown—like other Spaniards—except for the matter of landholding. The Pueblos' title to land was based on the royal cedula of 4 June 1687, which assured the Pueblos that they would have title to as much land as they needed. The cedula also restricted Spaniards from holding land in Pueblo areas, which protected the Indians from encroachment by white farmers and their wandering cattle. In the Southwest the Spanish also recognized the right of the Indians to sell their land if they so desired; but it had to be sold at auction in the presence of a Spanish official. This policy did not adequately protect Indian land title, and by the late eighteenth century, the Indians had lost or alienated much land through sales, loans, and leases that they may or may not have made voluntarily. As a result, on 23 February 1781, the royal *audiencia* in present-day Mexico City adopted an ordinance that prohibited the Indians of New Spain from selling their lands. This policy was necessary, because the Spanish government feared that the Indians soon would not have enough land to support themselves and would therefore become wards of the state. Consequently, this ordinance required the Spanish government, the general council of the Indians, or the royal *audiencia* to license all sales, leases, and mortgages. The acquisition of a license depended upon proving that the sale was necessary, and no one could obtain title to Indian lands without obtaining a license. This policy, which required governmental approval for Indian land sales, continued into the Mexican and American periods of the region's history. Spanish land policy theoretically dealt with Indian land titles on an equitable basis; but in reality, Spanish bureaucrats paid little attention to the protection of Indian property rights. The Spanish did pay for some of the lands that they seized in the name of the crown, but they never consistently abided by a policy that regulated the purchase of Indian lands.[4]

In contrast, the French made no pretense of acquiring title to Indian lands through formal proceedings. The French never recognized the right of Indian land title for agriculture or hunting, and they may not have recognized the right of occupancy either. Rather, governmental policy allowed the French to take possession of an area, then gain the consent of the Indian people by means of a treaty that placed those Indians under the absolute dominion of the crown. Instead of making an attempt to deal with the Indians on an equitable basis, as the Spanish did, the French merely claimed the right of possession by force; they never entertained the idea of purchasing Indian lands. The French were primarily interested in the fur trade, however, so they did not infringe upon Indian agricultural lands and thereby create disputes over title.[5]

The Dutch, who settled New Amsterdam, differed from the French. From the beginning, the Dutch recognized the Indians' right of possession, and the Dutch crown was the first European sovereign to grant lands to court favorites after Indian titles had been extinguished by purchase. The Dutch, for example, bought Manhattan Island in 1626. Three years later, the Dutch government declared, in the Charter of Freedoms and Exemptions, that its citizens could not settle beyond Manhattan Island until they had purchased the land from the Indians. In 1629 the government required its colony of New Amsterdam to purchase land from the Indians whenever new colonies were to be established. Soon thereafter, Connecticut, New Jersey, and Rhode Island adopted similar laws.[6]

The English policy for dealing with Indian land titles was, at first, similar to that of the French, because the English claimed land in the name of the crown by right of discovery. When the matter of alienation of title to Indian lands became a problem, the government turned such affairs over to the individual colonies for resolution. Usually, the colonies tried to purchase Indian lands to maintain peace and harmony between the two peoples. At first the colonial governments invariably allowed individuals to purchase land from the Indians. The Indians, however, did not understand or recognize the legitimacy of those land cessions or transfers, and the colonial legislatures soon prohibited private purchases of Indian lands. The new policy was designed to prevent animosity and bloodshed between the two groups and to give the colonial governments absolute authority to acquire Indian lands in an equitable and orderly fashion by purchase, treaty, or lease.[7]

This policy indicated that the British government recognized Indian title to certain lands based upon occupancy. Still, it did little to alleviate the problem the Indians had with the colonial farmers, who kept taking their lands. A high birth rate and a high immigration rate created great demand for more farmland—land that the Indians claimed as their own. These new Americans believed that the agricultural system of the Old World, which emphasized raising diverse crops and livestock, was so superior to Indian farming that it was beyond question. The new immigrants also maintained that since the Indians had not improved—that is, farmed—all of the land that they claimed, the Indians had no legitimate title to it. Actually, the Indians did use all of the lands that they claimed, because by hunting and fishing, they were able to provide the protein necessary to complement the vegetable foods that they raised. The white newcomers could forego those activities because they raised their own protein in the form of cattle and hogs. Thus, the Indians had not abandoned uncultivated lands, as the whites believed; instead, the Indians put those lands to what was for them the best possible use to meet their nutritional needs.[8]

Although many of the new farmers did not have any qualms about acquiring land from the Indians either by squatting or by purchase, the Puritans struggled with the moral implications of such actions in the tortuous manner in which they justified other matters necessary to save their souls. The Puritans needed some rationale for occupying the land that they found themselves on, because they did not hold it by right of a royal charter. Consequently, their right of occupancy depended upon the best justification that they could devise, and their faith served them in a superb fashion as they developed moral, Biblical, and legal arguments for taking Indian lands.[9]

The Puritans were particularly perplexed over whether a white man could occupy land that the Indians already claimed. The Puritans naturally answered in the affirmative, by basing their right on two important principles. First, they argued that the Indians did not possess the land by natural right; second, they turned to the word of God as revealed in the Bible, which commanded man to occupy the earth. These ideological and theological principles provided the basis for the Puritans' acquisition of Indian lands. For the Puritans, the land belonged to those who labored on it. No one could claim vast tracts simply because he hunted over it. Moreover, the will of God was at hand in the acquisition of Indian lands. In early January 1633, John Winthrop wrote that "if God were not pleased with our inheritinge these partes, why did he drive out the natives before us? . . . why dothe he still make roome for us, by diminishing them as we increase." In 1647, John Cotton argued that the Indians could not claim vast uninhabited and uncultivated lands simply for the pleasure of the hunt. Those lands were a *vacuum domicilium*, which was free for the taking. Certainly, the Puritans practiced a different work ethic from that of the Indians, which prompted Cotton to proclaim: "We do not conceive that it is just Title to so vast a Continent, to make no other improvement of millions of acres in it." For the Puritans, uncultivated land or "voyd places" were open to white settlers. In time, the land-utilization philosophy of the Puritans became accepted American thought, and Andrew Jackson would sound little different from John Cotton in repudiating Indian land claims. Thus the Puritans' right to the land was based undeniably upon the word of God. As a result, they had the moral obligation to claim all lands that they both needed and desired, no matter what the Indian claims might be.[10]

At first, however, the Puritans did not put this philosophy into practice, and they were careful to purchase lands in order to avoid any confrontation with the Indians. Moreover, the Indians often willingly sold lands to the Puritans. Because the Indians retained the right to hunt those lands and in some cases to plant it, few problems between the two civilizations developed over the wilderness. In fact, Indian-white difficulties usually involved boundary disputes, rather than the Puritans' right to purchase tribal lands. In time, however, white pressure for Indian lands, as well as international intrigue for the control of the continent, ended these relatively harmonious relations regarding the acquisition of Indian lands in New England.[11]

The problem of acquiring Indian lands was not as complex in the Middle Colonies, because the royal governors followed a policy of purchasing large tracts from the Indians before granting smaller tracts to the English settlers. The proprietary colonies of Pennsylvania and Maryland, however, were plagued with the problem of individual purchase, and William Penn felt compelled to proclaim a policy whereby individuals could not buy Indian lands. Only Penn or his commissioners could acquire Indian land, and anyone who did not abide by his ruling not only lost the land he had bought but also was subjected to a fine of ten shillings for every one hundred acres purchased from the Indians. After New York became an English colony, the governor's permission was needed before anyone could purchase Indian lands; the purpose was to slow, if not stop, the white farmers' uncontrolled acquisition of those lands.[12]

In the South the royal governors' slowness to grant land probably stimulated the tendency of the settlers to squat on Indian lands or to purchase lands from the Indians. This policy provoked enmity and bloodshed; and in July 1653 the Virginia House of Burgesses proclaimed that the approval of the legislature was necessary before anyone could purchase Indian lands. This law was designed to protect the Indians from loss and to give some order to the acquisition of Indian lands. Generally, such laws were inadequately enforced, because geographical conditions made those policies difficult, if not impossible, to enforce.[13]

By 1756 the British Board of Trade had become interested in gaining greater control over colonial affairs. In regard to Indian matters, it attempted to do so by appointing Sir William Johnson and Edmund Atkin as superintendents of the newly created northern and southern districts for Indian affairs. Their reports soon indicated that the colonists frequently and fraudulently purchased Indian lands or occupied those lands without benefit of title. As the colonists continued to acquire or to take Indian lands, relations between the two groups became more strained, armed conflicts increased, and the threat of a general Indian war became more ominous. Faced with these problems, in early December 1761, King George III approved a plan of the Board of Trade, which affirmed the crown's intention to protect the property rights of the Indians in order to keep the peace. This plan instructed the colonial governors to refrain from grant-ing Indian lands to white settlers, and it held null and void all fraudulent and unauthorized purchases. These instructions also required the gov-ernors to send all applications for the purchase of Indian lands to the Board of Trade in London. Then the board would make a recommendation to the crown concerning the sale, and the crown would make the final deci-sion about whether the sale could be executed. This long and bureaucratic procedure was designed to remove the matter of land purchases from colonial control and to centralize that authority with the British govern-ment. Even more than this, the crown's new directive essentially attempted to bring the colonials under greater British control while simultaneously attempting to deal with the alienation of Indian lands in a manner that would not only be equitable but would also help keep the peace.[14]

This policy did not go into effect immediately. Not until the autumn of 1763, when the crown faced the need to quell Pontiac's Rebellion and to shape a coherent administrative policy for controlling settlement west of the Appalachians, including the matter of acquiring Indian land, did the Board of Trade activate the policy that it had designed two years earlier. In King George's Proclamation of 1763, the colonists were not only pro-hibited from purchasing Indian lands; they were also forbidden from cross-ing the Appalachians without permission or a license. The British govern-ment hoped that in time it would effectively be able to deal with the Indians, but it could never do so on an equitable basis if the frontier farmers continued to take Indian lands. This new British policy did not matter in the slightest to the Americans. Settlers continued to cross the mountains and to stake claims to lands that they considered to be free for the taking.[15]

The complicated problem of acquiring and protecting Indian lands remained unresolved as the Americans drifted into war with Great Brit-ain. During the course of that conflict, the Americans, as they created a governmental structure based upon the Articles of Confederation, also tried to gain control of Indian-white relations, particularly regarding the problems of the ownership and acquisition of land. Article IX was an attempt to do this; it gave Congress the "sole and exclusive right and power of . . . regulating the trade and managing all affairs with the Indians not members of any of the states, provided that the legislative right of any state within its own limits be not infringed or violated." This article did not, however, provide an effective governmental structure for dealing with the Indian problem, because the central government did not enforce it in the area of land disputes. At best the Articles of Confederation merely continued the English precedent of recognizing the Indian people's posses-sory right to the soil. But Article IX did imply that Indian lands could be legally acquired as the result of negotiations between the national government and the tribes and that transfer of title or alienation was possible by means of the cession treaty.[16]

On 22 September 1783, Congress attempted to clarify its authority under the articles by proclaiming that it alone had the power to prohibit and forbid "all persons from making settlements on lands inhabited or claimed by Indians, without the limits or jurisdiction of any particular state, and from purchasing or receiving any gift or cession of such lands or claims without express authority and direction of the United States in Congress assembled." Purchases or settlements without this authority were declared "null and void." While this proclamation clarified the power of the central government over Indian lands beyond state boundaries, it left the impression that the states could deal with Indian land problems within their own borders on their own authority, no matter what the policy of the national government might be. This proclamation portended trouble for the years to come.[17]

Upon the termination of the War of Independence, the new American government began to treat the Indians like conquered nations. The commissioners, who made treaties with the various tribes, arbitrarily drew the boundary lines and did not offer compensation for the lands that the Indians were required to cede. Although governmental officials promised to protect the newly defined Indian lands from encroachment by white settlers, in reality the Indians continued to lose their lands to speculators and pioneer farmers alike. As the government worked to gain land cessions from the Indians so that it could, in turn, sell those lands in order to pay its debts and provide farms for soldiers who had fought for the nation's independence, the matter of Indian land cessions became more complicated than ever before.[18]

This new policy reflected a total lack of historical understanding concerning the merits of past colonial and crown policies. Instead of recognizing the right of the Indians to the soil that they occupied and cultivated, the American government denied the Indians any right whatsoever to their traditional lands on the premise that they, along with the British, were a conquered people and that the victors, while willing to be magnanimous, would dictate land matters as they saw fit. Those solutions involved land cessions from the Indians who had committed the sin of having fought on the losing side during the Revolutionary War. The major problem with this policy was that the Indians did not agree to the very principles on which it was based.[19]

The Indians who were forced to cede lands to the new government viewed the War of Independence as no more than a continuation of their long struggle to keep the American farmers off their lands. They had not sued for peace; and they did not consider themselves defeated. Moreover, they had no intention of ceding large tracts of land without receiving any compensation. Thus, because of conflicting Indian/white views over land, the new nation had no assurance of peace at a time when peace was desperately needed. Governmental coffers were not adequate for sustaining another war, and the nation's military resources were too limited to bring an Indian war to a successful conclusion.[20]

By 1786 the Indian policy of the central government under the Articles of Confederation was in shambles. The tribes were resisting the execution of the treaties that required them to cede substantial portions of their lands, and a general Indian war seemed likely. Fortunately, Henry Knox, secretary of war, foresaw the impending danger on the frontier and the implications it had for the success of the new nation. He believed the central government had to return to the British and colonial policy of purchasing land from the Indians, instead of taking it by force. He recommended that retroactive compensation be given for the lands that the Indians in the Northwest had already ceded and that future cessions be purchased. This suggestion was eminently sensible, because the Americans would surely want or need more lands. Knox believed the government had three options for shaping Indian policy. First, it could terminate all plans for future acquisitions. Second, it could take as much land as it desired, provided that it had the strength to do so. Third, it could acquire lands in a manner that would keep the peace and satisfy both parties to the negotiations. By 1786 the first two alternatives were not viable, but payment for Indian lands would keep the peace and at the same time achieve the desired end of facilitating American expansion. No viable alternative existed: necessity dictated a change in Indian policy concerning the alienation of Indian lands.[21]

By late summer of 1787, Congress had turned away from a policy of coercion for the acquisition of Indian lands to one of conciliation and purchase. That policy, acknowledged in the Northwest Ordinance, promised the Indians that the new government would not take land without their consent. This policy had an added advantage—the northern tribes were familiar with it, because the British and the northern colonies had long used this procedure for the acquisition of Indian lands. In the end, of course, the American government still acquired the lands that it wanted, but it did so by negotiating cession treaties, rather than by force. Difficulties remained, however, because this policy was based on the assumptions that the Indians would be willing, if not eager, to sell their lands and that the frontier farmers would not encroach upon Indian property.[22]

Two years later, in 1789, the Articles of Confederation were replaced by the Constitution, which caused a reorganization of the governmental structure. Under the Constitution, the federal government held broad powers to regulate commerce with the Indians and to treat with them.

Accordingly, in July 1790, Congress attempted to regulate the alienation of Indian lands by passing the first Indian Intercourse Act. The intent of that act was to make the concept of negotiated purchase by means of a cession treaty a workable policy. The act prohibited individuals or states from purchasing Indian lands, unless those lands had first passed to the federal government in a cession treaty. By so doing, the act recognized the right of Indian title by possession. Both President Washington and Secretary Knox hoped that this policy would keep the peace and permit the orderly sale of Indian lands to the government whenever needed. By purchasing large tracts of Indian lands, the policy makers also hoped to force the Indians to accept "conventional American customs"—that is, farming—and to become a part of white civilization.[23]

In 1793, Congress strengthened the Indian Intercourse Act when it authorized a $1,000 fine and a year in jail for private parties who tried to purchase land from the Indians or who settled on Indian land. It was not until the signing of the Treaty of Greenville in early August 1795 that the federal government succeeded in winning the Indians' acceptance of the cession treaty as an instrument for transferring land title. Once the treaty commitments could be enforced, the cession treaty was a viable instrument for alienating Indian land titles.[24]

In 1796, Congress authorized the president to use force to remove settlers from Indian lands. This protective policy remained in effect until 1802, but the federal government was not intent at that time on guaranteeing the boundaries of the various tribes in perpetuity. Rather, the underlying assumption of the policy makers, such as Washington and Knox, was that the Indian territories would be gradually pushed westward, because the Indians would willingly sell their lands and retreat as the white farmers advanced. The federal government did not have the resources to keep the white farmers from encroaching on Indian lands, and the federal, state, and local governments did not have the will for the strict enforcement of the removal of whites from those lands. At best, the federal government could only make the withdrawal of the tribes and the cession of Indian lands as orderly and as peaceful as possible. As a result of federal Indian policy, clashes between the two civilizations were inevitable, and the new policy proved to be scarcely more equitable to the Indian landholders than it had been during the Confederation period, when lands were taken by right of conquest. Consequently, many white Americans began to adopt the attitude that the Indian problem would never be solved until all of the tribes had been moved west of the Mississippi River.[25]

When Thomas Jefferson took office in 1801, the American frontiersmen were demanding additional land cessions from the Indians. Jefferson soon realized that the frontier farmers had no desire to incorporate Indian agriculturists into the white American cultural milieu. Still, he mistakenly thought that the Indians would be willing to accept the benefits of white civilization, sell their unneeded lands, and settle down to a peaceful life of agricultural endeavors. He was convinced that land held in fee simple would enable the Indians to become acculturated and assimilated into American society.[26]

This policy of cultural adaptation would free large tracts of Indian lands for white farmers; and in February 1803, Jefferson urged Benjamin Hawkins, agent to the Creek nation, to promote agriculture among his charges. Jefferson wrote: "This will enable them to live on much smaller portions of land. . . . While they are learning to do better with less land, our increasing numbers will be calling for more land, and thus a coincidence of interests will be produced between those who have lands to spare, and want other necessaries, and those who have such necessaries to spare, and want lands. This commerce then, will be for the good of both, and those who are friends to both ought to encourage it." With this reasoning, Jefferson could justify the acquisition of additional lands for the purpose of national expansion. The Indians would benefit incidently from the loss of their lands, because they would become small-scale farmers.[27]

About that same time, Jefferson expressed a more callous view to William Henry Harrison by noting that the policy of the federal government was to lure the Indians to agriculture and to encourage them to cede their lands. Jefferson wrote: "To promote this disposition to exchange lands which they have to spare and we want for necessaries, which we have to spare and they want, we shall push our trading houses, and be glad to see then good and influential individuals among them run in debt, because we observe that when these debts get beyond what the individual can pay, they become willing to lop them off by a cession of lands. . . . In this way our settlements will gradually circumscribe and approach the Indians, and they will in time either incorporate with us as citizens of the United States or remove beyond the Mississippi."[28]

Jefferson was invariably an optimist about his Indian policy, and he noted in his annual message to Congress in December 1805: "Our Indian neighbors are advancing, many of them with spirit, and others beginning to engage in the pursuits of agriculture and domestic manufacture. They are becoming sensible that the earth yields subsistence with less labor and more certainty than the forest, and find it in their interest from time to time to dispose parts of surplus and waste lands for the means of improving those they occupy and of subsisting their families while they are preparing farms." Jefferson, of course, was wrong in his assessment. Instead of amicably settling down to an agricultural life as white civilization knew it and instead of gratuitously ceding unwanted lands, the Indians became increasingly reluctant to part with the lands that they held for agriculture and hunting.[29]

An indication of the growing desire of the Indians to retain their lands came as early as the summer of 1801. At that time, members of a Cherokee delegation told Secretary of War Henry Dearborn that while they believed the United States government did not want to take their lands, they knew who did want to—the "frontier people." The Cherokee leaders poignantly asked how they could be expected to follow the wishes of the federal government and increase their agricultural activities if at the same time they lost their lands to white settlers or if they had to alienate more land to the government. Indeed, the Cherokee and the other southern tribes, such as the Creek and the Choctaw, had reason to worry, because the state governments in that region were impatient to seize control of Indian lands for the benefit of state coffers and settlers alike. The pressure that Georgia placed on the Creek, for example, was so great that this nation requested the federal government to post soldiers on its land to keep the frontier farmers from taking tribal property.[30]

Jefferson still maintained his belief that the Indians and whites could live in harmony, but he also believed that the Indians would willingly sell unneeded lands. By so doing, both Indians and whites would benefit, and harmonious relations would be established. In November 1805 he told a group of Creek chiefs who visited him in Washington: "You will certainly find your interest in selling from time to time, portions of your waste and useless lands, to enable you to procure stocks and utensils for your farms, to improve them, and in the meantime to maintain your families. . . . We are a growing people, therefore whenever you wish to sell lands, we shall be ready to buy; but only in compliance with your own free will." One may, however, doubt Jefferson's sincerity. As early as 1800, when France acquired the Louisiana Territory from Spain, Jefferson recognized that the gradual purchase of Indian lands and the assimilation of the Indian people was proceeding too slowly to meet the defensive needs of an expanding nation. As a result, Jefferson began to seek title to all Indian lands east of the Mississippi River for the United States.[31]

James Madison continued Jefferson's policy of assimilating the Indians into white civilization by encouraging the tribes to emphasize agriculture and by obtaining Indian lands whenever those properties were available for acquisition. During Madison's first year in office, he told members of the Creek National Council that they should increase their agricultural production and encourage their women to weave and spin so that they would have the food, clothing, and comforts of white farmers. The federal government also pressed for the cession of land whenever it believed that it needed to do so. As the nation moved toward war with Great Britain and as many of the tribes, particularly those in the Old Northwest, joined the British in the War of 1812, many governmental officials began to accept the frontiersman's solution for the Indian problem: not assimilation, but removal.[32]

Actually, the idea of removing the Indians to the west side of the Mississippi River originated with Thomas Jefferson in 1803. Once there, he believed, they would be far from the corrupting influences of white civilization, and they would have time to learn agriculture and to become educated and civilized before they had further contact with white society. The Louisiana Purchase, which gave the United States a vast domain wherein the Indians could be placed, made Jefferson's idea viable, because it enabled the federal government to trade or exchange western land for Indian land east of the Mississippi River.[33]

The War of 1812 postponed the further development or implementation of this idea, but with the conclusion of the war, the removal idea gained new supporters. In an October 1817 letter to Andrew Jackson, President James Monroe wrote: "The hunter or savage state, requires a greater extent of territory to sustain it, than is compatible with the progress and just claims of civilized life, and must yield to it. Nothing is more certain, than, if the Indian tribes do not abandon that state, and become civilized, that they will decline and become extinct." In his annual message to Congress two months later, Monroe proclaimed that "the earth was given to mankind to support the greatest number of which it is capable, and no tribe or people have a right to withhold from the wants of others more than is necessary for their own support and comfort." As a result of Monroe's belief, the War Department intensified its efforts to convince the Indians to exchange their lands for those in the West. Monroe, however, believed the Indians could be removed only upon their own consent, because the government still recognized their possessory right to the soil that they actually occupied.[34]

The federal government did recognize the Indians' right of possession; at least Chief Justice John Marshall said it did. In the 1823 case of *Johnson* v. *McIntosh,* he wrote: "It has never been contended that the Indian title amounted to nothing. Their right of possession has never been questioned." Marshall agreed, however, that the federal government had "complete ultimate title" by right of discovery and that Indian lands could only be ceded to the federal government. By so contending, he failed to equate the right of possession with full fee title in order to prevent the Indians from freely alienating their lands through freedom of contract. Marshall hoped that his ruling would help to strengthen and protect a still relatively new nation. If the Indians could not freely sell their lands, the British and Spanish could be prevented from purchasing those lands, and future international conflicts over land claims might be avoided. Marshall's decision, together with increased pressure from the southern frontiersmen, encouraged the Cherokee and the other southern tribes to refuse to make additional cessions. His decision also encouraged the state of Georgia to devise methods by which the Cherokee could be removed from the state so that Indian lands could be opened for white settlement and development.[35]

Although some Indian tribes had exchanged their lands for other property west of the Mississippi River by the time Monroe left office, few Indians had either agreed to removal or had actually left their homes for new ones. John Quincy Adams tried to follow the Indian policy of his predecessors by negotiating land-cession treaties with the various tribes, but throughout his administration he had to contend with adamant demands from the people of Georgia, the Old Northwest, and the states farther west that the Indians be removed from their eastern lands. Adams tried to rationalize that policy by arguing that the rapidly increasing American population no longer permitted permanent boundaries on Indian territories. Otherwise the American farmers could not feed a rapidly growing nation. Adams said: "To condemn vast regions of territory to perpetual barrenness and solitude that a few hundred savages might find wild beasts to hunt upon it, was a species of game law that a nation descended from Britons would never endure." This rationale denied the possibility that Indian farmers might be able to help meet the food needs of the American people. In November 1828, Peter B. Porter, secretary of war, spoke for the Adams administration in his annual report when he suggested that western Indian lands "be apportioned among families and individuals in severalty, to be held by the same tenures by which we hold ours, with perhaps some temporary and wholesome restraints on the power of alienation." For those Indians who refused to give up their land and move, Porter suggested that the states arrange to partition those lands in severalty so as to promote agriculture. He also recommended that a separate tract of land be reserved to support Indians who had lost their private holdings through "improvidence." Those lands also would be subject to state law.[36]

Well before Andrew Jackson became president in 1829, he had concluded that the Indian policy of the federal government, based on the land-cession treaty, was not only an absurdity; it was also detrimental to the well-being of whites and Indians alike. As early as 1817, Jackson argued that the Indian tribes should not be treated like independent nations with whom the United States dealt on a treaty basis. He did not believe that the Indian tribes possessed sovereignty, and he argued that Congress had the right to legislate for them just as it did for the other inhabitants of the nation. Given this premise, Congress did not need to rely on the slow and cumbersome cession-treaty process to define the boundaries of Indian lands. Instead, Congress could prescribe the Indian land boundaries according to the needs of the nation and of the Indian people. Essentially, Jackson contended that Congress should exercise the right of eminent domain over Indian lands. This meant that the Indians had to be fairly compensated for any lands that the government took. Jackson maintained that such a policy was not only right but was also expedient, because the federal government now commanded enough power to enforce that policy, much in contrast to previous Indian policy, which mandated that peace be kept by maintaining the good will of the tribes through the cession-treaty process.[37]

Jackson based his philosophy on the treaty of 1783, which ended the War of Independence and which transferred sovereignty over the Indians from Great Britain to the United States. For Jackson, the Indians, under the new government, occupied lands by the "grace of the conqueror." Accordingly, the Indian people, who thereafter had existed as independent nations within state boundaries, had no other choice but to maintain their independence by moving west or to become subordinate to state law. After becoming president, Jackson enunciated these beliefs in his message to Congress on 8 December 1829, which contained words reminiscent of John Cotton nearly two centuries earlier. Jackson maintained that the Indians had no claim to unused lands, because it was "visionary to suppose that . . . claims can be allowed on tracts of country on which they have neither dwelt nor made improvements, merely because they have seen them from the mountain, or passed them in the chase." Furthermore, removal would end the federal/state clashes concerning sovereignty over the Indians. Georgia moved quickly to take advantage of Jackson's attitude and in 1830 passed legislation incorporating Cherokee lands into the state's political system. The legislatures of Alabama and Mississippi did so as well.[38]

Jackson's view of the Indian land problem was in some respects Jeffersonian, because he believed a transition from a hunting to an agricultural society would make peaceful coexistence possible between both civilizations at some future date and because such a policy would enable the Indians to support themselves as small-scale farmers with less land. To accomplish this, Jackson held that "they must be first placed beyond the reach of our settlements with such checks upon their disposition to hostilities as may be found necessary, and with such aid, moral, intellectual, and pecuniary, as may teach them the value of our improvements, and the reality of our friendship." The pecuniary aid that he wrote about involved "assistance to all who may require it in opening farms, and in procuring domestic animals and instruments of agriculture."[39]

After Jackson's 1829 message to Congress, both houses began to debate Indian-removal bills that provided for an exchange of lands. During the next six months, supporters of the bills once again reiterated the old arguments for the seizure of Indian lands. At the same time the opposition, primarily composed of Quakers, Methodists, and Congregationalists, vigorously argued against the bills. Those who supported removal ultimately prevailed; Congress approved a bill, and Jackson signed it into law on 28

May 1830. After this, the Indians might become farmers, but they would practice their art in the West. This change in policy came at a time when the Cherokee had nearly fulfilled every aspect of Jefferson's dream.[40]

Although Jackson prevailed in winning congressional approval for his Indian policy, the Supreme Court proved less agreeable. Several years earlier, in 1827, the Georgia Legislature retaliated against the Cherokee within the confines of Georgia when it extended state law to cover all Cherokee lands. The Cherokee sought to restrain the state through the judiciary, and the result was the 1830 Supreme Court case of *Cherokee Nation* v. *Georgia.* At that time, Chief Justice John Marshall held that the Cherokee nation was neither a state in the union nor an independent nation; rather, it was a "domestic dependent nation" under the sovereignty of the United States government. Moreover, the Cherokee unquestionably had the right to possess their lands until they chose to alienate title by voluntary cession to the federal government. By so ruling, Marshall reaffirmed the Court's decision in the *McIntosh* case, which supported federal supremacy by right of discovery. The next year, in the case of *Worcester* v. *Georgia,* Marshall also held that Cherokee lands were a distinct political entity within which "the laws of Georgia can have no force, and which the citizens of Georgia have no right to enter but with the assent of the Cherokees themselves or in conformity with treaties and with the acts of Congress."[41]

John Marshall thought that the Indians had not only the right of possession and title to their lands but also the right of self-government. All cessions to the federal government had to be voluntary, and only the federal government could acquire Indian lands. Until such cessions occurred, the federal government had the obligation to protect the lands of the dependent nations from outside encroachment. Jackson, who did not agree, chose not to enforce Marshall's decision. He realized that to have enforced Marshall's decision would have required the use of the army to remove white settlers from Cherokee lands, and no one, least of all Andrew Jackson, wanted to see white soldiers shooting white farmers to ensure the right of Indian farmers to retain their lands.[42]

In retrospect, the acquisition of Indian lands by the removal process cannot be attributed to maliciousness on the part of Andrew Jackson. Jackson merely established the mechanics of a policy that many had advocated and that had been partially implemented long before his presidency. Although he had been an Indian fighter, he was concerned about the well-being of the Indians as well as the safety and security of the white population and the nation. Jackson had few viable alternatives on which to base an Indian policy that would accommodate the land needs of both white and Indian civilizations. He could have followed the policy of extermination, which the frontiersmen suggested; but few responsible officials, Jackson included, could countenance such a method to solve the Indian problem. Or he could have followed the advice of the humanitarians, who urged patience while the Indians became assimilated into white society; but such a policy had not yet come to fruition, and its success was very much in doubt because of the resistance on the part of both cultures. He might have, as the opponents of the removal bill suggested, provided reservations east of the Mississippi River, but Jackson did not believe the army had sufficient strength to keep the land-hungry whites from encroaching on Indian lands, even though those lands were guaranteed. Confronted with these choices, Jackson, like John Quincy Adams before him, saw removal as the only feasible policy. By removing the Indians from areas where they created jurisdictional problems with state governments and by granting them fee-simple title to lands far removed from white civilization, they could gradually adopt the skills and traits that would enable them to assimilate into white culture. Or if they chose to do so, they could retain their traditional way of life.[43]

This Indian policy was a paradox. Because most tribesmen did not adopt the social and economic trappings of a new civilization quickly enough, policy makers decided to send them into the trans-Mississippi West to enhance the process. This action meant that segregation became the method to achieve the goals of acculturation and assimilation. Whites coveted Indian lands more than they wanted integration with the tribesmen, and at the same time, federal officials argued that as long as the Indians retained communal control of their lands, they would remain in a state of savagery. Policy makers did not see removal as the abandonment of their goals to acculturate and assimilate the Indians on the basis of agriculture. Yet if the Indians stayed, the officials maintained, their fate at the hands of the frontiersmen would be far worse than removal and the cession of their lands. Consequently, from 1830 to 1834, Jackson persuaded many of the Indian tribes in the North as well as the South to accept federal policy and move west. Willingly or not, they began to withdraw from their eastern lands.[44]

Some of the Indian farmers among the Five Civilized Tribes absolutely refused to emigrate, and the federal government was compelled to accommodate them by granting or allotting them fee-simple title to lands where they were residing. By so doing, the government established the precedent of allotting land in severalty—a process that foreshadowed ominous consequences for Indian farmers late in the nineteenth century. Actually, the policy of allotting lands in severalty—that is, granting land to individual Indian owners—can be traced to a 1633 order of the Massachusetts General Court, which declared that the Indians who desired to live among the English in an orderly and civil manner were entitled to have land allotments on an equal basis with the English settlers. A little more than two decades later, in 1655, the Virginia Assembly also allotted land to the

Indians, but that policy involved assigning lands to groups or tribes, not to individual farmers.[45]

The allotment policy developed further in 1814, when Andrew Jackson negotiated the surrender of the Creek nation. For those Creek people who had remained loyal to the United States during the War of 1812, the federal government ceded a large tract of Georgia land so that any chief or warrior could have a section of land—that is, 640 acres—for agriculture. The land would belong to him and his descendents for as long as they occupied it. This grant was not one of fee-simple title, because upon abandonment, those lands would revert to the United States. Although in March 1817, legislation changed the inheritance provisions of this agreement, the allotment policy remained in effect. A year earlier, in 1816, William H. Crawford, secretary of war, had supported such a policy, because he believed that the Indian farmers who were willing to become citizens should be allowed to keep their lands and become assimilated into white society rather than being forced to move away. Admittedly, Crawford and others in government did not believe that many Indians would choose to become full-time farmers, but that is not the point. The development of an allotment policy signifies that the federal government was embarking upon a course that would lead to an expanded policy of forced allotment designed to speed the process of agricultural self-sufficiency and of adjustment to white civilization.[46]

The federal government applied the allotment policy to the Cherokee in the treaties of 1817 and 1819 and to the Choctaw in 1820. At that time, the government granted 640 acres in fee-simple title to each head of a household, provided he lived on the land and cultivated it. Moreover, on 29 September 1817 the federal government negotiated a cession treaty with the Wyandot, the Seneca, the Delaware, the Shawnee, the Potawatomi, the Ottawa, and the Chippewa in the Northwest, which granted allotments to those tribes. This treaty provided that the members of the Wyandot, the Seneca, and the Shawnee nations were to share equally in the allotted lands. Each person who received an allotment, upon the approval of the chief, could in turn sell it with the permission of the president of the United States. With this treaty, the federal government not only took another important step toward the full development of an allotment policy, but it also provided a method by which more land could be obtained from the very people who were to cultivate it like white farmers.[47]

The development of an allotment policy, particularly in relation to the Cherokee, the Creek, the Choctaw, and the Chickasaw, unified the two contradictory principles of removal and assimilation. First, allotment essentially gave the Indians homestead privileges within a specific area before that area became available for white settlement. Governmental officials determined the amount of land, based on need, that the Indians could claim. Second, it also provided for segregation by removal. As a result, the government did not solve the problem of acquiring Indian lands on an equitable basis while it was providing an opportunity for acculturation based on agricultural development. In fact, the opposite occurred. In the case of the Creek, for example, instead of providing that nation with a more secure land title, the allotment policy became a device to remove them entirely from their traditional lands. Because the Creek men were unaccustomed to holding land title in the manner of the white farmers, many soon lost their lands to speculators and other intruders. Some of the Creek leased their lands to white farmers, while others found that white settlers had staked claims and had planted crops on the very lands that the government had granted to the Indians in the cession treaty, and the federal government did not have the will or the resources to remove the settlers. When the Creek tried to claim their allotted lands, they found that the white settlers were prepared to use force to defend their new farms. The result was that the Creek had little choice but to accept removal rather than to assimilate with the white farming community.[48]

In retrospect, the most important consequence of the allotment policy was that the eastern Indians lost their lands. Once the white settlers had established themselves illegally on Indian lands, they were difficult, if not impossible, to oust. Speculators and traders quickly made loans to the new Indian property owners whom the federal and state governments now recognized as having the right to enter into contracts to alienate land titles. For those loans, speculators and traders received the owner's written promise to sell his allotted property as soon as the exact boundary of that land had been determined. Moreover, governmental officials urged the Indians to sell. As a result, speculators soon acquired from 80 to 90 percent of the lands that had been allotted to the southeastern Indians. Thus, faced with the alternatives of emigration or starvation, the southeastern tribesmen who held allotments had little choice but to accept the removal policy and to emigrate west of the Mississippi River.[49]

More allotments followed. By mid century, Indian policy was based on the premise that the common ownership of land would forever prevent the Indians from assimilating into white society and that the allotment of land in severalty was fundamentally the correct way to solve the Indian problem, or more correctly, the Indian land problem. This is not to suggest that the removal and allotment policies were based on malice. Rather, these policies were predicated on the belief that removal and private property, preferably in the trans-Mississippi West, if fairly arranged, were in the best interests of both Indian and white farmers. The allotment policy was designed to treat the Indians like white men, because it recognized the Indians' right to private property as well as their right to dispose of it as they pleased through freedom of contract. Thus, removal and allotment, not assimilation and acculturation, became the domineering features of early-nineteenth-century Indian land policy. Removal and allotment would continue, but the geography and the tribes involved would change as white farmers desired new lands in the trans-Mississippi West. While the federal government was developing and executing cession, removal, and allotment policies for the alienation of Indian lands, governmental officials, missionaries, and religious groups were seeking to teach the Indians how to farm on their new or allotted lands.[50]

7

Agricultural Aid and Education

By the late eighteenth century, many white Americans were convinced that red Americans had little choice but to give up vast tracts of land and become agriculturists. If the Indians ceded all but a portion of their lands, they could be taught to farm like white men. In the process, the Indians would see the advantages of settling on small tracts of land and would welcome the security of an agricultural life. During the last decade of the eighteenth century the federal government responded to this belief by implementing a policy designed to make farmers out of the American Indians and thereby to help assimilate them into white society.[1]

As early as 1789, Henry Knox, George Washington's secretary of war, recommended that the federal government provide farm tools and livestock to the missionaries who were working with the various tribes. Washington agreed with this advice and instructed the Indian comissioners, who handled treaty negotiations, to add stipulations that would allow the missionaries to reside among the Indians. Although those missionaries could not own land or engage in trade, they were permitted a reasonable amount of land for cultivation. By so doing, Washington hoped that the missionaries would show the Indians by example how to till the soil and raise livestock. The activities of the Quaker missionaries probably encouraged Washington to believe that religious groups could help the federal government civilize the Indians, make them self-sufficient farmers, and give them a new beginning in white society.[2]

On 26 May 1796 the Quakers began to work among the Six Nations in New York by distributing farm tools and providing agricultural instruction. Although the Quakers planned an incentive program for the Seneca, which involved awarding premiums to the best farmers for producing large quantities of wheat, rye, corn, and potatoes, as well as honoring women for making cloth from their own flax and wool, they made slow progress. Indeed, the Seneca men demonstrated more than a little reluctance to adopt the white man's system of agriculture, because it required them to plow, sow, cultivate, and reap—activities that brought ridicule from both sexes. The missionaries, however, operated on the premise that Indian males should learn to farm, because heavy field work was not suitable for women. Consequently, the missionaries emphasized the "gospel of soap," that is, they tried to teach the Indian women to spin, weave, wash clothes, and oversee housekeeping. Once the women were removed from the fields, the men could take over the tasks of farming.[3]

The missionaries also established demonstration farms to help explain such matters as crop rotation, grazing management, and livestock care. Usually, they offered this advice in the spirit of brotherly love, but when the Indians appeared reluctant to take advantage of this instruction, force was sometimes used. The American Board mission among the Sioux, for example, withheld all farming assistance unless the men agreed to plow. The early missionaries did not, however, quickly or dramatically change Indian life styles. In fact, the Seneca women in New York would not stop farming, and their chiefs informed the Quakers that the women's "habits have been long fixed" and that much time would pass before they would become housewives while their men became farmers, each thereby accepting their ordained place in white, agrarian society.[4]

In 1790 and 1791 the federal government made treaty arrangements with the Creek and the Cherokee respectively, which obligated the government to provide the Indians with agricultural implements and instruction. In April 1792, Secretary Knox made a similar offer to all the tribes in the Old Northwest when he wrote: "We would be greatly gratified with the opportunity of imparting to you all the blessings of civilized life, of teaching you to cultivate the earth, and raise corn; to raise oxen, sheep, and other domestic animals, to build comfortable houses, and to educate your children, so as ever to dwell upon the land."[5]

In 1793, Congress encouraged the acceptance of that offer when it appropriated $20,000 and authorized the president to provide tools, livestock, and seed to the tribes that wanted to expand their agricultural activities or learn to farm. This legislation, known as the Second Intercourse Act, also provided for the appointment of agents to live among the tribes. The agents' duties included providing agricultural advice and instruction to the Indians; the agents or their employees were directed to teach the Indians to plow, sow, harvest, and tend livestock. Congress periodically furnished the agents with funds for the purchase of "useful domestic animals" and the "implements of husbandry." Federal officials believed this aid would create a desire for private property in land so that these gifts could be used to the greatest advantage.[6]

George Washington, then serving his second term as president, recognized that the Indians might accept the white man's agricultural methods only with great difficulty, but the potential rewards were great. In a message to the Cherokee in late August 1796, Washington wrote: "Some among you already experience the advantage of keeping cattle and hogs; let all keep them and increase their numbers, and you will ever have plenty of meat. To these add sheep, and they will give you clothing as well as food. Your lands are good and of great extent. By proper management you can raise livestock, not only for your wants, but to sell to the White people. By using the plow you can vastly increase your crops of corn. You can also grow wheat, which makes the best of bread, as well as other useful grain. To these you will easily add flax, and cotton, which you may dispose of to the White people, or have it made up by your own women into clothing for yourselves. Your wives and daughters can soon learn to spin and weave." Washington promised medals for those Cherokee who excelled in raising cattle, cereal grains, cotton, and flax and who spun and wove cloth. About 1798, Silas Dinsmoor, the Cherokee agent, provided direct aid by distributing three hundred plows and a large number of carding combs to the Cherokee. By that time a mill had also been established at government expense to encourage the Cherokee to raise wheat.[7]

The federal government's support of Indian agriculture increased when Thomas Jefferson became president. On 15 May 1801, Secretary of War Henry Dearborn clearly defined federal Indian policy in a letter to Return Jonathan Meigs, agent to the Cherokee Nation. Dearborn wrote that "the labours of your predecessors have been directed in instructing the women of the Cherokee Nation in the arts of spinning and weaving, and in introducing among the men a taste for agriculture and raising of stock: in all of these, their success has justified a continuance in the same plan, and it is therefore desirable that the same course may be pursued." In December 1801, Jefferson reported to Congress that Agent Benjamin Hawkins had introduced sheep among the Creek in Georgia, and that these people also were raising cattle, goats, and horses. Hawkins was also making good progress in getting the Creek to fence their fields and to use the fifty plows that the government had provided since 1797. In addition, Hawkins had established peach orchards, and he had aided the endeavors of the Creek to grow cotton, some of which they had marketed as far north as Tennessee. He also had taught them to raise flax, rice, wheat, barley, rye, and oats, as well as apples, grapes, and raspberries. In 1802, Congress appropriated $15,000 to promote "civilization" among the friendly tribes. Although the Treasury Department did not keep precise records at that time, John C. Calhoun, who was serving as secretary of war in 1822, believed those funds had been used primarily to purchase agricultural implements and livestock for the Cherokee, Creek, Chickasaw, and Choctaw.[8]

Under Jefferson the federal government also encouraged missionaries to spend part of their time instructing the Indians in the art of agriculture, which the various missionary groups were agreeable to doing. Agriculture would encourage the acceptance of private property, and it would provide a secure foundation for a self-sustaining church and an orderly civil government. Farming would also help to prevent starvation and to make life less tenuous during the winter. Most important, farmers could become good Christians, while hunters were "unfavorable to the regular exercise of some duties essential to the Christian character." Usually, the agents and the missionaries attempted to introduce white agricultural practices by giving plows, livestock, oxen, and dairy cows to the chiefs who accepted agricultural aid and instruction. Gristmills were also provided, which the missionaries hoped would encourage the Indians to raise grain.[9]

When Jefferson met with various Indian delegations, he, too, encouraged them to give up the chase, accept the plow, and enjoy the secure life that came with it. In early January 1802, Jefferson said to a group of Miami, Potawatomi, and Weeauk Indians who visited him: "We shall, with great pleasure, see your people become disposed to cultivate the earth, to raise herds of useful animals and to spin and weave, for their food and clothing. These resources are certain; they will never disappoint you: while those of hunting may fail, and expose your women and children to the miseries of hunger and cold. We will with pleasure furnish you with implements for the most necessary arts, and with persons who may instruct you how to make use of them."[10]

Jefferson believed not only that whites could teach the Indians to farm but also that they were duty bound to do so. In 1805, in his second Inaugural Address, he voiced that belief when he said that "humanity enjoins us to teach them agriculture." Had most of the Indian tribesmen in the eastern United States heard that remark, they would have, no doubt, been perplexed, because most of them already knew a great deal about farming. In any event, Jefferson made a considerable effort to send agents among the tribes for the purpose of teaching them to cultivate the soil and to raise livestock. In 1806, Secretary of War Henry Dearborn's instructions to Nicholas Boilvin, the newly appointed assistant agent to the Sac in Wisconsin, indicates the application of that policy. Dearborn ordered Boilvin to plant a garden and fruit trees and thereby to teach the Indians the "arts of agriculture." Boilvin was to distribute the seeds from his crops among the chiefs who promised to plant them, and he was to introduce plows "as soon as any of the chiefs will consent to use them." Two years later, when Jefferson met with a delegation of Delaware, Mohican, and Munsee Indians, he said: "Let me entreat you therefore on the lands now given you to begin to give every man a farm, let him enclose it, cultivate it, build a warm house on it, and when he dies let it belong to his wife and children after him." For Jefferson, private property and the agricultural life would not only enable the Indians to have the best possible life; it would also enable the Indians and the whites to live together in peace.[11]

By 1810 the southern tribes had received plows, hoes, sickles, scythes, and cotton cards, as well as other tools in lieu of cash annuities. Six years later, Thomas L. McKenney, superintendent of Indian traders, asked each trader and agent to inform him about the "particular implements of husbandry" that the tribes needed. McKenney, too, was convinced that acculturation and assimilation would come only with the acceptance of an agricultural life. Indian agriculture would also encourage an appreciation of private property. The Chickasaw agent was the only one to reply, but he optimistically reported that his charges were "very desirous to get plenty of good Corn hoes, light barshear or shovel ploughs, axes, grubbing hoes, and every useful implement of husbandry." McKenney provided those implements as part of the annual annuity payment, but the Chickasaw refused to accept them, instead demanding payment in cash. McKenney still believed that the Indians really wanted farming equipment and seed, so he continued to send those supplies to his agents and traders with the urging that they plant demonstration gardens. McKenney thought the Indians had to be "anchored in the soil" before they could become citizens.[12]

Not until March 1819 did the federal government develop a limited but orderly program for civilizing the Indians—that is, for making farmers out of them. At that time Congress created a Civilization Fund of $10,000, which was to be expended annually to teach the Indians to farm and become civilized like white men. Congress did not, however, create any administrative machinery for the dispersement of the fund. Instead, the Madison administration invited churches and other charitable organizations to request funds for the establishment of schools among the Indians, where the children would learn to farm, among other things. In the words of the House committee that recommended the appropriation: "Put into the hands of their children the primer and the hoe, and they will naturally, in time, take hold of the plough . . . and they will grow up in habits of morality and industry, leave the chase to those whose minds are less cultivated, and become useful members of society." Unstated, but no doubt recognized, was the assumption that once Indian children had learned to farm, they would settle on their own lands; and this would be the end of the tribal life and the communal system of landownership. With such aid, officials believed the Indians would become civilized within a generation, a totally absurd expectation for such a revolutionary cultural change.[13]

The missionary schools that were established with the aid of the Civilization Fund were remarkably the same. Each boy or girl worked at tasks designed to integrate them into white society. The boys learned to plow, plant, and cultivate, as well as to milk cows and to feed hogs and beef cattle. The girls were taught to spin, weave, make candles, process meat, and keep house. The Cherokee mission school of Brainerd, near Chattanooga, Tennessee, was one of the most successful institutions that worked closely with the federal government to instill the values of white civilization and to teach the methods of white agriculture. Brainerd was founded in 1817 by the American Board of Commissioners for Foreign Missions, and by 1820, with the help of the Civilization Fund, it had become a model agricultural school. There, young Cherokee learned a new life style from resident farmers, blacksmiths, and mechanics. The white farmers had the responsibility for establishing a model farm and for providing instruction about crop procedures, livestock care, and grain milling. Other missions also created model farms. In 1824, McKenney reported that thirty-two church-related schools had an enrollment of 916 and that no insurmountable difficulty existed to a "complete reformation of the principles and pursuits of the American Indian." A decade later, in 1833 at Maumee, Ohio, Ottawa children attended schools and, upon leaving, most of them allegedly became farmers. With similar progress in agricultural advancement west of the Mississippi River, officials in the Office of Indian Affairs hoped that the tribesmen were turning their attention to agriculture. Funding soon became a pressing and enduring problem for the Indian mission schools. The $10,000 Civilization Fund was not enough to meet the educational needs—agricultural and otherwise—of the Indians; and Congress was reluctant to expand the fund. Nevertheless, by 1834, six religious organizations had established sixty schools, where 137 teachers were instructing about seventeen hundred children.[14]

In 1838, T. Hartley Crawford became commissioner of Indian Affairs and gave new support for agricultural education when he proposed creating local manual-labor schools, including model farms, for instruction among all the tribes. Crawford saw several important advantages in neighborhood schools, as opposed to the boarding schools then in existence. First, teachers could mingle with students and parents alike, and the parents could observe their children learning agricultural practices on the school farm. Crawford hoped the parents would emulate their children and thereby greatly enhance the civilization process. Moreover, the manuallabor schools would provide peers for those Indian students who returned home from the boarding schools.[15]

The Shawnee Methodist Mission at the Fort Leavenworth Agency in present-day Kansas became the model for Indian manual-labor schools. Within a year of its establishment in 1839, it had fifty students ranging in age from six to eighteen. In addition to learning how to read and write, the boys learned to plow, mow, milk, and feed chickens and hogs, while the girls studied the "various branches of housewifery." The boys, with the aid of a resident farmer, tended between five hundred and six hundred acres of fenced land. Because of the perceived success of the Shawnee Methodist Mission, Crawford believed manual-labor schools that taught agriculture offered "conclusive proof" that Indians could learn to live like white men. As a result, the Indian Office and the War Department worked to establish manual-labor schools whenever possible, and school officials, too, showed great pride in their accomplishments, as well as in their firmness in dealing with Indian children. J. B. Duerinck, superintendent of manual-labor schools, reported that at St. Marys in the Kansas Territory, as well as at the other schools, lazy Indians were told to either "root or die." Farming could be profitable, but it required the sweat of one's brow. Missionaries and governmental officials alike never questioned whether boys and girls from another culture would understand the rigorous Protestant ethic behind such advice.[16]

By the end of the Polk administration in 1848, the federal government was supporting sixteen manual-labor schools, with over eight hundred students, and eighty-seven boarding schools, with approximately three thousand pupils. Together with the missions, the Indian Office appeared to be disseminating the concepts, values, and techniques of white agriculture to a large number of young Indians. The results of this cultural transfer, however, were uneven among the tribes involved in the removal process. The federal government had not planned for removal to disrupt the process of giving the Indians a new way of life through agricultural aid and education, but the environmental conditions of the trans-Mississippi West differed greatly from those that the Indians had experienced in the East. Although they were accustomed to using the eastern environment to their advantage, the West was different, and successful agriculture required new techniques and technology. The Indians did not realize this at first. Indeed, white bureaucrats and farmers did not understand this either. Many of the tribes were also placed in temporary settings, or for one reason or another, they were required to move several times before they were assigned a permanent home. All of these factors retarded agricultural development among the tribes. While the Indians were struggling to adjust to a new environment, the agents of the Indian Office were continuing to encourage farming as a way of life by providing agricultural aid.[17]

The agricultural tribes from the Southeast adjusted the quickest to their new homeland beyond the Mississippi: they began to farm at once. By 1832 the Creek on the fringe of the southern Great Plains were producing twenty thousand bushels of corn annually, much of which they sold to other Indian emigrants. The Creek also raised other grains and vegetables, and they owned a large number of cattle. Five years later, they sold nearly $40,000 worth of corn, yet retained a large surplus to meet their own needs. Near Fort Gibson in present-day Oklahoma the Creek lived in comfortable houses and tended gardens, orchards, and fields, the produce of which, together with cattle, chickens, and eggs, they sold to the commissary. By the late 1830s the Creek were among the most advanced Indian farmers.[18]

During the 1830s the Cherokee, who resided near the Creek, once again became productive farmers. By 1837, some eight thousand Cherokee were living on an estimated one thousand to eleven hundred farms, and they had earned the reputation of being the best agriculturists in the trans-Mississippi West. The Cherokee raised cattle, hogs, sheep, cotton, and corn to a "considerable extent." Some Cherokee farmers contracted with the army at Fort Gibson to sell their produce, thereby earning an estimated combined income of as much as $60,000 annually—a considerable sum at that time—for the sale of potatoes, beans, peas, melons, and beef. Much of this money seems to have gone to buy whiskey from white traders, rather than for new seed, livestock, and equipment. Much of the corn that the Cherokee raised they sold to the newly arriving Indian immigrants so that they would have food while they, too, began farming.[19]

Not all of the new immigrant Indians were as successful as the Creek and the Cherokee. The Choctaw, who settled near them, did not have enough agricultural implements to become successful farmers immediately. The farm tools that the federal government promised to provide were slow in arriving. Five hundred hoes and plows, sent from Philadelphia in September 1831, did not arrive until after the planting season the following spring. In 1833 the Choctaw did not receive clearing and tillage implements until July, far too late in the growing season for them to begin farming with much chance of success. In spite of a chronic lack of governmental aid and support, the Choctaw proved that they were excellent farmers. In the autumn of 1833 they produced forty thousand bushels of corn which they sold to the federal government for redistribution to the newest immigrant arrivals.[20]

The Indian Office apparently had worked out the supply problem by 1835, when it procured one thousand plows for the Choctaw, who harvested a fifty-thousand-bushel corn crop in 1836 and developed a market for it at nearby Fort Towson. In addition, the Choctaw, who resided along the Red River, raised enough cotton to merit the erection of two gins by white traders who purchased all the fiber that these farmers wanted to sell. The commissioner of Indian Affairs estimated that five hundred bales of Choctaw cotton would be marketed in 1836. The Choctaw also raised a large number of beef cattle. They were so successful at livestock raising that they provided Creek contractors with beef cattle throughout the 1830s and furnished seed stock for the arriving Chickasaw as well. Some Choctaw farmers also grew wheat, but because they did not have enough bolting cloth to make grain sacks, they did not emphasize that crop. To have done so would have caused storage and handling problems.[21]

Nevertheless, for every example that the commissioner of Indian Affairs could point to as evidence that the Indians were making progress in becoming farmers, an example of failure could also be cited. The agents in the Upper Missouri Agency and at the Council Bluffs and Great Nemaha subagencies in Iowa and present-day Nebraska did not see much prospect of success in teaching the Indians to farm like white men. But one agent did believe that the tribes in the central prairies and the southern Great Plains could be taught to become good cattlemen, if given the opportunity, because the climate and soil conditions of those areas were best suited for grazing. Elsewhere, traders introduced whiskey, to the detriment of the Indians with whom they traded. Drunken Indians did not make good farmers, and many Indians spent their hard-earned money on liquor rather than investing it in farm tools, seed, and livestock.[22]

By 1840, many in the Indian Office were optimistic about the progress that the eastern Indians were making on their new lands. The Cherokee, Choctaw, Creek, Seminole, and Chickasaw had established prosperous farms in what is now eastern Oklahoma. Many of the Cherokee had planted orchards, which was an unusual agricultural activity among the Indians except for the Creek and the Navaho. During the 1840s, the Cherokee also displayed a great aptitude for handling the tools of white agriculture. Most of them could stock their own plows and helve their own hoes.[23]

Many of the Choctaw at that time had large fields located between the Red and the Arkansas rivers, and the productivity from these lands increased each year. Both the Choctaw and the Chickasaw, who lived among them, produced between seven hundred and one thousand bales of cotton for export annually. Many of the Chickasaw purchased large numbers of black slaves to cultivate the cotton. When the price of cotton dropped, however, some of the Choctaw quickly adjusted by planting more corn. Choctaw farms compared favorably with white men's farms throughout the nation. The Chickasaw also developed regional markets at Forts Washita, Towson, Smith, and Van Buren. In 1842 the Chickasaw sold twenty thousand bushels of corn, and in 1847 the agent reported that forty thousand bushels could be sold if suitable transportation to other markets were available. Eight to ten cotton gins were now located throughout the Choctaw and Chickasaw country to facilitate the processing and sale of the fiber. In 1846 the Creek raised enough corn to sell one hundred thousand bushels to white merchants, as well as to feed large numbers of hogs and cattle for commercial sale. Indeed, Creek beef cattle attracted drovers from Missouri, Illinois, and Indiana, to whom the Indians sold several hundred head that year.[24]

Much of the farming beyond the fields of the Five Civilized Tribes was conducted by white men whom the federal government hired and by Indian women. The Iowa in the Great Nemaha Subagency, for example, raised approximately fifteen thousand bushels of corn in 1842, while a white farmer provided technical advice about plowing and planting. Instead of merely showing the Indians how to plow, plant, and reap according to white methods, the agents and the white farmers usually did all of the plowing, planting, and harvesting themselves. As a result, the Indians received little practical experience, but as long as the farm produce was distributed among them and as long as the annuities arrived, they were content to remain idle. On still other occasions, the agents or farmers did a far better job of promoting the region to white farmers than they did of teaching agriculture to the Indians: this they did by drafting glowing reports about the fertility of the land and the great potential of their region for agriculture. At other times, the federal government spent too much money trying to establish blacksmiths among the Indians, for the purpose of making and repairing farm tools and implements. In 1849, for example, the government spent $5,615 for blacksmith services among the Chickasaw. This money would have furnished an estimated eight hundred plows (with three or four extra points), two thousand axes, and two thousand hoes—all of which were more than the blacksmiths assigned to the tribe presumably could make in five years. Waste such as this occurred because officials in the Indian Office had not clearly thought out the agricultural program of the nation's Indian policy.[25]

By the 1840s the resettled eastern tribes, as well as some of the tribes on the Great Plains, were making considerable progress in learning the white man's mode of agriculture from the missionaries and from governmental officials, though that progress was much less in reality than the official reports and production statistics indicate. Some of the problems that the Indian farmers faced were the results of nature and the environment, while others stemmed from the inefficient administration of Indian agricultural policy. During the mid 1840s, drought and flooding ruined corn crops for the Creek, and worms severely damaged the cotton crops of the Choctaw. The Pawnee, who lived along the Loup River in present-day Nebraska, were unable to make rapid agricultural progress, because the Sioux kept interrupting the Pawnee's farming operations by killing the field workers and destroying crops. At best, the Pawnee only raised a "tolerable crop."[26]

In addition, the agents or the white farmers who were hired to teach the Indians to farm frequently did not have the proper tools, sufficient amounts of seed, or adequate numbers of livestock. At the St. Peter's Subagency in present-day Minnesota the agent could not keep oxen in stock, because the Indians kept killing them for food. At the Fort Leavenworth Agency, the Indians killed their hogs and cattle before the animals could breed and thus build up the Indian herd. Serious abuses of the federal program also occurred when a farmer who was assigned to an agency did not live among his charges. Instead of plowing land for Indian fields, some farmers merely drew their salaries until they were dismissed from the Indian Service. When governmental farmers plowed fields and built fences, as they did among the Sioux, the Indian men were content to watch. In some cases the Indian men stopped performing their minimal farming tasks so that governmental farmers could complete the work. Ultimately, many governmental farmers found it "easier to do a thing *for* the Indian than to teach him how to do the same." By so doing, however, they made little progress in developing substantive agricultural endeavors among the Indians.[27]

All too frequently the implements that the Indians received were unsatisfactory for farming. If they were not broken or damaged, the tools often arrived late. The Indians also needed an outlet for their produce if they were to become fully integrated into the market economy of white agriculture. Only where they could capitalize on the needs of military posts, however, did they have an adequate market for their products. Although some tribes sold agricultural surpluses to other emigrant Indians, that market disappeared once the new arrivals had begun their own farming operations and the removal process had been completed. When military markets were available, the Indians, such as the Choctaw, soon found that white farmers, particularly Texans, tried to capture that outlet. Indeed, the Texans were chronically short of money, since little of it circulated in that republic during the 1840s; so they undersold the Indians at Fort Towson. For the Texans, low prices were better than no prices at all, but this competition damaged Choctaw productivity. As their market declined, some Choctaw farmers became careless, indifferent, and generally unenthusiastic about putting their efforts into the development of prosperous and productive farms.[28]

During the 1850s, mission schools and governmental agents and farmers continued to teach agricultural methods to Indian children and adults. At the St. Mary's Potawatomi Mission in present-day Kansas the missionaries were technologically advanced compared to those in other missions and agencies. In 1853, for example, they cut sixty acres of oats with a mowing machine. Whether this implement was a combination reaper-mower or just a hay mower is not known, but it enabled the Indians to employ one of the most advanced forms of agricultural technology. It was far superior to a sickle or cradle scythe, which most Indians and whites used to harvest grain crops. The Potawatomi, who observed the mower in operation, thought that it was the "wonder of the country," and they reportedly were "lost in admiration" as they watched it work. Even with this implement, which would have enabled them to expand hay and grain production and to capitalize on the ready market available not far to the west at Fort Riley, the Potawatomi were content to raise only enough agricultural produce to meet their own needs.[29]

Similar instruction could be obtained at the Chickasaw Manual Labor Academy near Tishomingo, Indian Territory, where the superintendent believed that Indian students should learn "scientific farming"—that is, they should receive training in the use of the most advanced technology such as reapers, threshing machines, and corn shellers. Whether they ever received training with those implements is not certain, but he did encourage his students to read the agricultural newspapers to keep abreast of the latest farming developments. The Civil War terminated this combination of practical experience and theoretical, or book, farming, but while the academy existed, it provided some of the most intensive agricultural education that Indian children obtained.[30]

While the tribesmen on the Great Plains were struggling to become farmers in the white man's tradition at mid century, the Navaho were cultivating about five thousand acres of corn as well as a small amount of wheat, potatoes, and cotton. They also grazed a large number of horses. About five thousand Pueblos also raised large flocks of sheep as well as wheat, but the lack of adequate tools severely limited their agricultural progress. In 1857, for example, the agricultural implements that had been sent from St. Louis were "entirely useless." The Indians at Nome Lackee Reservation west of Sacramento, California, also cultivated as much as one thousand acres and raised fifteen thousand bushels of wheat, in addition to an abundance of other vegetables. They, too, were plagued with a lack of implements, so they harvested their wheat with sickles rather than with reapers or cradle scythes. During the harvest they reportedly used those sickles with "extraordinary dexterity." At Mendocino the Indians in the California Agency also displayed their ability to manipulate the tools of white agriculture with great skill. Under the supervision of a white farmer, they used four large steel breaking plows to prepare their fields for planting. Each plow required from five to six yoke of oxen for draft, and three men—one to hold the plow and two to drive the oxen forward. In this way they could plow about one acre of virgin soil per day.[31]

During the 1850s, some of the tribes experienced major setbacks, which prevented rapid agricultural progress. In 1854, drought severely damaged the crops of the Cherokee, Iowa, Sac, and Fox, and the Creek and Seminole had difficulty in obtaining enough farming implements. The Chickasaw suffered from the twin plagues of drought and grasshoppers, while the Shoshone at the Utah Agency lost their corn crop to grasshoppers and crickets one year in the valley near Fillmore. Abolitionists in Indian

Territory were also upsetting the tranquility of the Five Civilized Tribes, while free-soil and proslavery forces were fighting among the Indians in Kansas Territory. Continued white migration was also presenting problems. The Kansas Reservation reportedly was an object of "filthy speculation" by whites who entered that reserve, opened farms, and cut timber. Late in the 1850s, smut destroyed the wheat crop at the Tejon Agency in California.[32]

Other problems existed as well. Many soil types and climatic conditions existed in the vast expanse of the trans-Mississippi West. In some places, federal officials did not know enough about the area to recommend the best agricultural practices. When the agents did encourage the Indians to cultivate their traditional crops more extensively and intensively, they sometimes discovered, as they did in the Puget Sound District, that the soils were nearly exhausted because of the continual cultivation of potatoes. There, even though production had decreased, the Indians were reluctant to break new lands, because they did not want to leave their traditional fields.[33]

In 1860, drought again was severe for the tribes living west of the Missouri River. Faced with starvation and the need for governmental aid, many tribesmen could see little future in agriculture. Indeed, the Oto and Missouri men, who had given up their summer hunt in north-central Kansas to stay home and farm, could see nothing but disaster resulting from a total reliance on agriculture. When their crops failed because of drought, they had few food resources to rely upon, and starvation became a grim possibility. For them and others like them, a wise balance between hunting, on the part of the men, and farming, by the women, provided the food resources needed to live comfortably and sensibly. Moreover, the Klamath along the West Coast had told federal authorities that their forefathers had managed to live adequately without agriculture and that they could as well, if whites would only leave them alone. These tribesmen thought it was much more sensible to follow their traditional cultural practices than to adopt new ones.[34]

Some agricultural progress certainly had been made among the Indians; nevertheless, by 1860 the most agrarian of these tribes were the very ones that had had the most agricultural experience before the removal era. Such groups as the Cherokee and the Creek proceeded to develop profitable farming operations on their own and in spite of governmental aid, which often came too late or in a condition that did not contribute to agricultural improvement. Among the less agricultural tribes, the federal government often ran into difficulty by trying to integrate subsistence farmers into a market economy. Instead of becoming commercial farmers, many tribesmen were content to practice subsistence agriculture and to rely upon food obtained from hunting. In order to make many of the Indians into commercial farmers, markets were also necessary. For some reason, federal bureaucrats never recognized the need to develop a market outlet for Indian agricultural produce. At best, officials in the Indian Office were content to praise the Indians for selling grain and livestock to other Indians or to the military posts. But Indian farmers needed more than this; they needed a fully developed market system, wherein they could easily ship their commodities to stable markets and receive fair prices. Perhaps the War Department could have purchased Indian agricultural products. The military needed large amounts of food and fiber, the purchase of which would have stimulated Indian commercial farming developments. This marketing and price-support concept, however, would not come to fruition until the twentieth century, and then it applied primarily to white farmers. Thus, in the absence of a market, many Indians could not see any advantage in producing large surpluses that would only spoil in storage: they were therefore content to continue subsistence agriculture.[35]

In the final analysis, progress in Indian agriculture was slow. By the Civil War, many tribesmen were struggling, or they had not yet been convinced that farming was the best way of life. The problem was not that removal upset the process and development of Indian agriculture. Rather, the problem was that many Indians had been placed in an environment where climatic and soil characteristics were far different from what they had known. The Indians needed time to adjust to that new environment and to the demands that it made upon them before they could develop successful operations. White policy makers, however, assumed that adjustment would come quickly, if only the Indians would learn to farm in the proper manner. The problem with this thinking was that whites did not have formal or ready access to agricultural instruction in the schools, nor had they developed the technology and scientific measures needed to make agriculture successful or profitable in the West. Indian children learned to farm by doing at the mission and at manual-labor schools, and their parents learned by observation and by helping the agents or farmers who developed "pattern" or demonstration farms at the agencies and reservations. This instruction merely passed on the best methods of eastern farmers, and these techniques were not always well suited to western agriculture. Thus, while federal officials tried to mold the Indians into "enlightened agriculturists," few whites claimed similar knowledge. White farmers would not learn to manage the Great Plains environment for agricultural purposes until the late nineteenth century. Thus, whites were demanding higher standards of agricultural progress for the Indians than the whites achieved for themselves.[36]

Progress in Indian agriculture also faltered when the tribes were assigned to reservations, such as Nisqually in Washington, which had poor soil or which were hemmed in by white settlers. At Nisqually and on the

Cherokee Neutral Land in present-day Kansas, whites grazed cattle without the permission of the Indians. Indeed, whites occupied the Neutral Land of the Cherokee Nation with such impunity that the settlers agreed to vacate those lands only after the army had burned several of their cabins. Moreover, when modern tools were available, those implements were not equally distributed among the farming tribes. As late as 1857 the plows that the Choctaw and the Chickasaw used were too light for more than superficial tillage. The students at the Chickasaw Manual Labor Academy were lucky indeed to have training with reapers, threshing machines, and an eight-horsepower Hoard and Son's portable steam engine. Most Indians were not so fortunate.[37]

Some tribes had difficulty in obtaining not only the necessary agricultural implements but also cattle. In 1833, for example, the Pawnee in present-day Nebraska had been promised oxen and other livestock valued at $1,000. Twenty years later, they had not yet received those cattle, because the government had placed the livestock with the farmers who were assigned to help the Indians, and in time those animals were sold and otherwise dispersed without passing to Pawnee control. In 1842, one observer reported that the cattle that contractors furnished to the tribes in the Indian Territory frequently were in such poor condition that they died before being delivered. Or unscrupulous whites stole the cattle, only to resell them to other tribes—all at government expense.[38]

Ultimately the federal government did not achieve rapid success and the desired results because it taught the wrong people to farm. By insisting that only men and boys learn farming methods, the federal government broke the Indian tradition, which relegated agriculture to women. By doing so, federal policy placed Indian men in a degrading position and immeasurably retarded the expansion and further development of Indian agriculture in the trans-Mississippi West. The officials in the federal government would have made much better and more rapid progress had they ordered their agents and the missionaries, who received federal funds for Indian education, to instruct the women as well as the men in the best agricultural techniques. Because white society required women to work in the home and not in the field, the federal government wasted the great agricultural abilities of the Indian women. Had tribal women been encouraged to take an active role in the farming process, Indian agriculture probably would have advanced more rapidly than it did. Some Indian women were better farmers than the agents or the governmental farmers who were sent to give them agricultural instruction. Indian women already knew how to grow corn, and the wisest federal policy would have been to encourage them to raise it as abundantly as possible, instead of trying to teach only their husbands how to farm as a new occupation. In retrospect, then, the federal goverment tried to impose an economic and cultural revolution upon the Indians by insisting that the males become farmers. That endeavor failed because policy makers did not provide adequate support. Most important, it failed because they did not understand the Indian cultures with which they were dealing or recognize the negative ramifications of their proposed changes.[39]

The federal government also erred by not emphasizing livestock raising among the Indian men. Since many of the Indian males were custombound to be hunters, it is not unreasonable to believe that federal officials could have devised an agricultural policy whereby Indian men would have been encouraged to assert their manliness by raising all forms of livestock. Had white officials been able to collect and disseminate the agricultural knowledge of the Indian men and women who farmed in the semiarid and arid West, the newcomers in the trans-Mississippi West probably could also have made easier adjustments to their new environment. But all of this was too much to expect from American culture and governmental organization at that time.

Finally, on the eve of the Civil War, the missionary and governmental manual-labor schools experienced problems that portended even greater difficulties to come. Indian parents became just as lonesome for their children at the boarding schools as their children were homesick. Some Indian parents petitioned headmasters and teachers to allow their children to return home before they had completed their education. Occasionally, when the schools prohibited Indian children from returning to the reservations, their parents spirited them away. One school official complained that few things were more trying for missionaries than dealing with "improvident ignorant parents" who took "useless" children home before the youngsters had completely learned the ways of white society.[40]

While agriculture was a central feature of Indian policy, Indian farming developed unevenly prior to the Civil War. Because that progress, as defined by white standards, often lagged, Luke Lea, commissioner of Indian Affairs, reached the conclusion that the Indians needed to be confined to reservations where they could be "compelled by stern necessity to resort to agricultural labor or starve." In 1852, Edward Fitzgerald Beal, superintendent of Indian Affairs for California and Nevada, urged that a reservation system be created. Congress supported his plan and, during the next two years, helped Beal establish the Tejon Reservation. For the next fifteen years, until 1869, the federal government entered into treaties with the Indians and assigned them to reserves where, among other things, they were taught to farm.[41]

Still, cultural change comes slowly, particularly when a people is forced to make a major adjustment in its way of life. On the eve of the Civil War, the nomadic hunting tribes of the Great Plains and the Far West were the most reluctant to become agriculturists. Before governmental officials could convince those tribes to accept the sedentary farming life, the Civil War disrupted Indian and white farming alike, changed the nature of agriculture in the Indian Territory, and prevented any reevaluation or reform of Indian agricultural policy.

8

From the Civil War
to Severalty

When Confederate batteries opened fire on Fort Sumter in April 1861, the nation tumbled into the Civil War. Indian-white relations were also thrown into confusion, and the further development of an Indian policy languished, because some of the tribes sided with the Confederacy and because the war caused federal officials to neglect many of the other tribes. In Indian Territory, for example, portions of the Cherokee, together with the Creek, Choctaw, Chickasaw, and Seminole, joined the Confederacy. On the Nome Lockee Reservation in California, agricultural development lagged as governmental officials became preoccupied with the war in the East. In Minnesota, the Civil War contributed to the great Sioux uprising, because the war caused intolerable delays in the distribution of annuities. And while nearly twenty thousand troops were serving in the West by 1865, the army was unable to protect peaceful tribes on the Great Plains, such as the Omaha, Pawnee, and Ponca, who were trying to become agriculturists, from the hostile Sioux and Cheyenne, who were not interested in adopting a sedentary life.[1]

Federal officials were particularly concerned with the response of the Five Civilized Tribes to secession, because these tribes occupied highly productive lands which would be an important source for grain and livestock. Indian Territory was also a potential staging area for attacks on either Texas or Kansas, depending upon which army held the region. When the war came, the Cherokee proclaimed neutrality, but some of them either felt the federal authorities had abandoned them or else they sympathized with the Confederacy and joined the secessionists. By the autumn of 1862 the Cherokee were split into two factions. The proslavery Cherokee held the Indian Territory, while the antislavery sympathizers, in the absence of federal protection, fled to free-soil Kansas.[2]

Although the Civil War divided the tribesmen in Indian Territory, no one is certain how many loyal members of the Five Civilized Tribes sought refuge in Kansas. By the fall of 1862, between fifteen hundred and two thousand Cherokee, mostly women and children, had made their way to the tribe's Neutral Lands in southeastern Kansas. Behind them, both white and red rebels were overrunning Cherokee country and destroying crops, livestock, and farm implements. Later, loyal whites, as well as Union soldiers, inflicted similar damage. Cherokee and Creek agriculture was severely damaged, because those tribes were the most divided during the war. Some of the Cherokee attempted to regain their farm lands in 1863, when they moved back into Indian Territory, but rebel forces once again drove them from their fields.[3]

By 1864, some of the Cherokee had returned to their abandoned farms. Others located near Fort Smith, Arkansas, and Fort Gibson, Indian Territory. Many would not, however, return to farming. When they were urged to do so, some Cherokee argued that Union officers would appropriate what they raised, as they had done in the past, and that teamsters, army-hangers-on, and rebels would take the rest. Until peace was firmly established, they intended to live upon the relief that the government provided. For the moment, at least, farming did not offer any remunerative advantage. Not all Cherokee were recalcitrant: some planted corn and potatoes in an attempt to return to a normal agricultural life, but their efforts were wasted. When autumn came, the grass near Fort Gibson had been overgrazed, and the military was in great need of grain. Because the Cherokee had raised a fair corn crop, it was the easiest to requisition. Unscrupulous whites also presented themselves as governmental officials and, under the guise of appropriation, took even more produce. The Cherokee did not resist, because they did not know who was authorized to take their crops, and they had experienced such losses before. At that time, the Cherokee received $1.00 per bushel for corn. Payment usually was made with vouchers worth $.90 on the dollar. Had the Cherokee received fair prices, they would have been paid $3.00 per bushel. In the end, the Cherokee were not only unfairly paid; they were also not reimbursed for three-fourths of the corn that was taken. The Cherokee who hid their corn crop were able to save it, while those who did not usually lost everything.[4]

At that same time, the Cherokee cattle supply was nearly exhausted, because thieves drove large numbers north while rebel troops continued to decimate the herds in the territory that they occupied. Federal troops at Fort Gibson were unable to stop the thefts, because they lacked adequate cavalry to do so. Two groups of loyal white civilians also stole cattle from Indian Territory during the Civil War years. The first were the rustlers, who entered the territory and drove the cattle to the southern border of Kansas, where the second group, euphemistically known as "cattle brokers," received the stolen animals. These cattle dealers purchased the animals at a nominal price, executed a bill of sale, and drove the cattle to other northern or western markets, where the dealers made enormous profits at the expense of the Indian cattle raisers. Kansas authorities ignored the illicit trade, because many cattle stealers were respected citizens. As a result, before the war ended, cattle rustling in Indian Territory became very nearly a "legitimate business operation," and robbery reached "unparalleled and astounding proportions." For those who were prepared to take the risk, the profits were high—$100 for a yoke of oxen, $12 per cow, and $10 for bulls and young cattle—the highest prices that had yet been paid for Indian stock. In this case, however, the Indian cattle raisers did not collect the benefits of their labors. When the agent tried to stop such thefts, a warrant was issued for his arrest, and his life was threatened. By such activity as this, Creek herds had been decimated by the end of 1864. By August 1865, between 200,000 and 300,000 cattle, valued at $4 million had been taken from Indian Territory without remuneration. When deputy marshals and the army were sent to end the practice, they frequently were guided into the territory by the most "arrant" of the cattle thieves, and little was accomplished, or the marshals were bribed and no arrests were made.[5]

Loyal Indians also participated in stealing and driving cattle northward in order to take advantage of wartime demands. Late in the war, at the Temporary Wichita Agency in Butler County, Kansas, the cattle trade assumed "enormous proportions." There, the Indians argued that if they did not sell all the cattle possible from the ranges in the territory, the rebels would steal the animals for their own use. Like the whites, these Indians were not particularly careful about whose cattle they rounded up for sale in Kansas. Nevertheless, William P. Dole, commissioner of Indian Affairs, did not object to the Indians' selling cattle to Union troops or sympathizers as long as the tribesmen received fair prices.[6]

In order to control the sale of Indian cattle, W. G. Coffin at Leavenworth, Kansas, who headed the Southern Superintendency, developed a policy whereby he granted purchasing permits to a few respected individuals. Cattle buyers were then required to issue bills of sale that described the stock, including the quantity, brands, and prices paid. Payment was to be made before the agent, chief, or other competent witness. Then the cattle could be driven to Kansas where the superintendent or the chief military officer would examine the bills of sale and compare the documents to the livestock in the possession of the cattle buyer. Then the superintendent would issue permits for sale. In this manner, federal officials hoped they could guarantee the Indians a fair price for their cattle and some compensation other than whiskey. This policy did not slow the drain of range cattle from Indian lands, but at least in theory, it offered some protection to the Indian cattle raisers whose enterprises the war had disrupted.[7]

While the Indian Territory was in chaos, the bureau struggled with the Indians elsewhere in its continued attempts to make them into farmers. In 1862, Commissioner Dole believed the federal government should continue the policy of confining the Indians to reservations where they could be taught to farm and could become accustomed to the idea of holding land in severalty. He thought the nomadic Indians had no choice, considering the rapid occupation of the West and, with it, the encroachment of white settlers on Indian lands. Governmental officials remained confident that the reservation policy was working among some of the tribes and that it could be successful for the others. The Omaha in Nebraska, for example, were "more or less instructed in agriculture" by the agency farmers. The majority of the males knew how to plow, harvest, and mow hay, and, in a "small way," they were becoming increasingly interested in raising cattle instead of horses. In 1862 the Omaha had nearly two thousand acres fenced for cultivation and pasture. The agency farmer was "constantly engaged in aiding their agricultural operations" by plowing the fields in the spring, by alloting small patches to the poorer tribesmen, and by providing whatever assistance was needed for planting and cultivating. Although the weather presented problems, the Omaha produced almost as much corn as they needed, together with modest harvests of wheat and potatoes and large quantities of garden vegetables. The Omaha also raised sorghum and used a steam engine to operate a mill, which pressed 13,050 gallons of juice that they then boiled into 1,350 gallons of syrup of "fair quality."[8]

Still, problems remained. Cutworms, chinch bugs, potato bugs, and other insects took a heavy toll of the crops. The governmental farmer complained that until the Omaha gave up their summer buffalo hunt, during which time they abandoned their fields with "bag and baggage," livestock would continue to damage their fields. Nevertheless, he was encouraged, because overall, more males were plowing, rather than relying upon their women to tend the fields. Since the Omaha, however, were positioned between the hostile Sioux, in Dakota Territory, and land-hungry whites in Nebraska, they farmed under constant apprehension.[9]

Among the Kickapoo in Kansas, the Indian Office could take pride that no governmental farmers were cultivating the tribe's land. The Indians tended the fields and raised wheat, corn, potatoes, beans, sorghum, and garden vegetables. Their harvests were abundant, and they urged their agent to procure even better seed. Nearly all of their horses, however, had been "Jayhawked"—that is, stolen by proslavery forces. This loss hindered the cultivation of their lands until the agent was able to purchase additional draft animals. Nearby, the Potawatomi also cultivated about two thousand acres in wheat, oats, vegetables, corn, potatoes, and buckwheat without the aid of a governmental farmer. In addition, they raised eight hundred cattle, a thousand hogs, and fifty sheep. Their wheat and corn averaged twenty bushels and thirty bushels per acre, respectively. In 1862 the Potawatomi received 260 plows, 184 hayforks, 125 scythes, and 10 harrows. They did not, however, receive modern farming equipment such as reapers, threshing machines, or hay mowers. Nevertheless, at the Potawatomi Manual Labor School at St. Marys, the instructors worked to make the boys into "industrious farmers" and thereby to provide them with the opportunity to gain "prosperity and happiness."[10]

In 1862, among the Ponca in the Dakota Superintendency, governmental farmers and hired breaking teams plowed 275 acres. The Ponca followed behind the plows and planted wheat, oats, and corn. Although the governmental farmer had requested a threshing machine, it had not yet arrived, and the Indians threshed by trampling the grain under the hooves of their horses. When insects attacked the potato plants, the Ponca spent three weeks picking them off by bushelfuls but were unable to save the crop. Drought and worms also destroyed the bean plants. In the Southwest, the Pima, the Papago, and the Maricopa continued to produce a wheat surplus, and in 1861 the army purchased a million pounds from them. Three years later, the Pima and the Maricopa also produced a good cotton crop, so the government encouraged them to raise even more by furnishing them with five hundred pounds of seed.[11]

Other tribes, however, did not fare so well in their agricultural endeavors. At the Grand Ronde Agency in Oregon, more than half of the oxen and most of the tools were useless, because they were old and worn-out. At Oregon's Siletz Agency, the Indians lacked plows, grain cradles, and scythes; and new tools and wagons were "very much needed." In 1862, frost destroyed the potato crop, and the reluctance of the Indians to become farmers made a dismal situation even worse. All of these factors, and perhaps personal prejudice as well, prompted a governmental farmer to report that the Indians could never be governed with kindness. "Fear of punishment," he wrote, "is the only means I use to control them, and I find it effective." A lack of adequate implements also plagued the agency farmer at the Umatilla Reservation in Oregon, where the Indians "earnestly" needed plows and harness as well as adequate draft horses and oxen. Although the Indians had been able to train their horses to pull the plow, these animals were unsuitable for breaking virgin soil. Nevertheless, the governmental farmer believed he would soon have "good farmers." Those who owned large herds of cattle and horses were reported to be "sharp traders" who usually managed to get the best of a bargain. In 1864 the Indians at the agency cultivated 726 acres without assistance, and the agent expected them to sell surplus grain and vegetables for as much as $1,000—a modest but potentially important entry into a market economy. Overall, the Indian farmers at the Umatilla Agency tried to manage their farms "just like white people." The lack of a mill, however, retarded the expansion of the wheat crop.[12]

When the Civil War ended, federal officials were free to give greater attention to Indian Affairs. In the peace settlement with the Five Civilized Tribes, the government fixed territorial boundaries and prohibited slave labor. Although many Indian farms had been abandoned, the Creek recovered rapidly. In 1865 they cultivated some 2,600 acres—more than double the area farmed prior to the Civil War. Two years later, they cultivated 6,000 acres; and in 1871 they tilled a remarkably large area of 28,600 acres. As their acreage increased, corn production rose dramatically. In 1866, the Creek harvested 125,000 bushels. With production as high as this, the federal government considered the Creek to have returned to self-sufficiency, so it stopped giving relief aid that year. In 1871 the Creek raised 625,000 bushels of corn and sold 500,000 bushels. They also increased their potato production from about 10,000 bushels to 100,000 bushels annually. In addition, the Creek raised peanuts, pecans, cotton, and tobacco valued at more than $7,400. The Creek trailed their prewar endeavors only in the area of livestock production, largely because whites continued to steal their cattle. In fact, the herds had become so depleted in 1866 that tribal leaders prohibited sales of cattle in order to rebuild the herds.[13]

The end of the Civil War did not, however, bring an immediate and widespread restoration of agriculture to Indian Territory. In fact, it brought more problems that influenced the nature of Indian farming in that region. Most important, peace brought a new flood of immigrants into the West. These settlers, together with the construction of transcontinental and local railroad systems, created an even greater demand for Indian lands. As a result, the federal government developed a policy of moving as many tribes as possible to Indian Territory. The peace treaties between the Five Civilized Tribes and the federal government contained provisions that granted considerable land to the government for the relocation of other Indians from outside the territory.[14]

During the next few years the Indian Office tried to resume business as usual—that is, to pursue the policy of making farmers out of all the Indians in the West. Still, the old problems remained. Some were environmental ones that federal officials had no control over, such as the loss of crops to grasshoppers at the Warm Springs Reservation in Oregon. Other problems were cultural, which the officials also could not control, such as the continued encroachment of whites on Indian lands or corruption among agents and contractors who furnished supplies to the tribes. In some areas, life continued as in the past. The Pima and the Maricopa cultivated their fields with "extraordinary diligence and success" and sold surplus grain to the military and to civilians at the "extraordinary prices" of several cents per pound. The Navaho, exiled to Bosque Redondo in eastern New Mexico, cultivated between two and three thousand acres and used a main irrigation canal, which extended over seven miles to water their crops. This water contained a large amount of alkali, however, and the soil was poor. They raised only one bushel of corn per acre in 1867— half the amount of the previous year. The Indians at Oregon's Umatilla Agency made rapid agricultural progress, and their crops would have been a "credit to any white farmer in the state." Among the Yankton Sioux in present-day South Dakota the governmental farmer spent more time lounging at the sutler's store than teaching his charges to cultivate the soil in the white man's way.[15]

At the California Agency, the officials gave much attention to teaching the young Indians to use agricultural implements, to handle teams, and to cultivate crops in a "careful and frugal manner." Although the agents and farmers who provided that instruction needed "patience and endurance," they made rapid progress with individualized instruction, so that some of the Indians became good farmers. Many federal officials believed that this kind of intensive agricultural training was vitally needed if the Indians were ever to become farmers. Properly trained, these officials argued, the Indians could learn to operate reapers, threshing machines, and grain mills, thereby saving the government the expense of hiring whites to do that work. Most of the tribes in the trans-Mississippi West, however, needed increased agricultural instruction and larger supplies of implements, seed, and livestock if they were to become self-sufficient agriculturists.[16]

In 1869, partly to help meet the need for increased agricultural education, President Ulysses S. Grant introduced a new "peace policy," which was founded on the optimistic belief that religious societies should be responsible for the formulation and administration of Indian policy, because those humanitarian groups would have the best interests of the Indians at heart. At the agency level, church-nominated employees would end corruption, provide the agricultural and academic education needed for acculturation and assimilation, and teach the benefits of Christian doctrines. This policy was not entirely new. The federal government had depended upon religious groups for a long time to help teach agriculture to the various tribes. Now, however, the government intended to emphasize the importance of the churches to a greater extent than ever before. First, this new direction in Indian policy involved creating the ten-man Board of Indian Commissioners whose responsibility was to exercise "joint control" with the secretary of the interior over the distribution of Indian funds and, by so doing, to help prevent the corruption that was notorious among the agents and contractors in the field. Congress formally provided for the creation of the new board on 10 April 1869, and in late May, its members met with Grant. The president assured the board that a basic aspect of the peace policy must be to guarantee the Indians land so that they could be "induced to cultivate," and the board agreed that, for that purpose, the tribes needed to be established on reservations.[17]

The second aspect of Grant's peace plan involved apportioning the agencies among religious groups so that these groups could nominate agents, farmers, and teachers and thereby remove the appointment process from spoils politics. The result would be, Grant hoped, an efficient and less corrupt administration of governmental policy and rapid progress in achieving the acculturation and assimilation of the tribes. This component of the peace policy was nothing less than an admission that the federal government had formerly failed in the administration of Indian policy. While Grant's peace policy did not radically depart from that of the past, governmental officials clearly hoped that the religious societies would accomplish the policy objectives that the officials had not been able to achieve. To develop a moral and ethical administration of Indian policy, no one seemed more trustworthy than missionaries from the various religious bodies. Major changes, however, other than structural or organizational reforms, did not occur immediately. The missionary boards had difficulty in finding people of high moral character who were willing to serve at isolated agencies. Moreover, many missionaries seemed to be more concerned about serving abroad than about working among the American Indians.[18]

The third focus of the new peace plan involved terminating the policy of dealing with the Indians by treaty. This aspect of Grant's plan originated in the House of Representatives and involved the most fundamental change with the past. Because the Indians no longer held the power that required the government to treat them like foreign nations, the time had come, many critics of past policy argued, to legislate for the Indians, rather than to negotiate with them. Thus the Indians essentially would become wards of the nation. Tribal loyalty would be broken down, and eventually, the Indians would be absorbed into white society as individuals, rather than remain members of their own nations. Furthermore, the House of Representatives wanted a voice in the determination of Indian affairs, and the Senate was willing to abrogate the treaty system, provided that the treaties that had already been negotiated would remain inviolate. Grant did not have a role in this development, but it became a component of his new peace policy in early March 1871. Although the tribes thereafter were not treated as independent nations, the government did not have any other suitable method for dealing with the Indians, so instead of making treaties with the tribes, federal officials used the euphemism of "agreements," which required the approval of both the House and the Senate. Thus, Andrew Jackson's earlier desire for the federal government to legislate for the Indians, as it did for everyone else, had not yet been achieved.[19]

In order to make Grant's peace policy work, the underlying assumption was that the Indians would have to be concentrated on reservations. With less land under their control, the Indians would give up the hunting life and become farmers. In time, these smaller domains could be subdivided and allotted—that is, apportioned to the Indians for individual farms. By segregating and isolating the Indians on reservations, the missionaries could conduct their work unhindered. On the reservations the missionary groups could implement their educational plans, which would, in part, make the Indians self-sufficient agriculturists, thereby reducing the cost of governmental support and opening the way for admission into white society.[20]

The good intentions of the federal bureaucrats and the high moral principles of the missionaries were, in the absence of adequate science, technology, and time, unable to overcome environmental obstacles on the Great Plains and in the Far West and the cultural resistance of the nonagricultural tribes in those regions. In the Medicine Lodge Treaty of 1867, for example, the federal government promised the Cheyenne and the Arapaho 320 acres for each head of a household, together with $100 worth of seeds and agricultural implements for their first year on the reservation, to help them begin farming. For the next two years, however, individual farmers received only a paltry $25 for seeds and tools. Moreover, the Cheyenne and Arapaho Reservation, located on the North Fork of the Canadian River in Indian Territory, was not the best area for agriculture as practiced with the techniques then known. This area of low precipitation and periodic drought would have tested the mettle of the best white farmers. To expect the Cheyenne and Arapaho, who heretofore had not been agriculturists, to begin farming and to become prosperous was more than a case of poor judgment on the part of policy makers; it was nothing less than cruelty softened under the guise of humanitarianism.[21]

By May 1870, about two hundred acres had been plowed and planted for the Cheyenne and the Arapaho, but the meager corn crop looked "very sickly" because of drought. In 1872, agent John D. Miles placed a governmental farmer with each Cheyenne and Arapaho band to encourage them to put more "heart" into their work. He believed this policy would not only increase instruction but would also stimulate a "spirit of emulation" or competition among them to "excel in results." The results were disappointing, however. White employees conducted almost all of the agricultural activities on the reservation. The Arapaho were somewhat more receptive to this new life style, and a few agreed to take the "corn road." But the Cheyenne were not: at the very best, they adopted a wait-and-see attitude, and they were not encouraged. In 1874, drought and grasshoppers destroyed the corn and vegetable crops. A year later, a cold, wet spring required them to reseed their crops. At best, only fifty acres had been planted with corn and various vegetables by the mid 1870s, even though some of the Cheyenne realized that the days of subsisting on the buffalo were drawing to an end. Still, some Cheyenne began to show more interest in agriculture; and in 1876 the agent divided seventy-five acres for each tribe into one- to five-acre family gardens. With this small beginning, he hoped that the Cheyenne and the Arapaho eventually would become full-time farmers.[22]

While the agent and the farmers tried to instruct Cheyenne and Arapaho adults about agriculture, the teacher at the agency school also taught the boys about farming. By the end of the 1870s, however, the Cheyenne and the Arapaho were still far more committed to the nomadic hunting life than to agriculture. A decade of agricultural aid and education was not enough to change the culture of these Plains Indians. Moreover, the agency always seemed to lack agricultural implements, or else the tools arrived too late to be of much use. Perhaps the Department of the Interior should have emphasized livestock raising instead of crop production for the nomadic Great Plains tribes. In 1872, Capt. Henry E. Alvord, special Indian commissioner, recommended, as others had years before, that the department "drop the corn talk" and stress cattle raising. Several years later, Bvt. Maj. Gen. Alfred Howe Terry supported that change of policy. The "pastoral state," he believed, would enable the most successful and rapid transition into white agricultural society, because such endeavors would involve "comparatively slight change of habits."[23]

In 1869 the Kiowa, the Comanche, and the Kiowa-Apache, who lived on a reservation in what is now southwestern Oklahoma, experienced problems similar to those of the Cheyenne and the Arapaho. They did not have an agricultural tradition, and the semiarid climate reduced their chances of agricultural success. By 1870 they had become "tired of being talked to forever about corn." The agent agreed with them and urged the department to begin work to make them into cattlemen, but he still employed local farmers to plow land, establish model farms, and teach the Indians to till the soil. He offered $500 in prizes to the ten Indians who were the most productive farmers. Only the Comanche, however, showed much interest in farming; with the aid of two white farmers, the Comanche raised 2,950 bushels of corn and 25 bushels of turnips on 154 acres, and cut twenty tons of hay. In 1871, an oppressively hot and dry summer discouraged the expansion of these meager efforts when it ruined most of the corn and wheat crops. When the government delivered farm equipment to the chiefs of the various bands for them to allocate, they quickly came to view the equipment as a symbol of wealth and kept it for themselves. These tribes also suffered from a lack of qualified personnel to teach farming under very difficult environmental and cultural circumstances. During the early 1870s the Kiowa agent estimated that he needed eighty farmers to instruct his more than three thousand charges, yet he only had one farmer for that work.[24]

During the 1870s, while the nomadic tribes on the southern Great Plains were struggling to become farmers, the Cherokee were being more successful with their agriculture. By 1872 they had made a remarkable recovery from the Civil War. In that year, they raised 2.9 million bushels of corn, 97,500 bushels of both wheat and oats, and 80,000 bushels of potatoes, while tending 160,000 hogs, 75,000 cattle, 16,000 horses, and 9,000 sheep. The agent was taking "great pains" to introduce fruit culture, and he hoped that every Cherokee family soon would have an orchard of grafted fruit trees. Although a few large-scale farmers cultivated cotton, small-scale farmers produced most of it, and the agent hoped that this crop would provide the basis for permanent tribal prosperity. He planned to emphasize the cultivation of cotton, because it was an important cash crop. Stock raising took too much time, usually several years, to return a profit, and the Cherokee needed money immediately. He also believed too many Cherokee were engaged in agriculture. Many needed to develop processing industries and thereby create a home market for Cherokee agricultural commodities. This activity would not only strengthen the Cherokee economy; it would improve the farm base as well. In 1874, despite drought, grasshoppers, and prairie fires, the Cherokee planted larger crops than ever before.[25]

For federal officials, the agricultural progress of the Five Civilized Tribes was both "complicated and embarrassing," because the Indians had more land than they needed. By the mid 1870s the Cherokee theoretically owned about 280 acres each; the Chickasaw, 775; the Creek, 247; and the Seminole, 82. Again, white farmers, land speculators, and governmental officials questioned the right of these Indians to hold land they could not cultivate. Although the commissioner of Indian Affairs recognized that the problem was a "difficult one," he also believed public policy would soon require these tribes to cede a large portion of land to the federal government: "There is a very general and growing opinion that observance of the strict letter of treaties with Indians is in many cases at variance both with their own best interests and with sound public policy. Public necessity must ultimately become supreme law; and in my opinion their highest good will require these people to take ample allotments of lands in severalty . . . and to surrender the remainder of their lands to the United States Government for a fair equivalent. Upon the land thus surrendered, other Indians should be located as rapidly as possible, and should be given allotment under the same restrictions." This policy would free other Indian lands for settlement by whites.[26]

Before the government could change its agricultural policy, however, whites in Indian Territory, particularly cattlemen, took advantage of the great expanse of grazing lands and the inability of the Indians and the federal officials to keep them off of it. Texas cattlemen grazed their herds on Quapaw land, for which they paid the Indians a nominal annual sum of ten cents per head, when they paid at all. One person grazed four hundred sheep on those lands, for which he paid three dollars for the entire season. Payment was made in the form of one sheep. Such flagrant violations of Indian property rights caused the Bureau of Indian Affairs to fix grazing fees on Indian lands at ten cents per head per month instead of ten cents per year, beginning with the 1880 season.[27]

Farmers, as well as cattlemen, were also intruding on Indian lands in the territory. Some fifteen hundred of the most "daring, intelligent and unscrupulous" whites had, by 1872, occupied Cherokee lands west of the ninety-sixth meridian. They argued that the Indians did not have the right to any land unless they cultivated it. As many as two thousand additional whites had claimed land in that same area by the time the Sixth Cavalry arrived to expel them. The intruders, who were greatly displeased that they had to leave, expressed much "wrath and indignation" against President Grant for having the army remove them; they threatened to "vote him down" for being an Indian sympathizer.[28]

Elsewhere on the Great Plains, environmental problems caused similar difficulties during the 1870s for those who were responsible for implementing the new peace plan, and agriculture remained a "precarious avocation," even for the best Indian farmers. Alkaline and windblown soils were unproductive or difficult to manage. Among the Pawnee, the Santee Sioux, the Winnebago, the Omaha, and the Oto on the Great Plains, grasshoppers and drought periodically ruined their crops and made "farmer" and "blanket" Indians question agricultural expansion or the acceptance of governmental policy. These tribesmen no doubt realized that white farmers were also subject to these misfortunes, thus agricultural life was not an absolute good. Misfortune could strike at any time; life was still tenuous on the Great Plains even if one became a farmer. The adoption of white agricultural methods for one's livelihood did not guarantee prosperity and freedom from hardship and want. Faced with this realization on the part of many Indians, federal officials, including the dedicated Quakers and other missionaries, had great difficulty in convincing the nomadic tribes that their future was tied to the plow.[29]

Governmental agents nevertheless tried to stimulate agricultural development, although those efforts were far too modest to meet departmental and Indian needs. Employees at the Sioux Whetstone Agency in Dakota Territory plowed two hundred acres and encouraged the tribesmen, with a great deal of difficulty, to plant seed. Unfortunately, drought ruined the efforts of both. Although the Pawnee in Nebraska used four mowing machines, three hay rakes, and many hayforks to put up a crop to winter their horses, they made little overall agricultural progress, because the Sioux still destroyed their farms, endangered their lives when they were in the fields, and stole their cattle. The Pawnee women continued to do most of the field work, while an "abundance of underdeveloped muscle" existed among the men. In 1873 the Indians at the Cheyenne River Indian Reservation in Dakota Territory received an ample supply of plows, harrows, wagons, harnesses, hayforks, hoes, and other implements; but because this was the first distribution of equipment among them, agricultural development proceeded slowly.[30]

Invariably, most of the tribes had too little seed, equipment, and wire fencing. At the Oto Agency in Nebraska, the Indians used butcher knives, tin cups, and their hands to dig post holes on the agency farm. At the Ponca Agency in Dakota Territory, many Indians harvested their wheat with butcher knives, because the one reaping machine was insufficient for that task. As a result, much of the crop shattered from the heads and was lost as it was being cut. Exceptions existed, of course. Among the Sisseton and the Wahpeton Sioux at the Devil's Lake Agency in Dakota Territory, the Indians had a steam-powered grist mill, mower, and reaper and a horse-powered threshing machine, all of which were very encouraging to the "industrious portion" of the tribes. The Winnebago in Nebraska also had a threshing machine. By 1878, several Sisseton in the Dakota Territory had purchased reapers from the sale of wheat. The nearby Yankton Sioux also used mowers and scythes with a "good deal of pluck" to harvest their rain-damaged wheat crop.[31]

In the Southwest, the Pima and the Maricopa quietly continued to go about the business of farming. In 1870 they cultivated 2,732 acres and raised a 40,850-bushel wheat crop, valued at $1 per bushel, as well as corn, beans, pumpkins, barley, and hay. The emphasis on wheat, by this time, stemmed largely from the encouragement of white traders, who had access to a flour mill located between Tucson and Prescott; but these traders invariably paid for the wheat in mercantile goods, rather than money, and at deflated prices as well. The Pima and the Maricopa also were beginning to experience a serious water-shortage problem, because white farmers above them on the Gila River were using increasing amounts for irrigation and mining. As a result, the Indians could raise only one wheat crop annually instead of two. In some cases, they did not have enough water for all of their crops, and land that had once been fertile stood idle. Friction also existed between Indians and whites when the cattle of one group ate the crops of the other because of inadequate fencing. The Indians in Arizona Territory still lagged technologically; they plowed with wooden moldboards hitched to the horns of their oxen in the Mexican fashion, instead of with iron or steel plows harnessed to doubletree hitches. Oxen were scarce, and these farmers usually had to take turns plowing their fields. If they had been supplied with light plows, horse collars, and harnesses and given the training to use their horses to cultivate the soil, their agricultural progress probably would have been much greater. Still, under very primitive conditions, the Pima managed to till between seven and eight thousand acres in the late 1870s.[32]

During the 1870s the Navaho in New Mexico were also struggling to become better agriculturists. Although late-spring frosts might injure their crops, they could rely on large flocks of sheep and goats for subsistence. Nevertheless, the government had to provide as much as two-thirds of their food. In 1872 they raised 130,000 sheep and goats, whose wool and hair they used for blankets, which they marketed for prices reportedly as high as $150 each. The wool would have had greater value had they been more efficient in removing it from their animals. Instead of shearing it off in one pelt, the Navaho hacked it off in clumps. Still, the agent was confident that the Navaho could become self-supporting, if only they could be given more agricultural education, seeds, implements, and livestock. They were, he reported, a "nation of workers" among whom the "drones" were very few.[33]

By the late 1870s, Indian agricultural policy was floundering at best. Grant's peace policy had not been able to keep patronage appointments from corrupting the Indian Bureau. By the early 1880s the Office of Indian Affairs was effectively circumventing the Board of Indian Commissioners. The board no longer controlled appointments; positions had become political spoils once again; and increasing demands for more Indian lands threatened to reduce Indian landholdings even more and to prevent the development of a firm agricultural economic base. Faced with these difficulties, policy makers were encouraged in 1879 when Capt. Richard Henry Pratt founded the Carlisle Indian School in Pennsylvania. The establishment of Carlisle began a renewed dedication to provide educational instruction that would include agricultural training. The primary purpose of Carlisle was to permit a select group of Indian children to be removed from their environment in order to teach them the rudiments of white civilization, after which they would be able to live and support themselves in white society.[34]

Although the Bureau of Indian Affairs continued to support the boarding and day schools on the reservations after Pratt established Carlisle, many officials believed that only off-reservation boarding schools could break the cultural bonds of Indian children and, in time, link them with white society. At Carlisle the boys and girls spent half the day in the classroom and half the day at work. This philosophy was not new; the missionary and government boarding and day schools on the reservations had long stressed the need for the Indian children to learn manual skills in addition to reading, writing, and arithmetic. In that tradition at Carlisle, the boys learned to cultivate the soil, to do blacksmith work, and to make harness. The girls received instruction in the domestic arts of cooking, sewing, and housecleaning. The "outing" system also was developed at Carlisle, whereby Indian children lived and worked with white employers for a period of time, in order to learn more about their trade as well as the nature of white society. Thus, Carlisle was essentially an apprenticeship for white civilization, and it quickly became the model for the development of additional off-reservation boarding schools. Three years later, Congress authorized the Department of Interior to utilize abandoned military posts for the development of similar schools. Congress was also willing to increase appropriations for the support of off-reservation boarding schools. In 1882, for example, it provided $50,000 to support the founding of Haskell Institute at Lawrence, Kansas, where, in addition to traditional studies, the students were taught farming.[35]

This new emphasis on the development of off-reservation boarding schools failed to provide the agricultural training that many hoped would enable the Indians to become peaceful, sedentary farmers. As had been true only a few years earlier in relation to the development of missionary and governmental boarding schools on the reservations, the parents became lonesome, the children became homesick, and parents sometimes abducted children. In addition, many schools tended to emphasize instruction in English and the mechanical arts, rather than agriculture. In some instances, the system of off-reservation boarding schools was used to hold hostage the children of hostile or recalcitrant Indians to make them bend to the dictates of bureau policy. This was particularly true at Carlisle, where Sioux children were sent in an attempt to subdue the people of Spotted Tail and Red Cloud, who had just been moved to the Rosebud and Pine Ridge agencies, where they suffered from the effects of severe winters and inadequate governmental rations. Parents were denied rations unless they sent their children to school.[36]

Ultimately, the boarding-school system failed to improve Indian agriculture, because only a small percentage of the children attended those schools and the ones who did attend emerged just as alienated from their past culture as they were from the white society that they were expected to enter as competent, if not equal, members. When the children returned from the boarding schools, they frequently faced ridicule or a feeling of isolation. They occupied a no man's land in which the ways of their past no longer satisfied them, yet the road to further advancement was blocked. The lands to which they returned frequently were unsuited for intensive or extensive agriculture, given the scientific and technological state of agricultural development at that time. As a result, these students had little choice but to return to more urban settings and ply a mechanical or domestic trade or to stay on the reservation and readopt their old culture or simply to waste their lives by idling their time away. In the end, agricultural education at the boarding schools proved to be a dismal failure, because federal officials did not have an agricultural policy that provided long-term technological, scientific, and financial support for the establishment of commercially oriented Indian farms.[37]

In 1880, while officials in the Indian Bureau were working to support at least modest agricultural training throughout the school system, most Indians in the trans-Mississippi West were facing the harsh reality of either making a living by agriculture or subsisting on governmental aid. Bureau officials took a degree of pleasure in the attempts of some Indians to raise livestock. At the San Carlos Agency in Arizona, for example, the Indians grazed 1,100 "highly prized" cattle, and some tribesmen milked their cows to supplement family diets. At the Umatilla Agency in Oregon, the Indians earned most of their income from stock raising. Primarily, they raised horses instead of cattle, earning an estimated $50,000 annually from their efforts. The Navaho grazed an estimated 700,000 sheep and 300,000 goats, as well as 40,000 horses. They marketed about 800,000 pounds of wool and wove another 100,000 pounds into blankets for their own use.[38]

Still, by the early 1880s, numerous problems hindered the further development of Indian livestock enterprises. In 1881, for example, inadequate congressional appropriations prevented the bureau from purchasing cattle for distribution to tribes that had not yet received livestock. The Kiowa and the Comanche in Indian Territory were forced to kill many of their breeding cattle, because the federal government did not provide enough rations to avoid the problem of acute hunger. Similarly, the Sioux on the Rosebud Reservation, who had received fifteen hundred cattle since 1879, did not have more than one-third of their herd still alive because of severe winters, insufficient fodder, and hungry tribesmen. When starvation did not compel the Indians to eat their stock cattle, some did so anyway. In 1881 the government issued four hundred head to the Pawnee in the Indian Territory to encourage them to become cattle raisers. Soon after that issue was made, however, some of the Pawnee began selling the hides to the agency trader.[39]

By the early 1880s, abundant grass and an adequate water supply in Indian Territory provided some of the best natural range on the continent for exploitation by Indians and whites alike. Although Texas fever had caused some Indian cattle raisers to abandon their efforts by 1882, 30 percent profits encouraged whites, particularly Texans, to reenter the territory and use Indian lands for their own purposes. The agent for the Kiowa and Comanche could not keep the white-owned cattle off unused reservation lands, and no one bothered to reimburse the Indians for the grass

In 1886 the San Carlos Apache labored, under orders from the United States government, to dig an irrigation canal on their agency lands. Although they became skilled cattlemen, they never developed an extensive irrigation system. National Archives.

that was eaten in the process. In 1883, white cattlemen made arrangements with some of the Cheyenne and Arapaho to rent reservation land at minimal annual fees of two cents per acre for a period of five to ten years. Half of the grazing fee was to be paid in cattle and half in currency. As a result, all but 465,651 of about 4.3 million acres came under white control. Elsewhere, cattlemen made similar arrangements or paid "grazing taxes." On 7 March 1883, for example, cattlemen organized the Cherokee Strip Livestock Association and leased 6 million acres from the Cherokee Nation in the Cherokee Outlet in the northwestern portion of Indian Territory. In September 1883 the Osage also began to lease tribal lands, and 380,000 acres of the reservation were under lease by the end of the year. Laben J. Miles, the Osage agent, justified the lessees because they provided needed revenue from unused lands. Because most of the leases were on the periphery of the reservation and because the lessees were required to fence their rented lands, the Osage Reservation would receive some protection from the cattlemen who illegally grazed their livestock on tribal lands. Here the lessees paid three cents per acre for a five-year agreement.[40]

The Bureau of Indian Affairs did not have a policy or enough power to prevent whites from exploiting Indian lands in this manner. Although federal officials did not approve of the leases, they permitted the Indians to make such arrangements with whites, because no formal land transfer was involved. The government treated those leases merely as licenses to enter the reservations for a specific purpose. In late April 1883, for example, Henry M. Teller, secretary of the interior, permitted the Cheyenne and Arapaho agent, upon tribal approval, to allow whites to graze cattle on the reservation for two cents per acre. Teller did not believe that a large number of whites should be permitted to use reservation lands in this manner, but a few "responsible" men could teach the Indians to become herders. "The Indians so employed," he wrote, "will soon become skillful herdsmen familiar with cattle raising. . .and become stock raisers themselves." He thought that any Indians who were employed as herdsmen should be paid in cattle. As a result, the Indians soon would have enough cattle to warrant the use of the entire grazing lands by themselves, and white cattlemen would be displaced.[41]

When the spring roundup came, more problems developed. White cattlemen descended upon their rented ranges and drove away all unbranded stock. If the Indians had not already marked their spring calves, they invariably lost those cattle. The agent for the Sac and the Fox reported that these whites "come at will, go at will, and do as they please, there being no law to intimidate them, nor force for local protection. Armed generally with two 45-caliber revolvers and a Winchester, they are 'monarchs of all they survey,' and a dispute is studiously avoided by the natives." In addition, ranchers drove large numbers of cattle north from Texas through reservation lands in Indian Territory on their way to Kansas markets. Many Indian cattle were either intentionally or unintentionally absorbed into those herds. Although the agent tried to locate and return some of those cattle, he did not have enough resources to regain much of the livestock.[42]

By 1885 the practice of leasing Indian lands for cattle grazing had become widespread, and although leases gave needed revenue to the various tribes, this policy divided the Indians, particularly the Cheyenne, some of whom threatened to renew hostilities to drive the white cattlemen from their lands. Leases also divided the whites, particularly the cattlemen and potential settlers who were not part of those agreements. Before violence occurred, however, Attorney General Augustus H. Garland, responding in July to a Senate inquiry, ruled that neither the federal government nor the Indians had the power to lease those lands. As a result, on 23 July, President Grover Cleveland declared the leases null and void. By early November the white cattlemen and some 210,000 livestock had left the Cheyenne and Arapaho reservation. Texas cattlemen continued to arrange leases with the Kiowa, the Comanche, and the Kiowa-Apache, and the Indian Office did not try to prevent them from doing so. The Cherokee Strip Livestock Association also continued to graze cattle in the outlet until 1890, when the government purchased the land to open it for settlement.

Similarly, in the Pacific Northwest, white cattlemen and sheepmen blatantly used unoccupied Indian lands for grazing and hay mowing, because their own ranges frequently were depleted or overgrazed. Poorly defined reservation boundaries contributed to the problem.[43]

While officials in the Department of the Interior vacillated over whether they could prevent whites from leasing Indian lands, they also gave increased attention to the wanton intrusion of whites on Indian lands for the purpose of permanent settlement. Part of the problem involved imprecise reservation boundaries as well as the vast tracts of unused lands. As a result, white settlers were not very "punctilious" about where they settled, and the Indians tried to extend their claims whenever they could. As a result, disputes arose between the two groups which developed into the "bitterest hostility." In many cases, the agents were helpless to resolve those disputes. The San Carlos agent at the White Mountain Reservation in Arizona clearly defined the problem: "It is a hard matter to take a crooked line 70 miles long, and ranging from peak to peak, and decide within a mile whether a ranch is off or on the reservation." Railroad pressures for land grants and rights of way through reservations compounded the problem, and unauthorized company surveyors sometimes entered Indian lands. The military also intruded on Indian reservations for the purpose of cutting timber.[44]

Those who settled illegally on Indian lands seldom faced the wrath of the federal government, and though they could be fined, those who had no property could not pay, and little effort was made to impose fines anyway. At best, federal officials could only remove the intruders, some of whom they forcibly evicted several times. Troops could not be stationed on every reservation because of the prohibitive expense. Consequently, when the illegal settlers were heavily armed, as they usually were, the Indians had little chance of removing them. When troops were present, "boomers" huddled along reservation borders, waiting for their opportunity to move across the line undetected and establish a claim to what they argued were public, not Indian, lands.[45]

Other misfortunes prevented the improvement and expansion of Indian agriculture. In 1880, the Zuñi in the Southwest produced a good wheat crop, but they did not have a fanning mill to clean the chaff from the threshed grain, and they only had one "elegant" steel plow for all of their tillage needs. At the Lower Brule Agency in Dakota Territory, the soil was "totally unfit" for cultivation because of the high concentration of alkali or because gullies and steep hills broke the land. In 1881 the temperature dropped to forty-one degrees below zero in Dakota Territory, and snow covered the ground, which made grazing difficult and contributed to the death of many cattle. During the summer months, "hot scorching winds" frequently shriveled the crops and stunted the grass. Although

In 1879, this Zuñi family had a pumpkin patch near its home. In addition to squash, the Zuñi raised corn and beans. National Archives.

white farmers who moved onto the plains in the 1880s experienced similar problems, officials in the Office of Indian Affairs did not realize the limitations of the climate in the trans-Mississippi West when they tried to make the Indians into white farmers. Some agents did warn their superiors about the hazards of the Great Plains climate. In 1881 the agent at Pine Ridge in Dakota Territory wrote: "The fact is, that by degrees the white man has taken from the Sioux pretty much all the land that can be considered arable. . . . White men well trained in farming, have tried to till the soil in this vicinity in Northern Nebraska and have lost all the money invested, and have not produced enough to pay for the seed. I can confidently venture to state that, if the experiment were tried of placing 7,000 white people on this land, with seed, agricultural implements, and one year's subsistence, at the end of that time they would die of starvation, if they had to depend on their crops for their sustenance." At that time, the nearby Rosebud agent wrote that "this country will not in our day become an agricultural country. Our Indians, if thrown upon their own resources and confined to this reservation, would soon starve to death." Governmental policy that was intended to make the Indians self-sufficient agriculturists as soon as possible was more than just optimistic; it was absurd. At best the Indians would need long-term aid until scientific and technological developments enabled them to have greater control or a manipulative influence over their environment.[46]

By 1883, some of the governmental agents and farmers among the Great Plains tribes had come to realize that agriculture was possible in the region if they produced wheat instead of corn. At the Crow Creek and Lower Brule Consolidated Agencies in Dakota Territory, the Sioux had "considerable" wheat for sale. Still, some of the tribes, such as the Northern Cheyenne and the Oglala Sioux at Pine Ridge, did not favor cereal grains, because they were accustomed to eating meat. They could not be made to change their cultural and dietary habits immediately and accept a new form of subsistence, to say nothing of a commercial economy. Nevertheless, by the mid 1880s the Indians at the Devil's Lake Agency in Dakota Territory were cultivating nearly twenty-five hundred acres and were using self-raking reapers and binders to harvest their small grain crops. Some of these machines were individually owned. The federal government purchased their surplus grain and gave it to the Chippewa in the form of flour.[47]

Technological progress had its drawbacks, however. Among the Sisseton Sioux in Dakota Territory, some of the Indians purchased expensive implements on credit at "ruinous" interest rates, just as their white counterparts did. As a result, a large portion of their crops went to pay the installments. Nevertheless, Indian farmers made some progress. In 1885 the Brule at the Rosebud Agency cultivated 2,286 acres. Some of these Sioux worked farms as large as eighty acres, the cultivation of which compared favorably with that of white settlers; this proved to the agent that farming was possible among the allegedly "self-willed," "stubborn," and "lazy Sioux." Even so, the agricultural life brought great cultural shock to the Indian farmers, who were "neither red nor white."[48]

Thus by the mid 1880s, Indian agricultural development was very uneven. Among the Five Civilized Tribes in Indian Territory, the most ambitious claimed the best lands. Some individual Indians worked farms ranging from five hundred to a thousand or more acres, exploited native workers, and earned large profits from their lands. In other words, they functioned like many white farmers. Most of the Indians were not that fortunate. Without sufficient agricultural training, because the Indian Office did not provide an adequate number of white farmers as teachers, the nonagricultural tribes had difficulty in learning to plow, plant, harvest, and tend livestock. Because of insufficient training, the Indians frequently wasted seed and neglected farm implements and livestock; and the implements, seed, and livestock often arrived too late or in too poor condition to use. In 1884 the farmers at the Warm Springs Reservation in Texas were still using cradle scythes to harvest their grain at a time when many white farmers were riding twine binders. The agent reported that farming there was "at best only a drudgery."[49]

Formal agricultural training at the boarding schools also failed to substantially aid the transition of the nonagricultural tribes. Too few Indian children received schooling, and not enough time was allotted for achieving the great cultural change that the training sought to achieve. Even though the Potawatomi boys in Kansas were "systematically trained in farm work" and even though they allegedly "begged" to be allowed to plow all day, most Indian boys did not receive adequate agricultural instruction. Furthermore, many of the elders in the nonagricultural tribes were openly resistant to so great a cultural change.[50]

Among the Southern Cheyenne in Indian Territory, the Dog Soldiers intimidated those who wanted to farm by destroying their crops and killing their cattle. At the Chiricahua Agency in Arizona, Cochise admitted in 1873 that his people would be better off if they learned to farm, but he thought that governmental officials would make the most progress with young people, because the older generation would not accept that life style. Nonetheless, while the Indians recognized the difficulty involved in cultural change, the agent enthusiastically professed that he would "learn them how" to farm.[51]

Where the Indians were successful farmers, a lack of markets hindered further development. The army usually provided the best outlet for Indian agricultural produce until a railroad entered their land. Thereafter, they usually lost that trade to white contractors. Local white contractors frequently purchased Indian commodities at low prices in order to supply the needs of the military. Even though Indian farmers usually were cheated in the prices that they received, without this limited outlet, they could not expand beyond self-sufficiency to become commercial farmers in the white tradition.[52]

Further agricultural development lagged because whites intruded on Indian lands, stole their livestock, or used their grazing lands. The slow progress of Indian agricultural development in the trans-Mississippi West, however, was not entirely the fault of incompetent or unscrupulous whites. The Indians themselves were often to blame. Lacking an agricultural tradition, the nomadic hunting tribes did not have the self-discipline to give all of their attention to the fields during the growing season. If they could, they would leave for their customary summer buffalo hunt, visit friends whenever the urge struck them, or simply refuse to tend their crops. Some recognized, as the Yankton Sioux did in 1871, that if they raised corn, they would have to eat it all winter, but that if they did not cultivate the soil, the federal government would give them beef.[53]

In the final analysis, governmental aid and education were inadequate, and not enough time had passed by the late 1880s for the nonagricultural tribes to undergo the cultural change necessary to enable the Indians to become farmers. In addition to cultural resistance, the environment was a great stumbling block to agricultural development on the Great Plains and in the Far West. Remarkably, at a time when white farmers on the Great Plains were beginning to organize politically to protest economic grievances that the environment in part had caused, the federal government was expecting the Indians to become self-sufficient and, ultimately, commercial farmers. Semiarid and arid climates, tough prairie/plains sod, bitter winters, grasshoppers, and other obstacles presented too great a challenge for the Indians who did not have an agricultural tradition. Indian culture could not yet triumph over the environment in regard to modern farming.

The Indians needed more time and more governmental support before they could become full-time farmers, integrated into a market economy. They had not yet reached that state by the late 1880s. Nevertheless, most federal officials believed that while Indian agricultural policy might be wanting in many respects, it was a success for the most part. The Indians had been concentrated on the reservations, and they had been taught to till the soil and raise livestock. The only thing that remained to be done before the Indians would be integrated into white agricultural society was to give them their own lands. The allotment of land in severalty would encourage recalcitrant Indians to take up the plow, once they had been guaranteed that the land they cultivated belonged to them. When the Indians realized that success depended upon working one's own land, private property would encourage them to make improvements and to become better farmers. The call for the allotment of Indian lands in severalty was not new. Indeed, governmental officials, missionaries, white farmers, and even some Indians had been advocating the development of a severalty policy for a long time. Some tribes had already been allotted lands. By 1887, however, the demand for the allotment of Indian lands in severalty had become a panacea—a panacea that would enable the Indians to become full-time farmers, elevate their standard of living, and allow them to assimilate into white society. It was a panacea that portended ominous consequences for Indian farmers.[54]

9

The White Man's Road

By the 1880s, most governmental officials, reformers, and western farmers had reached the opinion that severalty would solve the "Indian question." They reasoned that private property, if protected from speculators, would give the Indian a "sense of ownership" and provide him with the opportunity to "appreciate the results and advantages of labor" on the farm. By holding title to land, the Indians would become self-sufficient, ambitious farmers and therefore less reliant upon governmental annuities for subsistence. In time the Indians would become assimilated into the nation of white yeoman farmers. In a word, they would become civilized. Indeed, privately held farms were deemed to be the "door to civilization." Thus, severalty would begin a new era for Indian-white relations.[1]

This idea, of course, was not new. As early as 1633 the Massachusetts General Court had provided for land allotment to Indians. In the early years of the Republic, Secretary of War Henry Knox and, later, President Thomas Jefferson had favored severalty to encourage the Indians to become farmers. For Jefferson thought that individual ownership would prevent unscrupulous whites from obtaining lands held in common. During the removal years the concept gained increased support. By the 1850s, many treaties provided for the private holding of lands, particularly for chiefs and other tribal officials. Usually, however, these grants of private property were a form of bribery to win consent for land cessions, and they had little to do with the severalty policy developed during the 1880s. After the Civil War, reform groups, such as the Board of Indian Commissioners and the Indian Rights Association, continued to support efforts to grant lands in severalty as a means of pacifying the Indians and assimilating them into white agrarian society. In 1880, E. M. Marble, acting commissioner of Indian Affairs, held that all "true friends" of the Indian supported

136

allotment as a necessary feature of the "progressive age." On 6 March 1880 an important step on the road to a general allotment act came when the Utes agreed to a treaty that granted 160 acres to each head of a household, with additional acreage for grazing and for other tribal members. The federal government would purchase the unallotted Ute lands. Although the bill was bitterly debated, it became law on 15 June 1880.[2]

Because of the growing desire of whites for the allotment of Indian lands, on 19 May 1880, Texas's Senator Richard Coke introduced a general severalty bill which provided 160 acres for each head of household and 80 acres for single males. Orphans under eighteen would receive 40 acres, and other children, 20 acres. Allotment was to be made whenever the president believed that a reservation was suitable for division and the tribe was ready for agriculture. Allotment, however, depended upon a favorable two-thirds vote by tribal males who were over twenty-one years old. Upon allotment, the lands could not be sold for twenty-five years, so as to protect the Indians from land-hungry whites; but surplus lands would be sold.[3]

Carl Schurz, secretary of the interior, supported Coke's bill, because it would give the Indians permanent title to land, promote an agricultural life, end the need for annuities, and open more lands for white settlement and cultivation. Both the Board of Indian Commissioners and the Indian Rights Association supported the bill. Critics, however, charged that the western Indians could not be made into farmers simply by giving them 160 acres of land. Colorado's Senator Henry M. Teller led the fight against the bill, because he thought the benefits of severalty did not account for the various developmental stages among the tribes. He also believed that few Indians understood the concept of severalty whereby they would hold title to land but could not sell it without governmental approval. Senator Teller argued that the Indian had to be civilized before he would understand the concept and the value of private property and thereby be capable of wisely farming, leasing, or selling it. To grant the Indians land in severalty, Teller warned, was to ensure, not that they would become independent farmers, but that they would lose their properties within a generation.[4]

Although Coke's bill did not pass, Massachusetts's Senator Henry L. Dawes incorporated that plan into his own severalty bill, which he introduced on 8 December 1885. When President Cleveland signed that bill into law on 8 February 1887, it ushered in a new era for the American Indian. Like Coke's bill, the Dawes Act enabled the president to allot 160 acres to each head of a household on reservations where the land was suitable for agriculture. Single men over the age of eighteen were to receive 80 acres, while children under eighteen were allotted 40 acres. Where lands were unsuitable for cultivation, allotments were to be doubled to permit grazing and the development of a livestock industry. Those entitled to allotments were required to make their land selections within four years, after which the federal government would assign allotments to those who had not done so. The Dawes Act also enabled each qualified Indian to receive a patent for his land, but the federal government would hold it in trust for twenty-five years. During that time, the allotted lands could not be alienated. At the end of twenty-five years, the Indians would receive fee-simple title—that is, they could do what they wanted with it—sell, lease, or farm it. The federal government would purchase the surplus reservation lands and open them for white settlement. The money from the sale of surplus lands would be held in trust to support efforts at Indian education and civilization. Thus, the Dawes Act was designed to place the Indians on the white man's road to a peaceful agricultural life. Some reformers hailed it as the Indian Emancipation Act.[5]

Besides providing the opportunity for the Indians to become farmers on secured lands, the Dawes Act also offered major advantages for white farmers, because lands that were not needed for allotment were to be opened to white settlers. Advocates perceived several benefits. First, white farmers would have the opportunity to buy surplus Indian lands for their own use. Second, white families would provide examples of thrift, self-sufficiency, hard work, and the benefits of agricultural life to nearby Indian allottees. Third, severalty would aid the railroad companies by opening lands for new lines and by encouraging settlement, which would expand trade. Although the railroad companies did not actively lobby for the Dawes Act, company officials did nothing to jeopardize it, because severalty was in the railroads' best interests.[6]

To the extent that the Indians supported allotment, they did so grudgingly. For some, such as the Cherokee, who were exempt from that legislation, it was little more than an attempt to take lands that had been granted in perpetuity. To others, severalty offered the opportunity to secure at least a modest amount of land and freedom from the worry that the federal government again might move them elsewhere. Most of the western tribesmen wanted neither severalty nor the reservation system; they yearned to return to life before the arrival of the white men. The most troubling aspect of severalty, however, was the recognition among some whites and Indians that the western reservations were unsuitable for agriculture because of the scientific and technological limitations of that age, the absence of an agricultural tradition, and environmental limitations. In 1886, John D. C. Atkins, commissioner of Indian Affairs, recognized those problems when he wrote that "it would be a difficult matter for a first-class white farmer to make a living in the West."[7]

While the Indians awaited the implementation of the allotment program, some practiced agriculture. In 1887 the Bureau of Indian Affairs reported that Indian farms were more prosperous and better tended than at any time in the past. Where drought was not a problem, the Indians had made a "creditable" showing in agricultural productivity. Indeed, some Indian farmers, such as the Pima, gave officials reasons to be optimistic. Although the Pima cultivated only about ten acres per family, they were self-sufficient, even though "lethargy" prevented them from becoming commercial farmers. Even so, governmental farmers among them were successfully introducing alfalfa into their cultivation practices. Similarly, at the Round Valley Agency in California, Indian farmers were producing large quantities of vegetables, as well as some wheat, barley, and hops. At the Flathead Agency in Montana, the Indians were increasing their efforts to raise fruit trees. Potawatomi farmers in Kansas showed "calculation and thrift" by growing wheat, corn, oats, rye, and assorted vegetables. They used purebred Shorthorns to upgrade their herds, and they raised a "moderate" number of hogs. At the Cheyenne River Agency in South Dakota and at the Standing Rock Agency in North Dakota, Indian farmers also weathered the severe winter of 1886/87 better than did white cattlemen. Although these Indians had fewer livestock, cattle losses among the Indians averaged an estimated 15 to 30 percent, while figures for the white cattlemen reached as high as 75 percent. If these percentages are accurate, the great difference probably resulted because the Indians tended their livestock throughout the winter, while the white cattlemen let their animals fend for themselves. The Omaha farmers used money from the sale of hay, surplus lands, and cattle to purchase reapers and mowers.[8]

Still, problems remained. Unsurveyed reservations, for example, enabled white cattlemen to graze their animals on Indian lands and to plead innocence because they could not tell where the boundaries were located. Because reservation borders were not clearly defined, these cattlemen could not be prosecuted. Drought continued to plague the Great Plains and to discourage the Indians from becoming reliant upon agriculture. Wheat, which the agent at the Fort Berthold Agency in North Dakota had hoped would make the Indians into commercial farmers, returned only the amount of seed sown at harvest time. In 1887 at Pine Ridge, South Dakota, hail and drought destroyed the oat crop. Because these Sioux had not yet produced any crop for sale, the agent believed they needed to be "pushed and crowded" into doing so, but the lack of an adequate number of government farmers prevented him from applying that pressure. There, four farmers were responsible for teaching fifteen hundred widely scattered Sioux the principles of agriculture. Cultural tradition also prevented agricultural expansion. Among the nearby Yankton, many of the men had not yet made the adjustment from being hunters to being farmers. In addition, they tended to work in groups whereby they allegedly wasted time and effort and were less efficient than when they worked alone. Many preferred to survive on governmental annuities rather than to labor in the fields for their food.[9]

Farther to the south at the Kiowa, Comanche, and Wichita agencies in Indian Territory, the late arrival of seed wheat reduced the harvest in 1887, and no market existed for the wheat that was produced. The nearby Cheyenne and Arapaho did not have adequate markets for their corn either. Among the Five Civilized Tribes, cattlemen continued to infringe on Indian lands. By 1889, these lands were being occupied by more than thirty-five thousand intruders, mostly farmers, some of whom were "renegades and outlaws of the lowest class." The agent believed that a regiment of soldiers would be required in order to evict the intruders. White farmers in Kansas also grazed livestock on Potawatomi and Kickapoo lands and cut hay there as well. Cattle moved onto those reservations from as far away as seventy-five miles, and herds sometimes numbered from a thousand to fifteen hundred head. These Indians received only a dollar per head for the loss of their grass—a fee that the agent collected only with "infinite trouble and anxiety." In Nebraska a shortage of cultivators hindered the expansion of the Winnebago's corn crop. Frequently, a farmer would have to wait a week or ten days for his turn to use the few available implements. There were not enough reapers, and much of the wheat crop had to be cut with a mower and raked and stacked like hay, which made the heads shatter, thereby causing considerable loss. For the implements that were available, spare parts were difficult to obtain, or the wrong parts frequently were sent as replacements.[10]

Farther to the west during the late 1880s the Navaho continued to raise approximately 800,000 sheep and 300,000 goats. The wool clip, however, averaged only a pound and a half per fleece. Inbreeding and an insufficient number of new rams to improve their flocks caused the problem of low yield per head. The Navaho also were troubled with drought, which reduced the grass and made winter grazing difficult. They also lacked plows and other agricultural implements. At best, they were struggling to raise wheat, corn, and squash on approximately eight thousand acres. At the Colorado River Agency in Arizona, agriculture without irrigation was "simply out of the question" in the opinion of the agent. In order to keep the plants alive there, the Indians irrigated by using water jugs. This arduous work enabled them to cultivate about three hundred acres and to produce 250 bushels of wheat, 200 bushels of corn, 75 bushels of beans, and a "fair crop" of melons and pumpkins; but their ancient practice of killing the livestock of those who died continued to retard the development of a cattle industry.[11]

By 1890, most of the western tribes were still struggling to become agricultural. Thomas J. Morgan, commissioner of Indian Affairs, believed that Indian males over the age of forty did not make very "promising material" for becoming "enterprising farmers." Unaccustomed to tilling the soil, placed on infertile lands or lands that could not produce crops

Indian farmers in the Southwest developed hardy varieties of corn that could withstand heat and drought. This Navaho cornfield and hogan were located near Holbrook, Arizona, in 1889. National Archives.

without irrigation, and scattered far across the reservations, the Indians were not able to learn much from agency farmers. At best, the commissioner believed, the "large majority" of the Indians should become stock raisers to take advantage of the western grasslands and thereby become self-supporting. Yet federal policy not only prevented the Indians from becoming self-sufficient cash grain farmers; it also prevented them from becoming livestockmen because of its leasing policy for Indian lands.[12]

Although the attorney general had held on 21 July 1885 that the Indians, the secretary of the interior, and the president could not lease tribal lands, the Indian Office believed leasing should be allowed in order for the Indians to best use their lands. Officials in the Bureau of Indian Affairs held that a leasing policy was needed, because it would enable white cattlemen to profit from unused Indian land and would benefit the Indians with cash rents. In fact, the Department of the Interior had authorized several tribes to graze white-owned cattle on their lands, provided the owners paid the tribal agent a stipulated amount. Because the Indians of the respective tribes served as herdsmen, the department avoided the technicalities of the attorney general's ruling, but the tribes approved the grazing arrangements. Many of the tribes in Indian and Oklahoma territories continued to lease their grazing lands, despite the attorney general's ruling. As a result, numerous people, whom the agents could not control, entered reservation lands, and the federal government could not evict them immediately. To solve part of the leasing problem, on 17 February 1890, President Benjamin Harrison ordered all white-owned cattle to be out of the Cherokee Strip by 1 October 1890. This order was given, not so much to protect the Indians from the loss of their grass or to ensure a fair rental of those lands, as to guarantee that the cattlemen would not interfere with white settlers who would claim lands in the Cherokee Strip after that area had been opened for entry.[13]

Congress soon amended the Dawes Act to enable Indian landowners to lease their lands. In February 1891, Congress provided that an allottee could rent his land if "by reason of age or disability" he was unable to work. To do so, the Indian had to apply to the secretary of the interior for permission to lease his land. If approved, the Indian could lease it for farming or grazing purposes for a three-year period. After that time, the lease had to be renegotiated. This provision was designed to protect Indian landowners from renting their properties too cheaply for a long period of time. Pressure on the Department of the Interior to permit white farmers and cattlemen to lease Indian lands stemmed from the tendency of the Indians to allow their lands to remain idle, instead of putting them into agricultural production. Whites, who had more capital to invest in cattle and implements, thus sought the use of allotted lands for their own benefit. They rationalized that the Indians would make more profit from their lands by renting them, because the Indians were not cultivating them anyway. Moreover, lease money would enable the Indians to purchase implements, seed, and livestock for the land that they could efficiently farm. White farmers also would provide agricultural examples which the Indians should emulate. Some whites even argued that leasing was the best way to bring "wild lands" under cultivation and thereby prepare them for future occupation and use by Indian children.[14]

Not all cattlemen bothered to make leasing arrangements before they grazed their cattle on Indian lands. At the Unitah Agency in Utah, for example, the Indians asked the agent to remove the cattle that were illegally grazing on their lands. When he ordered the cattlemen to leave by 1 April 1891, the cattlemen petitioned the Bureau of Indian Affairs and received permission to continue grazing until 1 August 1891, thereby gaining the benefit of the entire growing season for grass. On many reservations, white cattlemen had been using Indian grazing lands for such a long time that they believed they had a "prerogative" to do so.[15]

Three years later, on 15 August 1894, Congress changed the wording of the leasing provision and made it possible for any Indian "by reason of age, disability, or inability" to lease his lands for five years. As a result, leasing increased, and by the mid 1890s, leases were relatively easy to obtain if one argued that an Indian allottee could not profitably farm the land. Moreover, the allottee simply could argue that he was unable to farm even if his problem only was "simple laziness." By leasing their lands, many Indians had little to do and quickly succumbed to idleness and intemperance. To combat that problem, the Board of Indian Commissioners recommended that leases be approved only in "rarely exceptional cases." In 1897, Congress changed the leasing system back to its original form by permitting the Indians to make three-year leases if because of "age or other disability" they could not farm their lands. These changes did not diminish the desire of whites to obtain Indian lands. Moreover, the inability of many Indians to make their allotments of 160 or fewer acres productive caused them to seek at least some monetary return from their lands. As a result, many Indians soon came to view land as a source of revenue, rather than as a foundation for economic independence based on their own agricultural labors. Most important, however, leasing was a major step toward the alienation of Indian lands. Once white farmers and cattlemen had leased Indian land, they either came to view it as their own or would not be satisfied until they had obtained title to it.[16]

When the Indians tried to improve the leasing arrangements, they sometimes were unable to do so. In 1892, at the Osage Agency in Oklahoma, for example, the Kaw agent allegedly joined members of the Arkansas City Cattle Company to maintain the old rent of four cents per acre, whereas the Kaw believed their lands merited a return of twice that amount. Moreover, by the end of the decade, the cattlemen had so overrun Osage lands that the tribesmen had difficulty in locating suitable farmland that was safe from roaming cattle. The problem was so serious, the agent claimed, that a "Texas steer has more rights upon the Osage reservation than the graduates of Indian schools." By 1895 the Winnebago and the Omaha in Nebraska also had leased almost all of the land—some fifty thousand acres, most of which had been leased to the Flournay Livestock and Real Estate Company at rates of between $0.15 and $0.50 per acre. The company in turn subleased the acreage for rates of $0.25 to $2.50 per acre. Similar problems existed among the Cheyenne, the Arapaho, and the Wichita in Oklahoma. Without federal protection from exploitation such as this, and subjected to agreements in which whiskey frequently was used to facilitate the proceedings, Indian lands that should have been used for their own agricultural purposes slipped away for the profit of others.[17]

Still, policy did not change to help the Indians retain their lands. In fact, on 31 May 1900, Congress restored the word *inability* to the leasing regulations and extended the terms once again to five years, provided the lands were used for "farming purposes only." Grazing leases remained limited to three years; thus, leasing became easier for Indians and whites alike. The Bureau of Indian Affairs held that the "age" reference meant that anyone could lease his allotments if he were under eighteen or too old to farm. The term *disabled* applied to unmarried women, the mentally impaired, the chronically ill, or widows without sons over eighteen. The term *inability* remained nebulous, but healthy allottees were not supposed to lease their lands.[18]

Not all whites believed the federal policy was beneficial for the Indians. Indeed, friends of the Indians were beginning to have second thoughts about the advantages of leasing. In 1900, even William A. Jones, commissioner of Indian Affairs, agreed that leasing was detrimental to the process of making the Indians into self-sufficient farmers. He argued: "By taking away the incentive to labor it defeats the very object for which the allotment system was devised, which was, by giving the Indian something tangible that he could call his own, to incite him to personal effort in his own be half." At that same time, the Indian Rights Association reported that the "practice of leasing is a growing evil." Frequently, those who desired to exploit Indian lands for agricultural purposes approached a tribal agent and offered a "bonus" if the official would only recommend approval of the lease. Those bribes could easily exceed the agent's meager salary. A year later, the Board of Indian Commissioners held that "leasing of immense tracts. . .to white men for terms of five years or more is likely to break up among several of our Indian tribes the most promising attempts which have been made at self support by cattle raising and grazing." Either one group or the other had to use those lands; the grasslands would not support both Indian and white cattlemen, particularly when the former group lacked the capital and managerial ability to develop large herds for sale in a market economy.[19]

In 1902, in spite of growing misgivings about the leasing policy, the Department of the Interior planned to lease approximately two-thirds of the Standing Rock Reservation in North and South Dakota, even though the tribesmen opposed this. The Indian Rights Association also objected to this plan, calling it a "backward step" in the efforts to make the Indians self-supporting farmers and a "complete reversal of government policy." Leases to white cattlemen would only guarantee that livestock would overrun native fields, because the Indians lacked fencing. Consequently, their crops would be ruined, and the Indians would be discouraged from making further efforts to become agriculturists. The department argued that the leases would ensure a monetary return of between $35,000 and $40,000—an amount that the Indians had never before received from white cattlemen. Even though the Indian Rights Association charged that the leasing scheme had an "unpleasant odor" about it, the agreements were made. In 1903 the Indian Office permitted nonresidents to graze cattle on the Blackfoot Reservation for one dollar per head. In 1905, white ranchers grazed an estimated ten thousand cattle on those lands, and perhaps grazed another ten thousand illegally, because the penalty for getting caught was only a fine of one dollar per head.[20]

The leasing policy continued, and on 1 July 1902, Congress permitted the Cherokee in Oklahoma to rent their allotments on an annual basis for grazing purposes and for five years for cultivation. The secretary of the interior had to approve leases for longer periods. Cattlemen were required to pay one dollar per head and fifteen cents per acre for grazing on unallotted Cherokee lands. On nearby Creek lands, leases would be approved for three years of grazing and for ten years of cultivation, with leases being available to the highest bidder. Even so, many of the Creek were persuaded to sign grazing leases at rates of twenty-five cents per acre annually, when fair remuneration would have granted them between one and three dollars per acre. Where Creek lands were leased through sealed bids, they received a "notable increase" in rents from white cattlemen. Generally, whites preferred to make informal leasing arrangements with the Indians, because they could obtain the use of the grasslands for less than when the Bureau of Indian Affairs supervised the leasing process.[21]

Large-scale cattlemen were not the only intruders on Indian lands during the last decade of the nineteenth century. In fact, small-scale operators caused most of the problems. At the Sac and Fox Agency in Oklahoma, the most troublesome intruders were "poor whites," who arrived with a few livestock, a wagon, and camping equipment and constantly moved about, grazing their cattle. White farmers also were a problem for the Cherokee, the Choctaw, and the Chickasaw. Intruders who were removed from tribal lands frequently returned, because there was no penalty to discourage them from settling on Indian lands. In the autumn of 1894, intruders posed such a threat to agricultural development on the Cheyenne and Arapaho lands that Secretary of the Interior Hoke Smith asked the secretary of war to furnish cavalry to keep the intruders off Indian lands. Intruders on Cherokee lands primarily settled in the cotton-producing sections of the Arkansas and the Canadian river valleys. By the spring of 1895, those lands had been almost entirely occupied. In the prairie section west and north of the Grand River, intruders cut hay that they marketed at the Kansas City stockyards, in the mining towns of southwestern Missouri, and to local cattlemen. By 1896, the number of intruders had allegedly almost reached the population of the Five Civilized Tribes.[22]

During the 1890s the tribes that were not plagued with the leasing problem continued to farm as best they could. In the Far West among the Pima and the Papago, the chief difficulty was insufficient irrigation water and poor-quality breeding stock. In 1893, drought caused the Navaho to lose not only their crops but also many sheep. Agricultural progress was slow, because one of the farmers assigned to the Navaho reservation was

During the late nineteenth century the San Carlos Apache earned some agricultural income by delivering hay to Fort Apache in Arizona. The hay, which was sold by the ton, was weighed on the scales at the left. National Archives.

charged with teaching agriculture to five thousand Indians, but he did not have a plow, a wagon, or a team to help convey his knowledge. By the mid 1890s, the agricultural position of the Navaho had improved, but their corn was "very light and chaffy." Thus, because of poor crop performance and because the Navaho preferred the "shepard life," one federal official recommended that they practice grazing to the "utmost limits." Nearby, the Pueblos still lacked threshing machines. As a result, they continued to tread grain from the heads by using cows, goats, and burros. Nevertheless, they had fenced fields and produced a grain surplus of which they marketed a "fair quantity." In the Pacific Northwest, at Oregon's Warm Springs Agency, Indian farmers raised less than before because much of the soil had been "worn out" from continuous cropping for fifteen to twenty years. At the Fort Belknap Agency in Montana, the Indians still raised so many horses, which were a sign of wealth, that they did not make much progress in cattle-raising.[23]

Drought also hindered agricultural development in the Great Plains. Insufficient precipitation frequently ruined the grain crops, and tribal agents continued to advise that the plains were suitable only for grazing livestock. With wheat averaging only five bushels per acre and corn returning only three bushels per acre, as it did at Pine Ridge in South Dakota, neither Indians nor white farmers could overcome the problems of a harsh environment, and agents contended that the time spent teaching Indians to farm was "virtually thrown away." The number of governmental farmers was too small to provide training, and this made matters worse. Moreover, while the Indians at the Rosebud Agency in South Dakota were seeking more oxen for draft and while Indians at the Devil's Lake Agency in North Dakota were struggling to farm with worn-out plows, white farmers on the Great Plains were beginning to use steam power to pull their plows and grain drills and to power their threshing machines. Yet, many Indian farmers, such as those at the Crow Creek Agency in South Dakota, could not get parts for their implements that needed repair. The agent for the Kiowa, the Comanche, and the Wichita in Oklahoma complained: "It was exceedingly hard to divide 115 plows among three hundred odd Indians who declared they want to use them." The chiefs made the situation worse by hoarding the farm equipment as status symbols. Thus, some Indians lagged so far behind technologically that they had little chance of catching up with their white neighbors and making their lands productive.[24]

Fortunately, the agricultural situation was not dismal everywhere. In the early 1890s the Flathead Reservation in Montana purchased three binders and several combined reaper-mowers. The agent expected those implements to help two hundred Indian farmers harvest a 45,000-bushel oat and a 40,000-bushel wheat crop from nine hundred acres. The Flathead also grazed about ten thousand cattle, which by 1895 they were trying to upgrade with purebred Holstein and Angus bulls. They also raised twelve hundred swine and as many as six thousand fowl. At the Santee Agency in Nebraska, Indian men were now farming, and they used binders—technology that placed them in the forefront of Indian farmers. The agent believed this was a sign that civilization was displacing "savagery." By mid decade the Blackfeet, who had received cattle from the federal government five years earlier, made their first shipment to the Chicago market. The Sioux at the Crow Creek Agency in South Dakota also furnished 150,000 pounds of beef for their own use. At that same time, the neighboring Rosebud Sioux sold more than fifteen hundred cattle at $30.84 per head to the federal government. The Iowa at the Great Nemaha Agency also developed into a "fairly successful community of farmers."[25]

By the turn of the twentieth century, then, the western Indians had made slow progress at best in becoming self-sufficient agriculturists. For the most part, they were struggling to become farmers or had abandoned entirely that white man's dream. On the Great Plains, drought, searing winds, and inadequate irrigation made crop raising difficult if not impossible for Indian and white farmers alike. Cultivated acreages were small. At the Devil's Lake Reservation in North Dakota, for example, the Sioux farmers averaged between twenty and eighty acres under cultivation. Most of that acreage was in potatoes, beans, corn, wheat, flax, oats, barley, and garden vegetables. Few raised livestock. At Fort Berthold in North Dakota, the reservation Indians still depended upon the federal government for 60 percent of their subsistence, though federal officials hoped that livestock raising eventually would make the Indians self-sufficient.[26]

At the Standing Rock Reservation in South and North Dakota, the Blackfeet, the Hunkpapa, and the Lower and Upper Yankton cultivated farms that usually contained two to ten acres. Although they raised cattle, their herds were not increasing rapidly, because they slaughtered many head each year for food. At the Cheyenne River Reservation in South Dakota, the Blackfeet, Miniconjou, San Arcs, and Two Kettle Sioux neglected their cattle by allowing them to range throughout the year without supplementary food or adequate care. Many of those Indians, together with those at the nearby Crow Creek reservation, did not own cattle; instead, they raised horses for prestige—an activity that provided scant monetary reward. Thus, those who would have been rich in their traditional cultural economy were poor in their new one. As a result, the Lower Brule depended upon the federal government for 60 percent of their subsistence needs. Most farms contained from five to twenty acres, which were largely planted in corn. The Loafer, Miniconjou, Northern Oglala, Two Kettle, Upper Brule, and Wazhazhe Sioux at the Rosebud Agency depended upon the government for 70 percent of their food needs.[27]

In Kansas, however, the Kickapoo, the Potawatomi, and the Sac and Fox held some of the best agricultural lands on the eastern fringe of the Great Plains. They had farmed for a long time and had the necessary implements and livestock to become increasingly productive; yet, they were "positively retrograding." These tribes failed to advance agriculturally because, like other Indian farmers on the Great Plains, many preferred to live off annuity payments and lease money. Their agent thought leasing had become a "mania." Those who did farm, however, were among the most productive Indian agriculturists in the American West. Each Kickapoo farmer cultivated between twenty and eighty acres of corn, potatoes, and other vegetables and raised livestock, such as beef and dairy cows. Nevertheless, those who farmed were far fewer than those who did not. Consequently, at the turn of the twentieth century, the Kickapoo were relying upon the government for 85 percent of their subsistence, while the Potawatomi needed the federal government for 75 percent of their food needs.[28]

Leasing and other pressures for the outright acquisition of Indian lands also retarded agricultural development in the southern Great Plains. At the Kiowa and Comanche reservation in Oklahoma, only a few Indians struggled to raise corn, wheat, oats, and livestock. As a result, the federal government met 92 percent of their subsistence needs. Although some cultivated more than a hundred acres, most planted from ten to seventy acres, part of which they seeded with drought-resistant kafir corn. Among the nearby Cheyenne and Arapaho, not more than 15 to 18 percent of the men farmed, and a "great portion" fed seed to their ponies instead of planting it. Although the Indians were to reserve at least 40 acres for their own cultivation, most of the Cheyenne and Arapaho informally rented all of their allotments to white farmers. Because of similar activities throughout Indian Territory in 1900, tribesmen were operating only 13 percent of the farms.[29]

In the Far West, Montana's Blackfeet attempted to develop irrigation with the aid of a white engineer, but their efforts brought little success. Many irrigation ditches needed repair or were worthless. Among the Pueblos, farmers still used sickles to harvest their wheat crops, because small plots did not merit reapers and because they did not have the means to purchase them. The nearby Pima and Maricopa lands were "practically worthless" without irrigation. Although irrigation agriculture had been important in the past, whites were using increased amounts of water upstream on the Salt and Gila rivers, and the Florence Dam had reduced the flow of the Gila to the extent that famine became a real threat to the Indian farmers who depended upon those waters.[30]

During the early years of the twentieth century, Indian farmers continued to flounder agriculturally. Most tribesmen on the Great Plains or in the Far West who did not have an agricultural tradition continued to lease their lands in preference to tilling the soil. Among the Sisseton in South Dakota, for example, a "great majority" were "anxious" to leave their lands. Drought discouraged them, and they preferred at least some return from their lands, rather than nothing at all. Yet, federal officials remained convinced that leases would benefit the Indians, because white farmers and cattlemen could provide examples of the best way to farm. Agents and farmers made little effort to obtain seed, to plow land, and to plant crops. In 1904 the Yankton Sioux planted three thousand fewer acres than in the previous year because of "promiscuous and indiscriminant" leasing by their agent. The federal government tried to make leasing beneficial to the Indians that same year when it required cattlemen to provide fencing on the lands that they leased for more than one year and required farmers to make "substantial improvements" on croplands that they occupied for more than two years. By so doing, officials hoped that the Indians could save the money earned from leasing and eventually move back onto their farms, where the sod would be broken and the fields fenced. Few realized, as did the Umatilla agent in Oregon, that once the whites had extensively developed Indian lands for agriculture, the Indians would not have the technology, capital, credit, or managerial ability to replace them at a later date. Furthermore, as was the case with the Cheyenne and the Arapaho in Oklahoma, improvements seldom were made, leases were broken, rents went unpaid, and trespass continued. In 1907 the Cheyenne and the Arapaho leased 471,000 of their holding of 518,240 acres. The few Cheyenne and Arapaho who did attempt to till the soil struggled against tremendous odds, because their small ponies were not strong enough to provide adequate draft power and because the government issued harness and collars that did not fit their horses. Kiowa farmers in Oklahoma also found the cost of implements prohibitive so that those who wanted to become farmers could not afford to do so.[31]

Inadequate irrigation facilities in the Southwest also continued to limit the expansion of Indian agriculture. Even so, the Pima and the Papago produced a surplus of wheat, barley, and hay which always found a "fair market" in Tucson. Among the Navaho and the Pueblos, the inbreeding of sheep continued to limit the wool clip to an average of two pounds per animal. Even so, the Navaho earned about $1 million per year from their livestock and from the sale of woolen blankets. The flocks, however, were suffering increased problems with scabies, lice, and ticks, and the Navaho needed a systematic dipping program. Where drought was a problem, grasshoppers sometimes made matters worse, as they did among the Zuñi. And where water was "king," white farmers continued to divert it for their own purposes, in spite of the prior rights of Indian farmers.[32]

On 27 May 1902, Indian agriculture received another setback when the secretary of the interior gained congressional authority to sell allotted lands of the deceased that were still in trust. Heirship land could not be willed, but when the allottee died, it was divided among the descendants in accordance with state law. As a result, many people might own an allotment or parts of several allotments. These small holdings further retarded the development of agriculture. Usually, heirs lacked capital or were too old or disabled to farm their inherited lands. Or the division created land parcels too small to merit agricultural development. Thus, many officials thought the sale of inherited land to white farmers seemed the best way to make it productive and, at the same time, to provide a monetary return to the heirs. As a result, Congress authorized the secretary to approve the sale of heirship lands. Before heirship lands could be sold, they had to be appraised; and then the Department of the Interior accepted sealed bids. If no bid met or exceeded the appraisal price, the land could not be sold. Frequently, land was reappraised at a lower value if the first bids were too low to permit a sale. Usually, other heirs or Indians wishing to add more acreage to their allotments could not afford to compete with white bidders. Thus, Indian farmers were effectively prevented from expanding and improving their farms. By 1905, Indian heirs had sold more than 250,000 acres valued at more than $2 million.[33]

More significant inroads on Indian lands came on 8 May 1906, when the Burke Act enabled the secretary of the interior to declare an allottee competent to manage his or her own affairs prior to the termination of the twenty-five-year trust period. This act enabled individuals to sell their allotments and enabled the secretary of the interior to grant a patent in fee simple to the heirs of an allottee who died before his trust period had ended. It also permitted the sale of heir allotments and the transfer of title to the purchaser, and it gave the secretary the authority to determine the legitimate heirs of an allotment. Francis Ellington Leupp, commissioner of Indian Affairs, believed that the Burke Act and previous legislation would solve the Indian problem, because Indians who held land in fee simple would not have legitimate claims upon the federal government for support. The Board of Indian Commissioners, however, believed that Indian lands of "great value" would soon be sold to whites.[34]

From the Dawes Act in 1887 to the Burke Act in 1906, the theoretical results of federal Indian policy were far different from the actual effects of federal legislation. Indeed, by the late nineteenth and early twentieth centuries, the best public lands had been homesteaded, and farmers, cattlemen, and land speculators were turning a covetous eye upon Indian lands. Once again, they raised the old argument that Indian lands were not being used and therefore were being wasted. White occupancy, through lease or purchase, for cultivation and grazing would make those lands productive. By 1906 the ability of Indians to alienate their lands with ease—lands that they could not farm because of environmental, technical, and financial limitations—eroded any possibility that they would become self-sufficient agriculturists, let alone commercial farmers, assimilated into the market economy of white agrarian society.[35]

By the early twentieth century, other problems were plaguing Indian farmers. The commissioner of Indian Affairs had not yet developed a comprehensive plan for agricultural training; no one else had either. The funds that Congress appropriated for hiring agricultural instructors were always inadequate for meeting the task at hand. With agency farmers receiving only $65 per month by the mid 1890s, few competent employees could be found. Agency farmers were often either incompetent and careless or far too concerned with politics to provide adequate agricultural training. Many could not speak the language of their charges. Without agricultural instruction, the western tribesmen could not become farmers in the short time that was expected of them. In addition, few teachers gave agricultural instruction. More emphasis was placed on learning English than on plowing and harvesting. In 1900, for example, only 320 farmers were employed to provide agricultural instruction to 185,790 Indians, not counting the Five Civilized Tribes, which were not provided with that service at all.[36]

Unless they had been provided with adequate agricultural instruction or other support in the form of implements, seed, and livestock, the Indians could not have become self-sufficient farmers, even had they been located on the most productive lands in a region with an ideal climate. Instead, they frequently lived on infertile lands in a hostile semiarid or arid environment and were cast to their own fate. When the Indians showed an inability to become farmers and thereby free the government of its obligation to provide them with rations, most whites assumed either that the Indians were uneducable or that they were too lazy to make the effort. In reality, however, the failure of the Indians to become independent farmers resulted from the government's failure to formulate and execute an appropriate agricultural policy. That failure stemmed from the inability of policy makers to understand the limitations of western farming, to recognize the poor location of Indian lands, and to comprehend both the need for intensive agricultural instruction and the difficulty of cultural change. This last factor was most significant in the failure of the old nomadic and hunting tribes to adopt a white man's agricultural way of life.

The Dawes Severalty Act also thrust the Indians into a monetary economy without providing them with the technological, educational, scientific, and monetary means to survive in it. As a result, the Indians sold or leased their land, because they could not farm it profitably or because the land was infertile or subject to a level of taxation that they could not pay. By 1906, heirship legislation and the Burke Act had subjected many Indian lands to alienation. Ironically, much of the reservation land in the trans-Mississippi West was suitable for grazing, and had severalty not deprived the Indians of their commonly held open-range lands, they might have been able to become self-sufficient cattlemen. Severalty, however, precluded this. Allotments of 160 or fewer acres were not large enough to graze cattle in areas where one animal might need thirty acres or more to survive. With the allotment process, the federal government also began to cut rations, in an effort to coerce the Indians to farm more intensively. This policy only forced the Indians to eat the cattle that might otherwise have contributed to building their herds. In addition, the nomadic tribes did not understand the value of land for sedentary agriculture. They understood that the land was basic to their cultural identity and that without it that identity would be lost, but they did not quickly grasp the white-held concept of land value in monetary terms.[37]

Given these policy and cultural problems, together with the environmental limitations of the American West for agriculture as it was then practiced, the Indians had little hope of becoming self-sufficient farmers within a generation, if ever. The Kiowa, the Comanche, and the Kiowa-Apache, for example, were located west of the 98th meridian in an area of Oklahoma known as the "famine district." Of the 155.6 million acres that the

Indians had held in 1881, by 1900, they retained only 77.8 million acres, of which 5.4 million acres were allotted. Although reformers had believed that severalty would help make the Indians become small farmers, it had not done so. Instead, severalty, cultural resistance, and the western environment, together with federal leasing and heirship policies and inadequate agricultural support, placed the Indians, not on the white man's road to self-sufficiency and civilization, but on the road to peonage. The twentieth century did not dawn bright with promise for Indian farmers.[38]

10

Good Intentions

A new century did not immediately change Indian agricultural policy. From the administration of Theodore Roosevelt to that of Herbert Hoover most policy makers and those who influenced Indian affairs remained convinced that an agricultural life for all tribesmen would solve the Indian problem. While this belief persisted, the old problems of allotment, leasing, heirship, and inadequate agricultural support remained. Despite those nagging obstacles to agricultural progress, Indian farmers continued to till their fields and to tend their livestock. Officials remained convinced that bureau policy would achieve economic emancipation and assimilation, if not acculturation, for the Indians within white society. Indeed, the evidence seemed clear.

In 1908 the Northern Cheyenne at Montana's Tongue River Agency grazed cattle valued at $75,000. By 1910 the Northern Cheyenne owned six thousand cattle, some of which they marketed for $38,000, and the Indian Office looked for "great progress" during the coming year. In addition, the Blackfeet in Montana grazed sixteen thousand cattle and received $160,000 from sales during fiscal 1911. By the end of December 1912, those sales totaled $292,161. On 15 April 1912 the Blackfeet also organized the Blackfoot Stock Protective Association. This organization required members to report thieves and cattle killers to the reservation's superintendent. A similar livestock protective association existed at the nearby Flathead Reservation, and federal officials continued to tout cattle raising as a key to economic self-sufficiency. In Oklahoma the Oto fields equaled those of their white neighbors, and the tribesmen planted fruit trees and alfalfa and built strong fences. In 1909, Cheyenne and Arapaho farmers in Oklahoma planted about four thousand acres, double the amount of the previous year. The agent for these tribes believed his charges could become self-supporting if they could retain possession of their land. Also in 1909 the Pima in Arizona increased wheat production by 50 percent with the aid of government farmers. Encouraging agricultural reports also came from the Yakima Reservation in Washington and from the Mission Reservations in southern California, where the Indians were farming more acreage and producing larger crops than ever before.[1]

Sheep herding remained important among the Navaho, the Pueblos, and the Mescalero in the Southwest, and with "persistent and intelligent encouragement," the Mescalero sold mutton on the Kansas City and Chicago markets for $5,000 to $9,000 annually and garnered another $5,000 to $6,000 from the sale of wool in St. Louis and Boston. In 1912 they sold 3,175 lambs for $2.75 per head, allegedly the highest prices paid for lambs in New Mexico that year. At that same time, the Navaho grazed approximately 1.4 million sheep which produced an estimated 3.4 million pounds of wool valued at $429,375. Merino grades produced another 293,463 pounds annually, worth $35,664. The Indian Office planned to improve the quantity and quality of that wool and mutton by up-breeding Navaho flocks with high-grade Rambouillet and Cotswold rams. At North Dakota's Fort Berthold Reservation, the Mandan, the Gros Ventre, and the Arickaree made a "good showing" in the production of small grains, and the Sisseton in South Dakota substantially reduced their seed costs by saving grain for future planting. By 1916, livestock production also had increased at most reservations, where 43,309 Indians tended an assortment of cattle, sheep, and goats valued at nearly $29 million.[2]

The Bureau of Indian Affairs reported in 1913 that it employed 249 farmers, 13 assistants, and 37 stockmen to teach the principles of agriculture on the various reservations. In addition to providing agricultural training, governmental farmers helped the Indians round up and destroy "worthless pony stallions," which many Indians still kept as status symbols. Governmental farmers and stockmen also tried to help the Indians up-breed small mares, which typified Indian horses, with larger stallions in order to produce stronger animals capable of pulling a plow or a reaper. The United States Department of Agriculture cooperated with the Bureau of Indian Affairs by sending technical literature and drought-resistant seeds to selected reservations, by giving advice concerning the eradication of grasshoppers and potato bugs, and by aiding the fight against livestock diseases, such as dourine, glanders, and mange.[3]

By 1914 the Bureau was also fostering a "spirit of rivalry" by organizing thirty-eight agricultural fairs on the reservations; fifty-eight were held in 1915. Officials encouraged Indian farmers to exhibit their produce and livestock at county and state fairs. These were not idle hopes, because several Indians in Oklahoma, Arizona, Nevada, and Wisconsin did so in 1915. Under the provisions of the Smith-Lever Act, Indian farm clubs also were

The Navaho began raising sheep during the late sixteenth or early seventeenth century. Sheep are still the most important livestock on the reservation. Here, three Navaho women are engaged in their annual task of shearing the sheep. National Archives.

organized on some reservations to encourage Indian children to learn more about agriculture. When a Papago boy won a $100 scholarship in an agricultural contest by producing fifty-one bushels of corn on an acre and when another boy won second prize for his Duroc-Jersey hog, bureau officials were pleased that Indian farmers were making progress. The Bureau of Indian Affairs also hosted farmers' institutes. Courses of instruction included gardening, dairying, and the raising of poultry, livestock, and fruit. Federal officials also established experimental and demonstration farms on several reservations. In 1908, for example, an experimental farm was established on the Pima reservation to determine the best grains, vegetables, fruits, and nuts and the best agricultural methods for that arid region. Agricultural researchers had high hopes that experiments with Bermuda onions, alfalfa, Egyptian cotton, drought-resistant olive trees, and pistachio nuts would help Indian farmers to develop new markets and new sources of income. Teachers at the Indian schools also taught agriculture "wherever practicable." Ideally, the goal of agricultural instruction was to achieve "practical results by simple means." Each school was to raise its own beef, fruits, and vegetables and to supply the needed milk, bacon, and poultry. Where the climate did not permit diversification, the superintendents were to emphasize the one crop that would grow the best or to graze all available land with cattle. In return, the Indian children would learn to become good farmers and cattlemen by following the examples of their teachers.[4]

Still, problems remained; one was that the bureau needed to increase the number and quality of governmental farmers. While some farmers were knowledgeable, many agencies had inadequate agricultural instruction. Governmental farmers often did not even live among their charges, but resided off the reservations. Tribal superintendents usually accepted the recommendations of their governmental farmers regarding the disposition of funds for agricultural work, and many governmental farmers connived with traders to profit from the sales of seed and equipment. With salaries ranging from $300 to $360 annually, the government did not offer sufficient remuneration to attract and keep good farmers. Agricultural instruction also was hindered because agents and superintendents frequently used their governmental farmer as a "man of all work." Consequently, governmental farmers spent most of their time doing work other than providing "systematic instruction" about agriculture. In 1910 the Board of Indian Commissioners expressed hope that a new system, which required competitive civil-service examinations for governmental farmers, would enable the bureau to put "expert farmers," rather than "college-boy" farmers, in the field. In 1916, however, preference for practical rather than theoretical experience changed when an appropriation act prohibited governmental farmers from receiving more than $50 per month unless they had a college education. This attempt to increase the qualifications was commendable, but it was too restrictive and therefore further retarded the improvement of agricultural instruction.[5]

The environment often posed an insurmountable obstacle for Indian and white farmers alike as crops perished under the blazing sun. Some tribes, such as the Jicarilla in New Mexico, also needed livestock to become agriculturists; but nearby Papago lands in Arizona were so barren that an estimated 140 acres were needed to support one cow. Adequate agricultural implements also continued to be in short supply. Among the Northern Cheyenne in Montana, for example, the lack of threshing machines caused a substantial loss of the grain crop in 1912. When the promised machines arrived, it was too late. By 1913 the federal government had also begun to phase out the practice of furnishing agricultural equipment, livestock, and seed in exchange for labor. Now the Indians had to pay cash for needed items or, as in the case of seed, return an equal amount after the harvest. While officials were pleased that most Indians who received such funding met their repayment deadlines, the $600-per-person annual limit of reimbursable funds kept Indian farmers chronically short of operating capital. Some Indians did not even use the reimbursable fund, because they wanted to avoid going further into debt. Yet, without equipment, seed, and livestock, Indian farmers had little chance of becoming self-sufficient agriculturists.[6]

Although these difficulties were serious, the greatest problems that restricted, if they did not prevent, further progress of Indian agriculture

Generally, Indian farmers have lacked modern technology for their commercial agricultural operations. This farmer at the Sacaton Agency in Arizona was more fortunate than most, because he had access to a McCormick binder to harvest his grain crop (ca. 1918). National Archives.

involved leasing, allotment, and heirship policies. In 1907, for example, Commissioner Francis Ellington Leupp still favored the continuation of the current leasing policy; he believed that the Indians were "embryo farmers" who, if permitted to lease their lands for short periods, would soon realize the value of those lands and be less likely to sell them in the future. Accordingly, the bureau planned to give the Indians more freedom by permitting the most "progressive" to collect rents without departmental supervision. As a result, leases expanded rapidly. At Fort Berthold in North Dakota, the Indians leased ten thousand acres to a white-owned sugar-beet company that used the Indians for planting, thinning, cultivating, irrigating, and harvesting.[7]

As the years passed, whites became increasingly interested in leasing Indian grazing lands, because most of the suitable public domain already had been taken. As demand grew, the rental fees increased by 25 to 100 percent. Whenever the bureau retained the right to approve the leases, prospective cattlemen had to submit sealed bids, and penalties theoretically were to be enforced if more than the allowed number of cattle were grazed on Indian lands. By 1911 the leasing of Indian lands raised the "gravest questions" of policy for the Indian Office, because officials now began to see it as a "positive detriment" to agricultural progress. As long as an Indian rented the land, he could not work it, increase his farming expertise, build capital, or advance on the road to self-sufficiency. In addition, competent Indians who were permitted to lease their lands did not show as much financial acumen as federal officials had hoped, though the policy makers did realize that the Indians often held more land than they had the capital to develop. If the Indians could not use those lands for cultivation or grazing, white farmers and cattlemen pressed for leasing privileges, and the rental fees theoretically would aid later improvement or expansion on the part of Indian farmers.[8]

The Indian Office continued to encourage the Indians to make better use of their grazing lands without resorting to leasing. In 1914, on the Sioux reservations, for example, white cattlemen no longer were permitted to lease large areas, and white-owned stock had to be removed from Indian lands. At that same time, the federal government also began furnishing tribal herds to upgrade reservation livestock and to encourage the Indians to give up the leasing tradition. More than $1.5 million was expended for the purchase of cattle, sheep, and horses. Government cattlemen were to oversee the herds and train the Indians to become stockmen; the natural increase of the herds would repay the federal government for the initial expense and would create a "nucleus" for building individual herds. The tribal herds also would reduce the expense that individuals incurred when they tried to purchase "first class" breeding stock to upgrade their own herds. The Blackfeet, among others, resented the tribal herd. They believed the federal government should furnish cattle without cost, and they preferred to sell their hay crops for cash rather than feed it to their own cattle.[9]

Even though the Indian Office hoped that the tribesmen would make better use of their grazing lands by becoming stockmen, economic change in the American West precluded that achievement. With grazing lands of the public domain rapidly diminishing, with the breakup of the larger Texas ranges into small ranches and farms, and with the increase in the number of Indians raising livestock, less land remained available for expansion by white agriculturists. Moreover, high prices for beef, mutton, and wool and increased demands for horses and mules, stimulated by the outbreak of a new European war, encouraged white cattlemen to look at reservation lands with jealous and greedy eyes. Although increased demand for the use of Indian lands boosted rental fees, it did not contribute to the expansion of Indian agricultural advancement.[10]

As the war in Europe intensified, both Indian and white farmers increased their agricultural activities. Driven by slogans that food would win the war and by the desire for great profits, whites pressed hard for the right to cultivate and graze unused Indian lands by arguing that the land was going to waste during a time of national crisis. While some whites believed the Indians should be "drafted" into farming, most simply wanted access to Indian lands. Commissioner Cato Sells was not prepared to force the Indians to farm, but he did believe the war would have a beneficial influence on Indian agriculture. It would enable the tribesmen to earn a comfortable living and, by so doing, to gain agricultural experience with a speed and intensity that had been lacking in the past. Because less federal

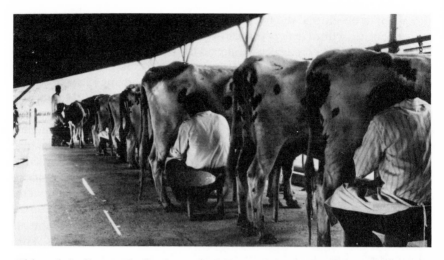

Although Indian agriculturists raised livestock in the Southwest, few tribes became noted for their dairy operations. This dairy at the Sacaton Agency in Arizona was an anomaly (ca. 1918). National Archives.

money would probably be available for Indian support as long as American troops were in the trenches, the Indians needed to take greater responsibility for feeding themselves. In 1917, with these thoughts in mind, Sells ordered his agents to take "aggressive action" to ensure that "every tillable acre on Indian reservations be intensively cultivated" to meet food demands at home and abroad.[11]

Although the acreage under cultivation increased from 5 to 100 percent on some reservations and while Indian farmers on seventy-three reservations cultivated 472,156 acres in 1917, up 11,360 acres from 1916, Indian farmers could not compete with their white counterparts. Environmental and cultural restraints, both red and white, prevented them from doing so. Because the Indians could not make maximum use of their lands, the Indian Office placed advertisements in newspapers and agricultural journals to encourage whites to lease those lands. High wartime prices and available land inspired whites to accept the patriotic call for help, but the new leasing regulations, established on 1 July 1916, seemed too restrictive to meet the needs at hand. Under those requirements, tribal lands could be leased for five years for dry-land agriculture and for ten years for irrigation farming. Allotted land could not be leased unless the owner was unable to farm it himself. The lands of minors could be leased only until the children reached adulthood, and the rental of heirship lands was restricted to an annual basis. The successful development or expansion of agriculture on Indian lands and the possibility of a protracted war demanded longer leases. As a result, by 1918, the bureau was approving leases for lengthier periods.[12]

Long-terms leases on liberal terms during the war years were not enough to still white agriculturists' demands for Indian lands, and the allotment policy only encouraged the further alienation of Indian property. Whites continued to insist that Indian lands be opened for settlement if the tribesmen would not or could not put them to profitable use, so the bureau hastened the process of issuing fee patents—that is, title—to Indian lands before the trust period had ended. As early as 1909 it was clear that the patenting process was a failure. On the Yankton Reservation in South Dakota it had a "disastrous" effect. Where the Yankton originally had controlled 230,000 acres, they now held about 87,000 acres. Similarly, more than 80 percent of the allotted lands of the White Earth Reservation in Minnesota had fallen into white hands, and by 1910, 60 to 75 percent of fee-patented land for the Omaha in Nebraska had passed into white control. A year later, an estimated 60 percent of all Indians who had received fee patents had sold their land. Although the Indian Rights Association "strongly opposed" patenting lands to incompetent Indians, Commissioner Leupp favored that policy, because it would force drunken Indians and those who did not send their children to school to become taxpayers and thereby bear their new social and economic responsibilities. When Robert Grosvenor Valentine became commissioner of Indian Affairs in 1909, he too encouraged competent Indians to apply for fee patents, because it would force them to "stand on their own two feet" and thereby make them independent of governmental aid. If the Indians sold their land, they would be forced to find employment in order to survive. Thus, clear land titles would solve the "Indian problem," and the faster allotted land could be patented, the better.[13]

Because the Indians were not applying for their patents quickly enough, Valentine organized a commission to judge the competency of the Indians and to issue patents to those deemed capable of managing their own affairs, no matter whether they wanted clear title to their lands. Valentine's first competency commission was assigned the task of evaluating the Omaha tribesmen, because more than 75 percent of the men were engaged in subsistence agriculture and were considered capable farmers. When the commission ended its work in March 1910, it made possible the patenting of more than twenty thousand acres to 244 Indians. Many of those Indians did not know English and were illiterate, and most of them were not capable of managing their own affairs, particularly when their lands soon would be subject to taxation and no longer were under the protection of the federal government. Of these patents, 107 were not wanted by the Indians. Quickly, white land-grabbers descended on the Omaha and offered them credit backed by land mortgages. The Indians used that credit to purchase horses, farming equipment, and other items, far beyond their means to pay, and soon between 60 and 75 percent of the land had to be sold to meet that indebtedness. In spite of the devastating affect that patenting had on Omaha agriculture, the Indian Office continued to grant fee patents as quickly as possible. New competency commissions were organized to visit the various agencies and determine whether the tribesmen were qualified to receive title before the termination of their trust periods. By the end of November 1920, when the use of competency commissions ceased, more than seventeen thousand patents covering approximately one million acres, had been granted; and the superintendents subsequently continued that work.[14]

In 1917 the patenting process quickened still further when, on 17 April, Commissioner Sells announced a "new declaration of policy." Thereafter, all Indians who had one-half or more white ancestry, who had reached the age of twenty-one, who had completed their education, and who were capable of managing their own affairs with the competency of an "average white man" were no longer considered to be under the supervision of the federal government. They were to be issued fee patents for their lands, even if they did not want them. The next year, Sells hailed this policy as the "beginning of the end of the Indian problem." Prior to this change in policy, the Indian Office had issued 9,894 patents between 1907 and 1917; but from 1917 to 1919 it issued 10,956 patents. In 1920, John Barton Payne, the new secretary of the interior, opposed the rapid patenting of Indian lands and returned to the policy of determining competency based on individual merit. For many Indians, however, this reversal of policy came too late; they had already lost their lands.[15]

While the Bureau of Indian Affairs was at work forcing fee patents on the Indians, it also continued to allot tribal lands. Although in 1907 the bureau pushed allotment as "rapidly as possible," allotment work had "decreased materially" by 1914. Only three allotment agents were in the field—a decrease of fifteen agents from only a few years before. The decline in activity and the reduction in manpower resulted because practically all lands "susceptible" to agriculture had been allotted by that time or were in the process of allotment. Although several large unallotted reservations still existed, particularly in the Southwest, the absence of water for irrigation prevented those lands from being allotted and relegated them to the grazing of tribal livestock. Nevertheless, by 1921, when most of the allotment work had been completed, 226,348 allotments had been granted for almost 37.2 million acres.[16]

While the allotment policy continued to retard Indian agriculture, the heirship problem did so as well. The sale of heirship lands further deprived the Indians of a land base. Because those lands frequently were divided into parcels too small to warrant cultivation, most heirs chose to sell their landed inheritance. In 1907, more than 106,359 acres of heirship lands were sold for an average price of $11.74 per acre; in 1908 an additional

91,302 acres sold for $14.27 per acre; and in 1909 another 102,708 acres sold for $12.86 per acre.[17]

On 25 June 1910, to expedite the sale of heirship lands, Congress authorized the secretary of the interior, through the bureau, to determine the legal heirs of deceased allottees in order to remove that responsibility from the state courts, whose determinations frequently were incorrect. Under this statute, whenever an heir petitioned for a patent or for the right to sell inherited lands, the supervising agent was authorized to take testimony, which he would then submit to the secretary for the determination of the legal heirs. Once the heirs had been determined, the competent heirs would receive fee patents and could sell their lands under departmental supervision. The secretary of the interior could even order the sale of heirship lands over the objections of the heirs. Soon the bureau became overburdened with heirship cases. With some allotments twenty years old, the settlement of contested estates became "extremely difficult." In 1911, for example, six hundred heirship cases on the Winnebago Reservation in Nebraska plagued the bureau, and this reservation was one of the smaller of the fifty-five reservations where allotments had been made. In order to speed the process of heirship sales, the bureau authorized the superintendents to put the buyer in possession of the land prior to the arrival of the patent from the General Land Office. In this way, the bureau could save two to four months between the time the sale was approved and the time the buyer assumed possession. By 1912 the average price of heirship lands had reached $20.30 per acre.[18]

Sales of heirship lands slowed in 1912 and 1913 because of drought; but in 1914, sales once again increased, and 90,768 acres were sold for an average price of seventeen dollars per acre. With sales of such volume at the discretion of the federal government, the Indians frequently suffered. Heirs often had lands sold against their will when officials connived with covetous whites to gain control of it. In 1919 the bureau estimated that heirs held only 50 percent of the lands allotted thus far. Commissioner Sells, who believed the Indians should be given as much independence as possible, encouraged his superintendents to issue patents to competent heirs, and he urged the superintendents to sell the lands of incompetent Indians. This policy, he argued, would put more Indian lands into white possession, and it would enable the monies derived from those sales to support the heirs. As a result of heirship policy, almost 1.4 million acres were sold between 1903 and 1920. During the 1920s, most of the heirs who had received patents continued to sell their lands. Pressure from land-hungry whites and the desire for quick and easy money encouraged them to do so; and thus more land passed from the domain of Indian agriculture.[19]

By 1920 the old problems that had restricted the expansion of Indian agriculture remained. Although high wartime prices and the bureau had

These Indian farmers are stacking hay at the Sacaton Agency near Pima, Arizona (ca. 1918). In the absence of irrigation, much of the arid Southwest was best suited for raising livestock. National Archives.

stimulated Indian farmers to expand production, drought in 1917 and 1918 had ruined many crops, as scorching winds "blasted" the wheat, corn, cotton, and hay crops in the Great Plains. Federal officials believed the war had stimulated the Indians to become more self-sufficient, but even so, the patriotic call for increased agricultural production did not bring a full harvest of Indian farmers into the economy of white America. By the end of the war, only 36,459 Indians were farming 762,126 acres, while 46,758 were raising livestock on 29 million acres. Although sales of farm commodities and livestock totaled $11.0 and $37.9 million respectively in 1919, Indian landowners were still leasing 9.8 million acres for a rental income of $807,503. In fact, despite wartime prices, renting still was easier and proportionately more profitable than farming. In addition, most Indian farmers still lagged technologically; therefore, extensive agriculture was impossible where many tribesmen cultivated at most 20 acres and where 160 acres or more were required for whites to farm successfully.[20]

Insufficient capital and knowledge about farming were still serious handicaps. Cash-crop farming meant single-crop agriculture in the West; and single-crop agriculture meant farming based on the production of wheat or cotton, large acreages, and mechanization. Without access to capital, Indian farmers could neither purchase new lands nor cultivate those they did own on a scale that would make a profit, and thereby move beyond subsistence agriculture. As a result, most Indian farmers were "doomed," and agricultural advancement was merely a dream. Only retrogression and the continued loss of land remained certain for most tribal farmers. Because banks would not make loans unless Indian lands were

Subsistence agriculture and insufficient credit have prevented most Indian farmers from purchasing large equipment such as tractors and disc plows. These at the Sacaton Agency near Pima, Arizona, were an exception. Tribal funds or credit with the agency probably were used to acquire these implements for communal use (ca. 1918). National Archives.

placed in collateral, these farmers had little choice but to mortgage their lands if they wanted to compete in a market economy. The risks were great, and the rewards were few.[21]

Indian farming also stagnated because the horses and mules that tribal farmers used for draft required substantial acreage for forage crops. With only limited acreage in cultivation anyway, the removal of additional land from cash-crop production restricted progress within a market economy. Indian farmers were caught in a vicious cycle. They needed capital to expand their operations beyond subsistence farming, and profitable commercial agriculture, in turn, would provide working capital for financing additional purchases of land, equipment, seed, and livestock. Without capital, however, Indian farmers could not progress beyond subsistence agriculture; furthermore, they needed capital even to reach a level of subsistence or self-sufficiency. Land was their only capital, but without experience in using it to enhance their agricultural endeavors, they were at the mercy of loan sharks and unscrupulous bankers.[22]

The bureau nevertheless tried to be optimistic, and federal officials believed they were acting in the best interests of Indian farmers. In 1920, 40,962 Indians produced $11.9 million of agricultural products, and 44,847 Indians raised livestock valued at $33.1 million. The nine thousand cattle at Montana's Tongue River Reservation were doing well on fenced ranges despite the drought. At the San Carlos Reservation in Arizona, the

Indians used registered Hereford bulls to upbreed the tribal herd, and the superintendent believed that cattle raising would be profitable in a "short time." The stockmen at the Fort Berthold Reservation in North Dakota shipped hogs weekly to St. Paul and made periodic sales of steers to Chicago through their cooperative shipping association. By the summer of 1924, they, too, were using purebred Herefords to improve the tribal herd. At South Dakota's Cheyenne River Reservation, the Indians organized a Farmers' Club, and six joined the Dewey County Farm Bureau.[23]

For the most part, however, the Indian agricultural landscape remained bleak. The postwar recession, coupled with drought, brought a collapse to the Blackfoot cattle industry; and stockmen on the other reservations suffered as well. At the Devil's Lake Reservation in North Dakota, the Indian farmers were reportedly "backward" and cultivated only 5,120 of 69,282 acres—one-third of which was tilled by four men and one woman. The "nomadic tendencies" of the Sisseton in South Dakota also caused them to neglect their crops and livestock, and the nearby Rosebud Sioux remained "rather prejudiced against extensive agriculture as a matter of principle." Small gardens and a few acres of corn satisfied their basic needs, while they drew rents from about 500,000 acres.[24]

Similar problems existed elsewhere. At the Crow Creek Reservation in South Dakota, patented land sold quickly, so the Lower Yankton Sioux, who lived there, showed little interest in farming. At the nearby Lower Brule Reservation, an estimated 55 percent of the original 300,000-acre reserve no longer remained in Indian possession, and the tribesmen were living in poverty. Officials complained that the Indians were not "industrious" and that they lacked "energy." Only a generation removed from a life of hunting and raiding, they were not yet capable of accepting the cultural life style of white Iowa farmers. At the Cantonment Agency in Oklahoma, the Cheyenne and the Arapaho were cultivating only 4,100 of 48,491 acres and were grazing cattle on only 4,869 of 28,344 acres of grassland. In the Southwest, the Navaho ranges remained overstocked, and the efforts to up-breed Navaho flocks with Rambouillet and Lincoln rams had failed, because those breeds were ill-suited to the hard desert environment. Many tribal ranges remained overstocked with "wild and worthless ponies," which were an "extravagance" at the expense of "remunerative stock."[25]

By the early 1920s, Indian policy continued unchanged. While the Bureau of Indian Affairs intended that all able-bodied males should cultivate a portion of their allotment, officials recognized that many reservations contained agricultural lands far beyond the ability of the Indians to cultivate or graze. To prevent the waste of unused land, the bureau encouraged the lease of surplus lands. While leases generally ran for five years, ten-year leases were permitted for irrigable lands. Allotments that were still held in trust could be rented by the allottees or heirs upon the approval of the tribal superintendent. Frequently, these leases brought no more than a dollar per acre. In 1925, whites leased more than 8 million acres of tribal lands for agricultural purposes.[26]

The allotment process continued during the early 1920s, although not on such a grand scale as in the past. In 1921, allotment work was completed on Arizona's Gila River Reservation, where each allottee received ten acres of dry land and ten acres of irrigable land. Allotment progressed on the Umatilla Reservation in Oregon, on the Bad River Reservation in Wisconsin, and on the Lower Brule and Cheyenne River Reservations in South Dakota. By mid 1925, allotments covered 37.9 million acres nationwide. The bureau also continued to issue patents to competent Indians as well as to those who had one-half or less Indian blood, whether they were competent or not. The sale of Indian lands also continued, although the postwar agricultural depression caused farm prices to drop and lessened the demand for Indian lands.[27]

Given the continuing problems of insufficient capital and governmental support, together with leasing, allotment, and heirship policies that restricted rather than enhanced Indian agriculture, the state of Indian farming was in worse condition than ever before by the mid 1920s. Most whites continued to view the Indians as dirty, lazy paupers or as picturesque figures, rather than as human beings who were still culturally primitive by white American standards and who were basically uneducated, frequently ill, usually hungry, and chronically poor. Clearly, whites needed to develop a new attitude toward the Indians that would enable the whites to provide the aid necessary to help the tribesmen adjust to twentieth-century white civilization. By ignoring the individual and collective qualities of the Indians, by exhibiting intolerance of their cultures, and by exposing the Indians—under the guise of federal policy designed for their protection and benefit—to some of the most greedy and unscrupulous land-grabbers in the nation, white Americans tended to treat the Indians as people in "perpetual childhood."[28]

By the mid 1920s, bureau policy did not substantially help the Indians to become self-sufficient, let alone commercial, farmers who could be acculturated and assimilated into white society. Not all whites, however, complacently accepted the Indians as second-class citizens; indeed, the critics of Indian policy became more vocal than ever before. Confronted with increased criticism of governmental policy, on 12 June 1926, Hubert Work, secretary of the interior, requested the Institute of Government Research to study the economic and social conditions of the American Indians and to provide recommendations for the improvement of federal policy. Under the direction of Lewis Meriam, specialists studied the matters of general Indian policy, health, education, economic conditions, law, family, community life, and the activities of women, migrants, and missionaries. When the commission issued its report in early February 1928, few could doubt that Indian agricultural policy had failed.[29]

In relation to Indian agriculture, the Meriam Report first attacked the use of governmental farmers. Instead of teaching the basic principles of agriculture, these bureau employees primarily served as field clerks and general laborers. Few taught agriculture; most of them were not qualified to do so. Low salaries precluded the employment of competent farmers to provide agricultural instruction. To rectify that problem, the commission recommended that agency farmers have the same qualifications and salaries of the agricultural-extension agents who served the white communities. A permanent agricultural-education program also was needed for each reservation, and the Indians particularly needed instruction about livestock raising.[30]

Agricultural education also needed major improvement at the Indian schools, because the teachers seldom provided the type of agricultural training that the children needed. Instead, school farms existed, first, to provide food for the students and thereby reduce expenditures, and, second, to provide agricultural instruction. Indian students tended dairy cattle, poultry, and gardens; but they were not taught to become dairy, poultry, or truck farmers in a market economy where the Department of Agriculture already blamed the problem of low commodity prices on an overabundance of agriculturists. Moreover, most instructors were as poorly prepared to teach agriculture as were the governmental farmers at the various agencies.[31]

The Meriam Report also recognized that the allotment policy failed to give the Indians a sense of responsibility and independence. Instead, the allotment program placed too much emphasis on individual landownership while it gave insufficient attention to teaching the Indians to use their new possessions wisely. Therefore, most Indians received allotments and fee patents before they adequately understood the responsibility of landownership. Heirship policy also complicated the process of encouraging agricultural advancement. When an allottee died, the bureau had to determine the legal heirs, regulate leasing agreements, and determine the feasibility of land sales. With this drain on staff time and finances, the sale of inherited lands was the simplest and cheapest course of action. Indian policy therefore depleted the land base and guaranteed that future generations would be landless. The acquisition of "surplus" reservation lands by whites after allotment only made matters worse, because it caused the Indian land base to shrink even more. As a result, several reservations could not support the population that lived on them.[32]

The allotment process had also proceeded so rapidly that much of the land, whether the Indians had selected it for themselves or whether it had been assigned to them, was unfit for cultivation in the white man's tradition. Even had the allottees been excellent farmers, they would have had great difficulty cultivating those lands for self-support, to say nothing of earning a profit from them. When good land was allotted, the Indians seldom understood its value or how to use it properly, and it soon passed into the possession of whites. If fee-patented or heirship land was not sold, it frequently was rented to whites, so the Indian owners did not use it for their own farming enterprises. Indeed, many superintendents preferred to lease Indian lands, because the fragmented allotments and heirship holdings were simpler to administer in this fashion, rather than to provide the proper agricultural training for their charges.[33]

Allotments with fee-simple patents also created serious tax problems for the Indians. Once an Indian had clear title to his land and the trust period had terminated, his land was subject to taxation; but without an income from the land, he was not able to pay, and his lands were subject to seizure for tax delinquency. Whites who had settled in Indian country demanded that the Indians pay their fair share of the tax burden for the support of schools, local government, and road building. But Indian land values were greater proportionately than their incomes, so property taxes were difficult to pay. If they were to pay taxes at all, the Meriam Report recommended a graduated income tax, which would be more appropriate to their financial situation.[34]

By the mid 1920s, most friends of the Indians and most policy makers still believed that independence depended upon agricultural progress. They assumed that the Indians were capable of becoming self-sufficient and, eventually, commercial farmers if they received the proper training and support. In fact, many Indians were engaged in some form of agricultural endeavor in spite of the problems that confronted them. Therefore the Meriam Report recommended that the bureau emphasize self-sufficient agriculture, because one-crop farming required commercial expertise— that is, the ability to buy, sell, and transact business at the proper time, as well as the skill to marshal capital for extensive farming operations. Because the Indians still were in the process of transition from a hunting-and-raiding to a sedentary-agriculture culture, they needed governmental aid and support. Because subsistence farming is a family-oriented endeavor, the Meriam Report also recommended that Indian women be given greater training for tending gardens, raising poultry, and preserving fruits and vegetables. In contrast to the past, Indian males now knew more about agriculture than did the women.[35]

Within the overall context of agriculture, the Meriam Report also favored emphasis on livestock raising, because the Indians had an aptitude for it and because so much of their land was fit only for grazing. The cancellation of leasing agreements and the issuance of tribal herds under the

These Indians at the Sacaton Agency raised vegetables under the direction of a federal agent (ca. 1918). National Archives.

supervision of a competent stockman would do much to stimulate subsistence agriculture. Tribal herds could help build individually-owned herds. In time, Indian-owned livestock would graze the reservation range lands, leasing would end, and the monetary returns would be greater than ever before.[36]

The lack of irrigation, however, was a major problem which the bureau had to resolve if it was to aid the Indians in the development of a self-sufficient agricultural economy. Without irrigation, many Indian lands could not be utilized for agricultural production; but irrigation systems were expensive to build and operate, and irrigation farming required an expertise that most Indians lacked. In addition, governmental officials estimated that white farmers needed at least two thousand dollars and two years' experience to succeed in irrigation farming in the West; and even so, many white farmers failed when engaged in irrigation agriculture. The Indian Irrigation Service not only failed to formulate regulations for the operation and maintenance of irrigation projects on Indian lands; it also changed the methods by which those projects were capitalized. Until 1914, tribal gratuity funds financed many irrigation projects, but on 1 August 1914, Congress required those costs to be repaid. Thereafter, the individual landowners who would benefit from the water were to bear those costs. Many Indians saw this action as a betrayal of the government's promise to develop irrigation facilities for their benefit. Moreover, some projects were not feasible from the beginning; many systems were built without engineering advice or supervision; and some irrigation canals could not supply the quantity of water needed. The Indians invariably could not pay for the projects, and the government frequently placed a lien on the land and required that payment be satisfied when the land was sold. Much of the irrigation expansion on Indian lands had occurred merely to impress Congress and to gain further funding. Irrigation projects on reservation lands became ends in themselves, and their benefits to

Indian farmers were only of secondary importance to the Irrigation Service.[37]

The Irrigation Service also remained poorly organized and under-staffed. Its employees received the lowest salaries of any in governmental service, and its work suffered from shoddy planning and lack of coordination and standardization among the reservations. Worst of all, the Irrigation Service served white agriculturists more than it served Indian farmers. Between 1911 and 1926, for example, 3 million acres of patented lands were sold to whites for $46 million, while an additional 2 million acres of heirship lands passed into similar hands for approximately $37 million. When those sales were made, the water rights were transferred as well. Consequently, where irrigation facilities existed, the Irrigation Service frequently was responsible for maintaining those systems for the whites who possessed the land.[38]

The development of irrigation projects also brought the federal government new and incredibly complicated problems concerning water rights. With 55 percent of the 241 reservations located in the semiarid or arid West, irrigation water became an important factor in the successful expansion of Indian agriculture. In 1908 the United States Supreme Court ruled, in *Winters* v. *United States*, that Indian water rights were reserved from the time the reservations were created. Yet the court left many issues concerning water use unresolved, and the bureau acted on the assumption that beneficial use had to be made within a specific period in order for the Indians to maintain their water rights. Because beneficial use would dictate future claims to the western waters, the bureau sought to lease irrigated land to whites who would make good use of it and thereby guarantee the water rights to Indian land. Water, however, differed from other property because it could be interfered with far from the lands of an Indian irrigator; therefore water rights were subject to lengthy state and federal litigation before the rightful users could be determined.[39]

By autumn 1927, 150 irrigation projects covered 692,057 acres of reservation lands, only 362,018 acres of which were irrigable. Whites controlled 244,829 acres, and Indians occupied only 117,189 acres of those potentially irrigable lands. Nearly 40 percent of irrigated Indian acreage was located on Arizona's Pima Reservation and on the Unitah Reservation in Utah. Clearly, project managers favored white irrigation farmers who could best use the water to reap high yields from intensive agriculture and thereby meet project assessments. For the most part, the Indians lacked the necessary experience, capital, and tools to begin irrigation farming, so their acreages within an irrigation system remained dry. By mid July 1932, irrigation ditches were serving approximately 190,000 acres of Indian land, but water did not flow through them.[40]

Thus, as the nation moved closer to economic collapse, Indian farmers were destitute. Among the Northern Cheyenne on the Tongue River Reservation in Montana, drought, grasshoppers, and hail destroyed the 1925 grain crop, and reservation officials withheld funds from tribal cattle sales to help repay old "reimbursable fund" debts. Without crops or funds, these Indians were near starvation. In 1928, drought and grasshoppers also destroyed the crops at the Crow Creek, the Lower Brule, the Pine Ridge, and the Rosebud Reservations in South Dakota. At Pine Ridge and elsewhere, hoes, rakes, and walking plows remained the basic agricultural tools. Leasing continued to be a "great evil," as it did on the Cheyenne and Arapaho Reservation in Oklahoma. The Navaho range in the Southwest still was overstocked with approximately 1.3 million sheep and goats—more than double the carrying capacity—and far too many horses grazed tribal lands.[41]

The onset of the Great Depression made a bad situation worse for Indian as well as white farmers. Prior to the stock-market crash in 1929, cotton from the Gila River Reservation in Arizona sold for $0.17, but in the autumn of 1931 it brought only four cents per pound. At the same time, wheat prices fell from $2.25 to $0.85 per hundred pounds, and alfalfa prices fell from $18 to $6 per ton—all economic setbacks that were "discouraging" to Pima farmers. During the winter of 1931/32, privation and want were so great among the Indians that the Department of the Interior distributed 6.2 million pounds of flour for human consumption and 5.6 million pounds of wheat for livestock feed, which the Federal Farm Board made available through the Red Cross. Moreover, drought returned to the Great Plains in 1932 with a severity that led to the dust bowl in the southern plains during the remainder of the decade.[42]

The Hoover administration had only a slight interest in Indian problems, other than to reorganize the Bureau of Indian Affairs so as to simplify management and make it more responsive to the "needs of the field." Indian agricultural progress was symbolic, rather than a mark by which to gauge the future. In 1930, for example, New York's Mohawk farmers formed a dairymen's league to promote knowledge about this endeavor. Two years later the extension agent for the Lac du Flambeau Reservation in Wisconsin convinced almost every family to cultivate a garden. At the Colville Reservation in Washington, Indian children showed renewed interest in 4-H work, particularly in relation to sheep raising. By 1931, 343 clubs with 3,377 Indian members had been organized nationwide. The Hopi produced an estimated 750,000 bushels of corn annually on about six hundred acres, which they irrigated with well water; and they used Hereford cattle to improve their livestock. By the early 1930s, however, the policy of forced acculturation had failed. Between the passage of the Dawes Act and the embarkation on a new course of policy in 1934, the Indians had lost an estimated 83 million acres, or about 60 percent of their original 138 million acres of reservation lands. Nearly half of the remaining acreage was located in the semiarid or arid West, where, without irrigation, agriculture was a high-risk business.[43]

Thus, during the first thirty years of this century, Indian policy remained firmly grounded on the assumption that the Indians should become self-sufficient farmers. Often, however, the Indians could not cultivate their lands, because of the harsh environment, inadequate capital, and insufficient training. Consequently, Indian landowners chose to lease their land or to sell it. Leasing provided a steady though limited income, without the risk involved with actual farming. In addition, land sales prevented the Indians from becoming self-supporting, industrious agriculturists. To a great extent, however, the failure of Indian agricultural policy had been caused by errors of judgment, rather than malicious intent. Few recognized that cultural change comes slowly. Even by the early 1930s the Indians were far from the elusive goals of acculturation and assimilation. Indeed, the Indians were near cultural disintegration. Even so, the critics of Indian policy were becoming more vocal, organized, and effective. As the nation sank into economic depression and as a new administration came to power, American Indian policy was on the verge of dramatic change.[44]

11

The Indian New Deal

On 21 April 1933, John Collier became commissioner of Indian Affairs. His appointment clearly placed the bureau on a new course. With Collier at the helm, governmental officials had the opportunity to steer Indian policy on a revolutionary course that would change Indian-white relations for all time. Collier's view of the Indian world and its place within white society was far different from that of other friends of the Indians. He believed that all efforts to acculturate and assimilate the Indians into white society had failed. Previous agricultural policy, with its accrual of allotment, heirship, and leasing problems, had served only to destroy the inherent strengths of Indian tribal life. Henceforth the Indians might become farmers, but they would not do so within the context of white culture. Instead, the Indians would retain their cultural uniqueness; no longer would they be forced to emulate white farmers. Cultural pluralism, not acculturation and assimilation, was the key to the Indian problem. Tribal communalism, rather than private individualism, Collier maintained, would engender viable social alternatives to life based on the precepts of white society. In order to achieve that goal, Collier asserted, the federal government would help the Indians become self-supporting within their own ethnocentric world.[1]

Collier brought a new sense of mission to the bureau. He had a messianic drive to aid the underprivileged and abused, which he had developed during more than a quarter-century of social work. Collier had become associated with a new breed of reformers during the previous decade. After visiting New Mexico's Taos Pueblo in 1920, he became convinced that the cultural unity of the Pueblos provided the answer to the Indian problem. Soon he was crusading to protect tribal institutions and property

174

Although most Indian farmers in the twentieth century emphasized grain and livestock production, some, such as those on the Morongo Reservation in California, tended orchards. During the 1930s the Morongo had an extensive drying yard for apricots. National Archives.

rights. As a research agent for the Indian Welfare Committee in 1922 and as executive secretary of the American Indian Defense Association in 1933, he actively opposed the principles of the Dawes Act. Instead, he championed the preservation of Indian civilization by the creation of a "Red Atlantis." Only by so doing, he argued, could the Indians live self-sufficiently and with dignity.[2]

Collier's attacks on the Bureau of Indian Affairs during the 1920s helped convince Secretary of the Interior Hubert Work to commission the Brookings Institution to investigate the bureau; the result was the highly critical Meriam Report. Thus, by the time of Franklin Delano Roosevelt's election to the presidency in 1932, Collier had gained a reputation as the most outspoken champion for dramatic reform within the Bureau of Indian Affairs. As such, he was a leading candidate for the position of commissioner of Indian Affairs. When he accepted the position, many Indians and reformers, along with Collier, believed the time had arrived for an "Indian New Deal."[3]

Upon assuming office, Collier clearly perceived the problems that confronted the Indians, and he acted quickly. His reform plan was broad, but agricultural policy remained the cornerstone of his program. Collier's agricultural plan involved guaranteeing the land base for future generations, preserving the soil, and extending federal aid in a multiplicity of forms to create a sound economic base that would make self-sufficiency possible within the tribal community. To achieve these goals, Collier first pushed for the creation of the Indian Civilian Conservation Corps, which would not only provide needed work relief by creating thousands of federally supported jobs but would also help prevent soil erosion and restore reservation lands to full agricultural productivity. President Roosevelt supported Collier's plan, and the new agency took the field in July 1933. Officially known as the Indian Emergency Conservation Work, it became known unofficially as the Indian CCC or CCCID. On the premise that reservation problems differed from those found elsewhere, the bureau, rather than the CCC, maintained control of this organization. Work camps, for example, were established for single and married men and for married men with families. If the project enabled them to do so, married men could live at home or take their families with them. Moreover, the work of the CCCID was directed not toward improving the public domain (as was that of its parent organization) but toward conserving and restoring reservation lands and training the Indians to use their lands wisely.[4]

By July 1933 the CCCID had planned seventy-two camps for work on thirty-three reservations. With a congressional appropriation of almost $5.9 million for the first six months of work, the CCCID turned its attention to soil conservation, range management, and forestation. Most of the work was scheduled for reservations in Arizona, Montana, New Mexico, Oklahoma, South Dakota, and Washington, where "relative poverty" and "relative need" determined the first expenditure of funds. Collier believed the erosion problem was "acute" on the reservation grazing lands in these states, and he called upon the Indians to diversify their crops and to institute better range-management practices. In addition, Collier obtained $6 million from the Public Works Administration for the Irrigation Service to begin work on lands of the Navaho, Pueblos, San Carlos, Wind River, Yakima, Duck Valley, and Walker River reservations to provide water for subsistence farming.[5]

By autumn, the CCCID was employing 14,400 Indians on sixty-nine reservations. A presidential allotment of $4 million enabled Indian conservation work to continue into 1934. During that time, the CCCID built stock-watering ponds and terraces that checked runoff and prevented gullying; eliminated rodents, such as prairie dogs, flickertail gophers, and ground squirrels, to save forage; drilled wells for irrigation; and developed springs, so that livestock would not have to travel more than two miles for water. The CCCID also seeded crested wheatgrass on wind-eroded lands and built fences to make possible the rotation of range lands for livestock grazing and to promote natural reseeding. During the first year, the CCCID built 746 reservoirs and developed 719 springs. CCCID workers also constructed 12,931 check dams or stock-watering ponds, which protected an estimated 59,480 acres from erosion. The CCCID also eliminated 44,052 head of "useless range stock," mostly wild mustangs on reservations in the Southwest, to improve grazing for remunerative sheep, cattle, and goats. This aspect of the conservation program encountered stiff resistance, however, particularly from the Navaho and the Hopi, because these animals represented cultural as well as economic security.[6]

While the bureau became increasingly involved with soil conservation, Collier also made it responsible for the administration of a new land policy. He had long believed that the allotment policy of the previous administrations was destructive to the well-being of the Indians. Allotment, together with heirship and leasing policies, continually eroded the Indian landed estate, and he was determined to reverse that policy. On 12 August 1933, Collier ordered his superintendents to halt the sale of all Indian lands that the federal government held in trust or that were otherwise restricted. This decree covered both allotted and inherited lands. Moreover, the superintendents were not to issue fee patents, certificates of competency, or requests for the removal of the restrictions that prevented the sale of Indian lands except in cases of severe individual distress or emergency.[7]

This order was merely a stopgap procedure, designed to halt the sale of Indian lands. The Indian lands that remained, Collier believed, had to be brought under a new system of Indian ownership. The approximately 17 million acres of allotted lands and the 7 million acres of heirship lands had to be "salvaged" for productive use by the Indians themselves and "permanently safeguarded against voluntary or forced alienation." To achieve that end, "deep readjustments," voluntary on the part of the Indians but supported by legislation to ensure success, were needed. He also argued that lands that the Indians had lost because of previous agricultural policy had to be "recaptured" to enable them to live full and productive lives as subsistence farmers. Henceforth, all Indian lands not under individual control also were to be subjected to communal use. Tribal rather than individual development of those lands, he believed, would guarantee the preservation of their landed estate. To accomplish this, however, the Indians necessarily would become "laboratories and pioneers" in the new efforts to restructure the tribal world.[8]

Indeed, a reversal of Indian land policy was essential if the tribesmen were to avoid generations of, if not permanent, poverty. No allotments, for example, had been made at the Sisseton Agency in South Dakota since 1892, yet Fred A. Parker, agency superintendent, reported that all land subject to allotment was gone and that within a generation all reservation land would be in heirship status and the Sisseton would be relegated to "wanderers over the face of the earth," with no future but to "perish miserably in dire poverty." On the Round Valley Reservation in California, out of 50,000 allotted acres, only 2,400 cultivated and 13,000 grazing acres remained in Indian possession. Of that total, 10,000 acres were inherited lands, and most heirships were too small to merit individual development. Some heirs leased their lands for only one cent per acre annually.[9]

Because the allotment policy had turned the Indians into "paupers," instead of making them responsible, self-sufficient farmers, and because many whites still supported past policies, Collier believed legislation would be necessary to guarantee protection for his land-policy decree. By early 1934, Collier had developed an extensive plan to redirect the course of American Indian policy. Submitted to Congress in mid February, it became known as the Wheeler-Howard bill. It called for governmental organization on the tribal level, improved education, the establishment of an Indian court, the abolishment of the allotment system, the restoration of surplus lands to the tribes, and the acquisition of new lands with a $2 million annual fund. It also authorized the secretary of the interior to transfer privately held lands to the tribe if he deemed such action necessary for the consolidation of Indian lands. The bill also enabled the tribes to assume control of restricted lands upon the death of the owner, rather than having the land be divided among the heirs. The heirs, however, would maintain a proportional interest in tribal lands. In addition, the bill authorized the secretary to sell current heirship lands to the tribe, and it provided for an Indian agricultural and industrial credit fund. Collier believed the Wheeler-Howard bill would ultimately enable the Indians to gain control of their own affairs, prevent future allotments, and consolidate and expand Indian lands.[10]

In sharp contrast to past policy, the policy that Collier urged was communal, rather than individual, ownership of property. While many Indians supported the Wheeler-Howard bill, others feared that the government would now force them to give up their allotments and their heirship rights to the tribe. Some, particularly those who were the most acculturated and assimilated, such as the Indians in Oklahoma and New York, objected to the renewal of tribal sovereignty in daily affairs. Friends of the Indians, such as the American Indian Rights Association, also objected to the Wheeler-Howard bill, because it would segregate rather than integrate Indians into the white community by stressing tribal self-development and self-government. The friends thought the Wheeler-Howard bill would prevent the Indians from becoming self-sufficient farmers on the model and scale of white men.[11]

In order to win congressional approval for his program, Collier was willing to compromise. Even so, opposition within the Indian community, as well as in Congress, remained strong. Before the bill became law on 18 June 1934, it had been redrafted in the House and heavily amended in the Senate. The compromise bill extended the trust period indefinitely for Indian lands, restored unallotted and surplus lands to tribal ownership, made possible the voluntary return of allotments to communal control, and provided for the addition of lands with an annual fund of $2 million. The Wheeler-Howard Act, also known as the Indian Reorganization Act

(IRA), directed the secretary of the interior to safeguard Indian range lands by preventing overgrazing, by limiting the number of livestock on the reservations, and by implementing soil-conservation practices. It also authorized a $10 million revolving credit fund for economic development.[12]

After the act was approved, Collier solicited support from the various tribes, which were to vote whether to accept or reject it, because the IRA was not to be forced upon them. Collier hoped that an educational campaign would convince the tribesmen that the acceptance of the IRA would mean both protection for their property and cultural and economic independence, whereas rejection would mean a continuation of the old problems of allotment, heirships, and inadequate credit. As a result of voting during the next two years, 181 tribes were placed under the IRA, another 14 tribes were included because they did not hold elections, and 77 tribes rejected the new policy. Collier was upset that the returns were far less than unanimous, particularly because the Navaho, who formed the largest tribe, rejected the legislation. The Navaho did so largely because Collier's opponents had convinced them that the act would permit the federal government to take their sheep and goats in a stock-reduction program, and the Navaho believed that such a policy would make them destitute. Although Collier argued loud and long that the stock-reduction program and the IRA were separate components of bureau policy, the Navaho rejected the IRA rather than face the prospect of starvation in the effort to improve their ranges. Still, Collier generally was pleased with the provisions of the IRA, and he believed that 18 June 1934, the day on which President Roosevelt signed the act, would become known as the Independence Day of the American Indians.[13]

While the Indian Reorganization Act became the heart of the Indian New Deal, the livestock-reduction program became the hand by which the bureau intended to lead many of the tribes to agricultural self-sufficiency. On the Navaho Reservation, for example, overgrazing was the most serious problem that prevented further agricultural progress. By 1933, Navaho sheep and goats had denuded more than 2 million acres in Arizona, Utah, and New Mexico. Because of overgrazing, together with drought, much of the land did not have enough vegetation to prevent serious erosion. With the Navaho ranges overstocked by as many as half a million head of livestock, the agricultural situation was critical in the Southwest. To rectify this problem, Collier supported a stock-reduction program that would not only remove four hundred thousand "excess" sheep and goats from the ranges but would also enable the institution of a soil-conservation program. The Papago ranges in Arizona were also being overgrazed during the mid 1930s; sixty thousand horses and cattle were grazing lands that had a carrying capacity of only twelve thousand cattle.[14]

A stock-reduction program, Collier believed, would increase both the quantity and the quality of the range lands in the Southwest. This improvement would make possible the production of healthier lambs with more meat and sheep with heavier fleeces. Any immediate decrease in income would be offset by increased earnings over the years ahead. Mistakenly, however, Collier assumed that the Navaho would readily see the advantage of stock reduction and agricultural diversification. Yet, for the Navaho, the program represented a venture into the unknown, because it denied them the security that their livestock provided. Thus, when Collier presented a detailed stock-reduction plan to the Navaho Tribal Council in October 1933, its members balked. They contended that the previous generation and drought had caused the soil-erosion problems, not their livestock. While most Navaho grazed a few sheep, goats, and horses, which provided food, clothing, money, and social status, few large-scale livestockmen existed. Therefore, instead of reducing the livestock of small-scale herdsmen, the Navaho believed the overgrazing problem should be solved by expanding the size of the reservation to provide more grazing lands and by developing irrigation systems that would make it possible to produce additional forage for their animals. Language and cultural differences further aggravated the buruea's inability to win immediate or total acceptance of a stock-reduction program. Bureau officials, speaking through an interpreter, had difficulty getting the Navaho to understand such concepts as carrying capacity, maximum holding, sheep unit, grazing district, and nonproductive stock.[15]

Although Collier convinced the Navaho to accept his plan to remove a hundred thousand sheep and goats from their herds, he unwisely called for an across-the-board reduction based on the average dipping records for the period 1929-34. Consequently, the large-scale herders were able to meet their reduction quota by merely culling unwanted animals, while the small-scale stockmen were forced to eliminate productive animals that were essential to their livelihoods. As a result, many small-scale herders were brought to the verge of starvation. Moreover, while the Tribal Council had agreed to the stock-reduction program, the individual Navaho herdsmen had not, and most of them obstructed the full implementation of the program. When the time came for the Navaho to approve the Wheeler-Howard Act, they were embittered by the stock-reduction program and the failure of the government to extend the Navaho boundary to the east. The result was the rejection of the Indian Reorganization Act.[16]

In March 1934, Collier again went before the Navaho Tribal Council to win approval for the reduction of 150,000 head of goats and 50,000 sheep. This plan also called upon the Navaho to sell 80 percent of their annual lamb crop. Faced with bureau threats to impose stock reduction even if the Tribal Council did not approve it and recognizing that the sales would provide desperately needed income, the Tribal Council accepted the program on 2 July 1934. According to this plan, the Federal Surplus Relief Corporation (FSRC) would purchase the livestock with a $250,000 fund. All male goats that were not sold and that were not from milk-producing strains were to be castrated. The government also promised to can the goat meat and use it to stock the pantries of reservation schools. This plan failed because of logistical problems with the packers, so some goats were shot and allowed to rot. This action, though limited, completely discredited the stock-reduction program among the Navaho. The bureau achieved the reduction of 50,000 sheep and 148,300 goats for nearly complete success in this phase of the program, but the Navaho saw failure instead of success. Navaho sheep and goats provided status and wealth for their owners, and herding patterns helped to bond families. To see their livestock shot and wasted while the herdsmen were going hungry violated Navaho cultural values and what they believed to be good sense. In 1935, when the FSRC pledged $400,000 for the purchase of 200,000 sheep and goats, few Navaho participated in this extension of the reduction program, and the government was able to purchase only 16,225 sheep and 14,716 goats. As a result, the remaining funds were transferred to a similar stock-reduction program among the Laguna and the Acoma Pueblos, who agreed to a 50-percent reduction of their herds.[17]

While the Navaho stock-reduction program was not a total failure, it was nearly so. By 1936 the federal government had spent $476,000 to purchase 315,000 sheep and goats; yet Navaho herds had been reduced by only 212,000 head because of natural increases. After 1935 the Navaho refused to make further voluntary reductions, and the bureau resorted to forced reductions and the issuance of grazing permits to control the numbers of livestock, particularly horses, that Navaho herdsmen could graze. Sheep prices on the open market also exceeded governmental purchase payments, and the bureau had a constant fight to keep herd levels within the forage limits of the reservation. Increased Navaho resistance to the stock-reduction program, which included the opposition of the Tribal Council after 1942, led to the termination of the project six years later. By that time, however, the program was beginning to show positive results, because the number of sheep on the reservation had been reduced from approximately 1.3 million in 1930 to 465,000 head in 1948.[18]

While the Bureau of Indian Affairs was moving to reduce surplus livestock among the Indians of the Southwest, it was also working to improve the quality of the animals that henceforth would provide the basis for a viable livestock economy. Although Navaho sheep produced wool with tough fibers that were well suited for weaving blankets and carpets, the clip was light and the meat was of low quality. Continuous inbreeding exacerbated the problem, and previous attempts to introduce new rams and breeding practices had failed to bring about any substantial improvement in the Navaho flocks. In the absence of a controlled breeding program using the best rams available, lambs were not only born from November to July, with a resulting high loss due to winter weather, but the 300,000 lambs that were produced each year were inferior in quality. In addition, the 80,000 to 100,000 feeder lambs sold annually were not uniform in size and age. Without adequate forage, the lambs also arrived at market in poor condition, rather than as well-finished livestock produced under the best range-management practices. When Indian stockmen sold their feeder lambs to post traders in the autumn, those that fell below minimum weight and size were rejected and returned to the range, where they became part of the breeding stock. Moreover, male lambs that were stunted from insufficient feed or by genetic deficiencies were seldom castrated. As a result, the Navaho maintained flocks with breeding stock that was unsuitable for commercial feeding and slaughter. For commercial purposes, then, Navaho herds were in a state of degeneration.[19]

Collier realized that the quality of the southwestern flocks was inferior and that improvement was mandatory if the Indians were to become commercially viable livestock producers. As a result, he requested the Department of Agriculture to make recommendations for improving the sheep in the Southwest. In June 1934 the Department of Agriculture responded by recommending that an experimental laboratory be established. There, the type of sheep best suited for the ranges could be determined, and the government could teach the herd and range-management practices needed to improve meat and wool production and to conserve the soil. Collier agreed, and the Bureau of Indian Affairs, the Bureau of Animal Industry, and the Soil Conservation Service pooled their resources to establish the Southwestern Range and Sheep Breeding Laboratory at Fort Wingate, New Mexico. Construction began on 10 February 1936, and scientists soon were experimenting with Romney, Corriedale, and Cotswold rams for up-breeding. Scientists at the experiment station hoped to produce a hardy crossbreed that would adapt to the arid climate, that would produce wool acceptable to the commercial and home-weaving industries, and that would sire market lambs of good quality. From these experiments, crossbred rams were available to the Navaho for controlled breeding programs. The results of this experimental work were less than the officials desired. Although breeding experiments improved the uniformity and the color of the fiber, the wool was not as suitable for the carpet or native rug-weaving industries, and the weights of the fleeces were not adequately increased. The introduction of Rambouillet rams by the Navaho Service had also encouraged the herdsmen to breed for meat rather than for improvement of the wool. Nevertheless, the joint efforts of the experiment station, the agency officials, and the Navaho herdsmen contributed to the development of larger, healthier, and more marketable sheep. In addition, they substantially reduced the mortality of newborn lambs by controlling the lambing season, and the wool clip increased by 58 percent in little more than a decade.[20]

In contrast to the livestock problems of the Indians in the Southwest, those of the Great Plains tribes stemmed from an inadequate number of cattle, which prevented the development of a viable livestock economy. Insufficient credit, deficient annuities, and inadequate breeding stock in the past had kept the plains tribes from building individual or tribal herds. Despite these problems, the great drought of the 1930s actually benefited the livestock-raising programs of the Indians on the Great Plains. The drought, the searing heat, and the dust-bowl conditions had dried, burned, and covered up much of the pasture lands on the Great Plains, so white cattlemen would face severe financial losses if their livestock died. By 1935, many white cattlemen who leased Indian grazing lands at the Red Shirt Table Community on the Pine Ridge Reservation in South Dakota had liquidated their herds. In June 1934, however, the federal government instituted an emergency drought cattle-purchase program which enabled cattlemen to cull their herds, receive a check from the government, and make the adjustments needed to ensure a viable operation as soon as normal precipitation had returned to the Great Plains.[21]

While most of the cattle purchases from the emergency drought area contributed beef to the Federal Surplus Relief Corporation, the Bureau of Indian Affairs used the cattle-purchase program as an opportunity to acquire livestock for the reservations that needed cattle. In late summer 1934 the Department of Agriculture allocated $800,000 to the Department of the Interior for the purchase of "well-bred" cattle in the drought-stricken area of the Great Plains. All purchases were to be made in cooperation with the Aberdeen-Angus, Shorthorn, and Hereford Breeder's associations, thereby benefiting white and Indian cattlemen alike. Although title to the cattle remained with the Indian Service, the Indians were permitted to select the favored breed for their reservations. In order to obtain the cattle, however, the Indians had to sign an agreement pledging that within three years they would return one calf for every cow they received. Upon return of the calf, title to the cow would be transferred to the Indian owner.[22]

Although administrative delays slowed the implementation of the program, the bureau estimated that it would acquire 25,000 cattle through the drought cattle-purchase program. Bureau officials also anticipated the purchase of an additional fourteen to fifteen thousand head of purebred cattle in the drought area. This program, bureau officials believed, would give the Indians the "greatest opportunity" they had ever had to become successful livestock producers. Reservations that were not included in the original distribution program were to receive cattle from the repayment, or "Revolving Fund," of cattle at the end of three years. But because the calves used for repayment would not immediately be suitable for breeding stock, the bureau intended to allow four years for repayments in kind among these Indians.[23]

Cattle that would be brought to the reservations under this program could be distributed either to organized corporations and associations or to individual cattlemen. All heifers thus acquired were to be between twelve and twenty-four months of age and were to be bred prior to arrival on the reservations. In addition, all cattle had to be either purebreds or grades, and only a limited number of bulls were to be acquired. The cattle were also to be branded with the letters ID, to signify that they belonged to the Department of the Interior, thereby ensuring proper accounting of the annual increase and title to the livestock.

Although this cattle-distribution program held great promise, it was not without serious problems. On the Fort Totten Reservation in North Dakota, for example, the cattle that arrived from New Mexico were infected with brucellosis, and some of the livestock had to be destroyed at governmental expense. In addition, the Indians needed more cattle than they received through the program. In spite of problems such as these, however, bureau officials believed the drought cattle-purchase program was "very encouraging," and the most optimistic officials contended that the Indians who participated could not help but become successful cattlemen. Indeed, this aid helped Indian cattlemen and their livestock to increase from 8,627 owners with 167,313 head to 13,787 owners with 229,343 head by the end of 1934. By February 1936 the drought cattle delivered to the Pine Ridge Reservation in South Dakota were doing well, although the Sioux remained in "desperate need" of more livestock to increase the economic return from their land, and leasing also remained a serious problem. With rents at only ten cents per acre, white stockmen grazed thirty thousand sheep on reservation lands.[24]

The Sioux were beginning to organize livestock associations, a development that bureau officials in particular were glad to see. They believed that Indian-organized cattle associations not only would improve breeding and range-management practices but also would help educate other tribesmen about proper livestock-management procedures and enable them to make improvements in marketing, disease prevention, and fencing. By the spring of 1937, some of the best livestock associations were located on the reservations at Fort Hall, Idaho; Fort Belknap, Montana; San Carlos and Mescalero, Arizona; and Yakima, Washington. Indeed, the Klickitat River Cattle Association on the Yakima Reservation was responsible for overcoming the reluctance of Indian cattlemen to test their animals for tuberculosis. With 100 percent participation by association members in a testing program, the nonmembers fell into line. Ultimately, all reservation cattle received the test. The cattle associations could market their livestock in carload lots, thus increasing sales efficiency, reducing expenses, and thereby improving profits. At the Blackfoot Agency in Montana, a cooperative marketing association earned prices from 12 to 30 percent higher than those received by nonmembers. These associations made the best use of reservation lands that were checkerboarded with allotted, tribal, and white holdings. By fencing a large tract of land, association members could command a larger amount of range land than individual operators could muster; thus small-scale cattlemen could receive the same economies of scale that the larger operators enjoyed.[25]

Indian cattle associations also flourished in the Southwest. In June 1938 the San Carlos Tribal Council approved the creation of nine cattle associations. When a tribal member wanted to enter the cattle business, he had to apply to the board of directors of a specific association, and his name was placed on a waiting list. When his name came up for admittance, he was allowed to select twenty heifers from either the tribe's or the association's herds. These were considered loan cattle, which the new operator had to replace with two other cattle within seven or eight years. If a member died, the association rounded up the deceased person's cattle for its own herd. Each member was responsible for participating in the roundups, branding, and other work. If a member failed to meet his obligations, the association levied a fine. In addition to maintaining a tribal herd of two thousand head, half of which were registered Herefords for up-breeding individual herds, the San Carlos Apache also created a "Social Security" herd on 6 December 1938. Sales of purebred Herefords from this herd were to aid all tribal members who were without support, who were too old or ill, or who suffered from a physical handicap and were unable to work. The Jicarilla Apache in New Mexico also maintained an "Old People's Herd," to support the aged, infirm, and those in need of help.[26]

The clan-based cattle associations of the San Carlos had achieved a remarkable degree of success by the late 1930s. In 1937, for example, the associations marketed more than ten thousand head of "fine Hereford stock," which brought the owners an average return of about $32 per head and the association members more than $3.5 million. These associations were so strong and knowledgeable about the cattle business that on two occasions the Indians rejected all bids until they received the prices that they desired. On another occasion, they rejected all bids and shipped their cattle directly to a Los Angeles market, thereby obtaining prices $4.13 per head higher than the best offer at the reservation. San Carlos cattle were desirable, because buyers came to the reservation each year from Oklahoma, Texas, Colorado, California, and Iowa, as well as Arizona. Moreover, by the end of the decade, the six hundred registered Herefords that the federal government had furnished to the tribe during the drought of 1934 had increased to twelve hundred, all of which contributed to the further up-breeding of the San Carlos herds. All registered animals remained the property of the tribe.[27]

The Pueblos in New Mexico also organized successful cattle associations which provided for better management of the range lands, supported vaccination and dipping programs, organized roundups, graded cattle, and held sales. In Arizona the Papago resisted such organization of their cattle enterprises, because cattle served them as a reserve fund, or the Papago Bank, to which they could turn in time of need. Thus, whenever the Papago were unemployed or suffered a crop failure, they sold a few cattle for needed cash. This freedom of action and the willingness to forgo greater profits from an orderly semiannual roundup and sales system prevented better organization of the Papago cattle industry through an association system. Social, religious, and cultural demands also required the Papago to barter, give, or butcher a cow at any time. Cattle associations prevented such freedom of action, thus violating cultural traditions.[28]

While the federal government was simultaneously encouraging some tribesmen to reduce their livestock and encouraging others to expand their operations, Indian farmers who cultivated crops were continuing to struggle under the burdens of drought, dust, and heat, without access to adequate credit, technology, or scientific assistance. Heirship lands continued to plague Indian farmers by preventing them from controlling tracts that were large enough to permit successful agriculture on a commercial basis. With these problems, Indian farmers reaped "heartbreaking experiences," rather than bountiful crops, during the 1930s; but in Kansas, inadequate credit and equipment stimulated the Potawatomi to organize the Potawatomi Harvesting Association. This cooperative, consisting of seven families, purchased and shared essential agricultural equipment with the aid of a Resettlement Administration loan. Most Indian farmers on the Great Plains were not so fortunate, however. Subsistence garden plots withered under the burning sun, while irrigation wells and streams went dry. In mid decade, moreover, grasshoppers descended upon the few crops that remained on many Great Plains reservations.[29]

In 1936, federal aid became available for drought-stricken Indians when the Resettlement Administration made its relief program available to them on the same basis as whites. The Works Progress Administration also ordered field representatives to employ Indians as well as whites in various work-relief programs. In addition, the Farm Credit Administration (FCA) authorized its Emergency Crop and Feed Loan Division to grant loans of $200 per farmer in the drought area, while the Production Credit Division of the FCA also provided loans for feed, seed, fertilizer, and livestock to aid Indian farmers in the drought area. Congress, too, extended funds to Indian farmers through the Agricultural Adjustment Act of 1938. In 1939, for example, the Pueblos received funds from the Agricultural Adjustment Administration for the withdrawal of some land from commercial production in order to plant soil-building and conserving crops such as alfalfa. Although this relief helped alleviate a "distressing" agricultural situation, it did not provide Indian farmers with all the needed aid. Because state and local authorities, together with agency superintendents, who administered these programs, frequently did not know about what types of agricultural aid were available or how to obtain them, many Indian farmers received delayed or inadequate aid.[30]

Many Indians, such as the Pima in Arizona, also continued to lack essential farm tools, and many made a better living with a $2.00-per-day job than by farming. The most successful Pima farmers raised alfalfa on fields that they then rented to white cattlemen for pasture. Alfalfa did not require special attention, except for irrigation, and the crop returns were almost entirely net profits. This type of farming enabled the Pima to hold nonfarm jobs yet receive an income from their croplands without spending a great deal of time planting, cultivating, and harvesting. And the cattle companies that rented the pasturelands often hired Indians to tend the herds, thereby providing additional agricultural income. When the "pasture program" began in 1932, only 478 white-owned cattle were grazing on Pima lands, but in 1935, 6,900 head—the maximum carrying capacity of the pastures—grazed those lands and provided an estimated $70,000 for approximately 450 Pima farm families. A "pasture committee" and an extension worker administered the "pasture project," which made possible the orderly marketing of an important agricultural commodity. The Pima also solicited buyers for their alfalfa hay crop. With 6,610 tons baled and sold for $5 per ton in 1935, this agricultural enterprise contributed significant income to the Pima economy. Bureau officials had hoped to encourage the Pima to cultivate cotton instead of alfalfa pasture, but without adequate credit to purchase equipment and lacking the technical skills to raise and market this cash crop, the Pima preferred to earn most of their agricultural income from their pasture lands.[31]

To improve crop production among the Pueblos and the Navaho, as well as among other Indian farmers in the West, irrigation was required. Collier recognized that the need for irrigated land was "terrific," if the Indians ever were to become self-sufficient farmers; but he also recognized that the problem was "largely insoluble." With the aid of $6 million from the Public Works Administration, however, the Indian Irrigation Service was at work by late 1933 developing plans for the expansion of irrigation systems on reservation lands. This funding was designated for the construction of reservoirs and the building or rehabilitation of canals and laterals. Irrigation projects among the Navaho especially were intended to help them diversify and become less dependent upon livestock for subsistence and income. Ultimately, governmental officials believed these projects would boost subsistence cultivation and make the Indians less reliant upon governmental relief.[32]

To expand agricultural production further and thereby ensure self-sufficient, if not commercial, agriculture among the Indians, the bureau joined the Federal Surplus Relief Corporation and, later, the Resettlement and Farm Security Administrations in an attempt to expand the Indian land base. Federal officials intended to do this by purchasing submarginal acreage from white farmers who had been unable to use it productively and by transferring it to the tribal estate for agricultural development. Collier and other bureau officials strongly emphasized that the word *submarginal* did not mean that useless lands were being acquired for the Indians. Rather, most of the cultivated submarginal lands would be returned to grass to supplement reservation grazing lands, thereby expanding the Indian livestock industry. The program to purchase submarginal land also would make it possible to consolidate the reservation lands that white ownership had checkerboarded, as well as restore many landless Indians to the tribal domain; and it would help to protect Indian watersheds. Although original plans called for the acquisition of 3.5 million acres of submarginal lands in ninety-eight purchase projects at a cost of more than $11 million, the program had been drastically reduced by late 1935. Funding for this, as well as most other New Deal projects, always was less than projected or required, and difficulties in obtaining title slowed the acquisition process. Nevertheless, between 1934 and 1940 the federal government spent $2.4 million to acquire 993,673 acres for Indian development.[33]

In addition to the program to purchase submarginal land, the Indian Reorganization Act provided for the restoration of surplus lands to tribal control and for the acquisition of other lands. Bureau officials estimated that 3 million acres could be acquired, but Congress provided only $1 million for the purchase of new lands during fiscal 1936. This funding hardly contributed to a needed land-acquisition program. Bureau officials estimated that the Indians needed an additional 15.9 million acres to ensure a "modest" standard of living from agriculture, while 9.7 million acres more were required to bring the Indians up to the subsistence standard of living of an "average rural white family."[34]

Nevertheless, the original appropriation contributed to land purchases for thirty-four reservations in eight states, the largest being in Minnesota and South Dakota. By December 1936 surplus lands had also been returned to tribal domain on reservations in five states. Although the Indian Reorganization Act authorized $2 million annually for the acquisition of new lands, after 1937 Congress reduced that funding by more than 50 percent, and appropriations had nearly ceased by 1940, because many congressmen believed the expenditures were wasteful. Moreover, budget recommendations for fiscal 1942, 1944, and 1945 did not include provisions for the purchase of additional lands for the Indians. Because the House and the Senate Indian Affairs committees did not oversee the expenditures of the Resettlement Administration, the program to purchase submarginal land ultimately made a greater contribution to the expansion of the Indian land base than did the provisions of the Indian Reorganization Act.[35]

During the 1930s the drift towards a new European war essentially ended the agricultural progress of the Indian New Deal. As the Roosevelt administration devoted more and more attention to international affairs, even transferring the Bureau of Indian Affairs to Chicago in 1942 to make room for war-related office personnel, Indian farmers were all but forgotten. In 1940 the Sioux still desperately needed cattle in order to develop a viable livestock industry. Without adequate livestock, they continued the old practice of leasing their lands to white farmers and cattlemen. Still, friends of the Indians remained optimistic that the tribesmen in the West could become self-sufficient cattlemen. Reports that Indian cattle sold for more than $3.1 million in 1939, up from sales of $263,095 in 1933, heartened them. Although 16,624 Indians raised 262,551 cattle in 1939, insufficient credit, despite the "repayment cattle pool," prevented the adequate expansion of the industry. In addition, the leasing system kept many potential cattlemen in a state of lethargy and poverty.[36]

This is not to suggest that progress had not been made by the end of the 1930s. In 1935, most of the goats were sold from the Acoma and the Laguna herds, and the Zuñi also made considerable reductions among their goat herds. By the end of the decade, the Pueblos' livestock nearly met the carrying capacity of the range. Most Pueblo stockmen now recognized that overgrazing destroyed the grassland and caused soil erosion. Stock ponds, corrals, and a better understanding of livestock raising had improved range-management practices among them. In May 1939 the Navaho agreed to the reduction of more than ten thousand head of "nonproductive stock," that is, "useless" horses, from their reservation. This reduction was the result of many years of coaxing by the bureau. Although the horses marked for removal were "decrepit, bony, half-starved" creatures, the horse-reduction program still was an emotional issue, because the animals had a "high prestige value" for their owners. Horses provided the Navaho with recreation, enabled them to make gift exchanges for marriages, and served for ceremonial purposes. Although Navaho stockmen who owned fewer sheep than the bureau permitted could replace each horse with five sheep, social status based on horses was hard to change. Many Navaho owned more than one hundred horses, and some owned three hundred; yet, the government demanded reduction to ten. Even though the Navaho disliked parting with their horses, some now recognized the damage that these animals caused to the range land.[37]

Navaho, Acoma, and Laguna flocks also were being improved with the use of Rambouillet rams, while Corriedale rams were up-breeding Zuñi flocks. By 1940, Acoma and Laguna herds had been reduced by 48,548 sheep units, mostly in the form of horses, goats, burros, and aged steers. With better range and breeding-management practices, the birth weights of lambs and calves and the production of wool increased by 20 percent, and the number of calves rose 30 percent. Better shearing and cooperative selling also increased prices by 10 percent. Among the Navaho, the large-scale stockmen now were bearing the brunt of the stock-reduction program for sheep and goats. Officials also had divided the Navaho range into eighteen land-management districts and had calculated the carrying capacity of each.[38]

Moreover, by 1940, 700 men and 35 women owned 25,000 cattle in twelve cooperative livestock associations on the Fort Apache Reservation in Arizona. They had repaid the 600 purebred Hereford cattle that they had borrowed from the revolving cattle pool, and they had developed a high-grade herd. Papago cattlemen had increased to 1,325, up 133 percent since 1934. The San Carlos also were well established in the cattle business. With nearly 28,000 head grazing tribal lands, the San Carlos had increased their cattle 1,000 percent since 1923. Moreover, no tribal lands were being leased to white cattlemen. One-fourth of the 581 cattle owners who participated in twelve livestock associations were women. One-third of the 1,600 purebreds that the San Carlos had borrowed from the revolving cattle pool had been repaid. Throughout the West, Indian cattle associations also had expanded, from 53 in 1933 to 150 in 1939. Members totaled 9,017—all of whom helped to improve the raising and the marketing of Indian-owned cattle.[39]

In many respects, however, agricultural life was little changed from what it had been in the immediate past. In 1940, less than 10 percent of the Yuma in Arizona earned their entire living by farming, and only 50 percent were part-time farmers. They raised a few cows, goats, and chickens but did not graze sheep. In general, the Yuma preferred to rent their lands to white farmers for the production of cotton, alfalfa, and truck crops. The nearby Maricopa were little more advanced. They cultivated some wheat, barley, corn, and beans on about fifteen hundred acres and raised a few cattle, but they did not cultivate their crops adequately, and they preferred to rent their lands to white farmers and cattlemen. Navaho livestock owners still held "extreme bitterness" toward bureau officials for the forceful, even coercive, manner in which the livestock-reduction programs had been thrust upon them. Although progress had been made with the Navaho concerning soil conservation and proper range management, far too many tribesmen continued to keep horses for prestige, all of which placed a burden on the Navaho range and prevented the expansion or the improvement of the commercial livestock industry. The Navaho also needed to give increased attention to up-breeding, to controlling lambing time, to wool grading and handling, to cooperative sales of fleeces and lambs, and to the development of better storage facilities for their crops to make systematic marketing possible. In Oklahoma, the Kiowa, the Comanche, the Kiowa-Apache, and the Wichita were also struggling agriculturally. They, too, preferred to lease their lands if they still retained possession of any. The continued division of estates among their heirs also undermined any hope of agricultural progress. During the 1930s, land sales slowed down but never ceased, as the tribesmen alienated their farms to acquire money for subsistence. By the end of the decade, most of these Indians had quit farming. The Cherokee also chronically lacked draft animals, equipment, capital, and credit.[40]

Although World War II caused increases in commodity prices, which benefited those Indian farmers who had crops to sell, the war years did not profit the Indians on a scale similar to that of World War I. High livestock prices, which encouraged whites to seek Indian lands for leasing, together with fragmented holdings as a result of the heirship problem, prevented the Indians from consolidating their lands to take the greatest advantage of the wartime economy. By 1942, every acre of Indian land that was not being used by the tribesmen themselves was under lease. As a result, the Indians were left to their own devices, but without adequate seed, feed, livestock, and equipment and without the ability to acquire the financing to obtain those needed items, Indian farmers bordered on extinction. In addition to these problems during the early 1940s, the Indian New Deal itself was in trouble. After 1937, Collier was on the defensive. His antagonists continually charged that bureau programs designed to foster tribal self-determination, restore religious traditions, and reassert cultural uniqueness promoted segregation rather than assimilation, to the detriment of all Indian people. As Collier's opposition gained congressional support, funding for the bureau decreased, particularly for land purchases, credit, and construction projects. By early 1945, Collier no longer could protect the bureau from his detractors, and on 19 January 1945 he resigned.[41]

While publicly proclaiming the achievements of his tenure, Collier probably left office disappointed that the Indian Reorganization Act had not provided for all of the reforms that he had sought. Certainly, he was not pleased that many of the tribes refused to participate in his plans for the creation of tribal governments and of political and economic organizations that were foreign to them. Still, one should not be so critical of the Indian New Deal as to overlook its achievements, which indeed were substantial. If Indian farmers did not progress under it as far as bureau officials and others had hoped, it was because they had so far to go. In terms of capital, technology, and education, the distance between white and Indian farmers was so great that substantive reform, even under the best of circumstances, could not possibly have placed the tribesmen in a competitive position within the agricultural economy of white America. Moreover, Collier was seeking only to bring the Indian farmers up to a subsistence level; he did not stress the goal of commercial production because he did not want Indian farmers to compete with white agriculturists. At the same time, the harsh environment of the trans-Mississippi West prevented even subsistence agriculture, and the land base was also inadequate for farming.[42]

The Indian Reorganization Act did halt the allotment process, however; it also helped to increase the Indian land base by 7 million acres by the cancellation of leases, and it provided for the purchase of 398,189 acres of new lands with IRA funds. Tribal funds secured another 265,852 acres, and the Indians regained more than 621,000 acres of surplus lands that the Dawes Act had created. Provisions of the Taylor Grazing Act of 1934 also made possible the acquisition of approximately 4 million acres, and soil-conservation and irrigation projects made many reservation lands more productive than ever before. By 1940 the tribesmen owned approximately 40 percent of the livestock that grazed on Indian lands; they also had substantially improved their cattle and sheep herds, ranges, and marketing practices. The cattle-pool program continued to help many Indians enter the livestock business and to earn a greater income from agriculture. Federal encouragement of cattlemen's associations aided the drive to foster improved range management and facilitated better marketing agreements and procedures. The Indian New Deal also provided a revolutionary change for Indian-white relations, and the Indian Reorganization Act remained a foundation for the formulation of Indian policy. The IRA fostered the "cultural pluralism" that Collier wanted to ensure. In many respects, then, the Indian New Deal changed life for the American Indians. Some, however, profited from those changes more than others did, and for most Indian agriculturists, the promises of the Indian New Deal were greater than the reality.[43]

In fact, only twelve thousand Indians were able to live on the newly acquired and rehabilitated lands, much of which were unsuitable for subsistence production, and many Indians were reluctant to cultivate tribal lands, because they did not have clear title to their fields. They hesitated to invest the money, labor, and time required to make permanent improvements, such as homes and buildings, if they could not absolutely control the use of that property. Heirship lands, the unwillingness of the Indians to return their inheritance to tribal control, and the continued lack of credit for agricultural purposes—all combined to encourage the Indians to lease their land rather than to cultivate or graze it for their own purposes. By the end of World War II, 2.5 million of 4.5 million acres of Indian allotted and tribal lands were being leased to whites at the abysmally low figure of one dollar per acre. Another 11.5 million of 47.5 million acres of grazing and timber lands also were being leased to whites for ten to twenty-five cents per acre. As a result, net income for Indian farm families averaged only $501 in 1945, compared to $2,541 for white farmers. Nor did the Indian New Deal provide relief for those Indians who had lost their lands prior to the passage of the Wheeler-Howard Act, other than to offer the hope that they might regain the use of other lands acquired through the land-acquisition program.[44]

Congress's failure to provide funds for a significant land-acquisition program was a major flaw in the Indian New Deal, and it prevented any possibility of making either subsistence or commercial farmers out of the tribesmen. Although in 1934 Congress had authorized the expenditure of $2 million annually for land purchases, it had not appropriated much more than $5 million before World War II. This failure stemmed largely from the opposition of western congressmen, who resisted the bureau's efforts to purchase the land of white farmers. And because some funding for land acquisitions came from sources such as the Resettlement Administration, Congress hesitated to make additional allocations to the bureau for that purpose. Nor did Collier pursue the matter of land acquisitions with the diligence needed to solve the problem because of the expense and administrative time required. As a result, white-owned land continued to checkerboard the reservations.[45]

In relation to the most popular program of the Indian New Deal, Congress substantially reduced funding for the CCCID in 1937. The Roosevelt administration's desire to reduce federal spending and the coming of a new European war also made funding increasingly difficult for the Indian CCC. And on 10 July 1942 the program terminated when Congress completely cut its funding. During the nine-year existence of the CCCID, more than 85,000 Indians participated in this work-relief and conservation program, and $72 million had been expended to restore lands on more than seventy reservations. Because of CCCID work, reservation grazing and croplands were in better condition than ever before; but while the work of the CCCID was beneficial, not all Indians profited from the results. Because many of the Indian grazing lands still were leased, white stockmen frequently benefited most from the construction of ponds, terraces, fencing, pest and weed control, and other conservation measures designed to restore and conserve the grasslands. As long as heirship policy continued to fragment Indian lands, the work of the CCCID could be of no more than cosmetic aid. The surface of the land might be restored, but it would not profit many Indian farmers. Probably few enrollees remained on the land to raise crops and livestock as a result of their experience and training in the Indian CCC, and the war years also caused many of them to leave the reservations, never to return.[46]

By the end of World War II, many problems remained for Indian farmers. In addition to the need for more land, they also still required better agricultural training through demonstration and extension work. Many did not know how to make the best use of irrigation systems. Others failed to rotate crops, apply fertilizer and insecticides, up-breed their livestock, or follow the best range-management practices. Many who wanted to improve their farming practices lacked the capital and credit necessary for success. After 1945 the use of new and improved forms of science and technology accelerated. Only those farmers who commanded the necessary capital, credit, and managerial ability to apply that science and technology on increasingly large acreages were able to expand production, reduce unit costs, and increase efficiency while being confronted with surplus production and chronically low prices. Few Indian farmers could do so. As a result, fragmented lands, leasing, inadequate access to capital, and technological backwardness continued to plague Indian farmers. As white farmers began to use their lands more intensively than ever before, the matter of water rights also emerged as a critical issue for the tribesmen. Although water rights had posed problems in the past, the debate over who owned and controlled the western waters elevated to an unprecedented emotional level during the postwar years. The stakes were high, because without water, many Indian farmers had little chance of survival. While the battle lines were drawn in state and federal courts concerning the matter of the western waters, the Indians also had to contend with an old concept in a new disguise—termination. It, too, boded ill for Indian farmers.[47]

12

The Termination Era

When the guns of World War II fell silent and peace was finally restored, the Bureau of Indian Affairs professed that Indian farmers had made great improvement in agricultural operations during the course of that conflict. Even so, Commissioner William A. Brophy recognized that American agriculture had entered a new age and that successful farming in the future would require larger farms, the intensive application of science and technology, and expanded capital investment. Indian farmers had little opportunity for expansion or for acquiring new machinery, chemical fertilizers, and pesticides; and they did not have adequate credit. Indeed, Commissioner Brophy noted that the land available to the Indians for agriculture was "insufficient in quantity and quality" to support them. Certainly, the shortage of agricultural land was critical. Moreover, after the war began, Congress did not apropriate funds under the Indian Reorganization Act for the acquisition of additional land until fiscal 1947. Although Congress provided $350,000 for land purchases at that time, the amount was insufficient to meet Indian needs.[1]

The Indians wanted to reorganize their land holdings in order to make them more manageable and productive, and the congressional appropriation offered them the chance to do so. Those funds were to be used for the acquisition of heirship lands belonging to old or incapacitated Indians and trust lands belonging to those who were permanently living beyond the reservations. Bureau officials reasoned that these purchases and subsequent consolidation would simultaneously benefit Indian agriculture and provide support for the needy. This plan, however, offered too little too late, because land and commodity prices remained high after the war. As a result, many white farmers sought Indian lands in order to expand production and to capitalize on high postwar prices. As the demand for Indian lands increased, prices for them became more inflated than ever before. Many Indian landowners, who had lived in poverty with their allotments and who had slim chance to turn their lands into productive farms, now sent to the Department of the Interior a "flood" of applications for the sale of their trust lands and for patents in fee, which would enable them to alienate their lands at will.[2]

In addition, the heirship tangle further complicated the land problem by removing approximately 7 million acres of the "best lands" from use by Indian farmers. Still, bureau officials saw improvement in the development of Indian agriculture. By the end of the war, 12,718 Indians had participated in the Revolving Cattle Pool, and $6.6 million had been loaned from the revolving credit fund, from which the bureau acknowledged the loss of only $2,746. Moreover, cattle raising, the most important part of Indian agriculture, grossed more than $16.3 million in sales during 1945—a substantial increase from less than $1.3 million in 1932. The tribesmen also consumed an estimated $9.2 million worth of beef at home. Cash receipts from the lease of 10 million acres of grazing land contributed an additional income of $2 million. Approximately 535,000 acres of Indian-operated lands were irrigated, and the bureau was at work improving Indian farmland through various conservation measures, such as contouring, terracing, strip cropping, and the reseeding of grazing lands. In addition, bureau officials hoped that the 597 Indian 4-H clubs would improve Indian farming in the years ahead by teaching and promoting the best agricultural methods. The bureau also provided "on-the-farm" training in cooperation with the Veteran's Administration, which enabled veterans who owned or leased farms to receive classroom instruction for agricultural improvement. This program was most successful in North Carolina, where sixty-nine Cherokee veterans received aid in developing plans for raising crops and livestock and for training in soil conservation. By 1947, Indian agricultural income reached $49 million, up from $1.8 million in 1932. The first figure, however, included an estimated $12 million in farm products consumed at home. A year later, Indian cattlemen owned 363,000 head, valued at $45 million, and sales of live and dressed beef totaled $17.8 million.[3]

In spite of these reasons for optimism, serious problems remained for Indian agriculturists in the postwar world. In the Southwest, inadequate range land and the vestiges of the stock-reduction program continued to limit Navaho well-being. Overgrazing and soil erosion prevented most of the Navaho from maintaining the 250 head of sheep per family that were required for subsistence. With the additional factor of drought influencing Navaho stock raising, adequate grazing lands simply did not exist to support the sheep in numbers required to provide a minimal standard of living for those tribesmen.[4]

In Arizona's Colorado River Agency, the Yuma showed little eagerness to farm; they preferred to lease their lands to white operators. Moreover, while federal officials provided extension services designed to help transform the Indians of the Pima Agency from subsistence to commercial farmers, partially with the help of the experiment station, which was instrumental in developing a long-staple "Pima cotton" and other plants, an insufficient water supply for irrigation hindered agricultural expansion. Only 14,600 acre-feet of water was available for the 100,000 irrigable acres of the agency's Gila River, Salt River, Maricopa, and Fort McDowell reservations. In the absence of enough water for ditch irrigation, the Papago practiced their age-old method of floodwater farming.[5]

Most Indian agriculturists in the Southwest depended upon cattle raising for their farm income. The San Carlos Apache in Arizona were among the most successful of these stockmen. In 1947, they still maintained eleven cattlemen's associations and the "social security herd" of 4,450 head. The registered breeding herd totaled 1,800 animals, while 37,500 cattle grazed San Carlos lands, and cattle sales reached $250,000 annually. In addition, San Carlos cattlemen utilized all reservation grazing lands; whites no longer leased that area. In many respects, however, the San Carlos were the exception. Although the Indians at the nearby Fort Apache Agency depended to a great extent on cattle raising for their income, their range was not stocked to carrying capacity, and livestock raising had not been developed to the extent that it had been on the San Carlos Reservation. And while the Navaho and the Hopi had reduced their sheep to 388,000 head by the end of the war, they still grazed 59,000 goats and 35,000 horses—the latter being more important to their cultural tradition than to their economic well-being and to the proper use of the grazing lands.[6]

As in the past, the environment remained a limiting factor, in part preventing the success of Indian agriculture. Of the approximately 56 million acres under the bureau's jurisdiction, two-thirds were located in the semiarid or arid West. By midsummer 1947, only slightly more than 8 percent of Indian lands had been treated with the proper soil-conservation techniques. This slow process caused Commissioner Brophy to lament that at that rate, at leasty sixty-six more years would be needed to complete the soil-conservation measures required on Indian lands. By 1949, soil erosion had made 20 percent of Indian lands unfit for cultivation, and conservationists estimated that the tribesmen were losing approximately 36,000 acres to soil erosion annually. Half of the grazing land was of "poor quality." Some areas required forty-five acres to support one cow; half of the cropland also needed irrigation to be productive.[7]

Because of these problems, the average Indian family's income from agriculture was only $918 in 1947. Even at this low rate of earning, how-

ever, two-thirds of the Indian families on the reservations relied upon agriculture for "slightly more" than half of their annual income. By 1949 the
average agricultural income for Indian farmers had dropped to about $500,
compared to $2,500 for white farmers. With low returns such as these
and with land prices still high from wartime inflation and postwar expansion, more acres were fee patented than at any time since 1932. In 1948
alone, 67,000 acres of Indian lands were fee patented, and although the
bureau did not record the amount of acreage that the owners subsequently
sold to white farmers, the lure of cash no doubt prompted many Indian
landowners to take advantage of a quick profit. The next year, fee patents
were issued for 146,655 acres, and 10,000 acres were alienated through
the approval of deeds, the issuance of certificates of competency, and
special orders by the secretary of the interior which removed restrictions
prohibiting the sale of Indian lands. Moreover, Congress made it possible
to alienate an additional 75,000 acres by approving special laws that per-
tained to the Five Civilized Tribes. This grim state of Indian agricultural
affairs prompted John Ralph Nichols, commissioner of Indian Affairs, to
remark that the Indian "problem" could only be solved by the use of
"men, money, and imaginative and patient management." The situation
was far beyond the stage where anyone could expect results from "pana-
ceas or 'overnight' solutions." Despite these problems, a far-greater difficulty threatened future progress among Indian farmers at midpoint in the
twentieth century. This policy had a frightening sound and a final implica-
tion: governmental officials called it "termination."[8]

With the resignation of John Collier and the conclusion of World War
II, governmental officials once again embarked on a major effort to assimilate the Indians into white society. Assimilation would mean that the Indians had achieved equality with their white neighbors. It also meant that
the federal government would be freed from the responsibility of protecting and supporting the tribesmen and could withdraw from the Indian
business. In short, the relationship between the Indians and the federal
government would be ended for all time.[9]

Although termination had its roots in the nineteenth century, it also
was a product of the postwar economic boom which mandated the full
utilization of the nation's resources, such as land, timber, and minerals,
and the reduction of federal spending and regulation. Termination also
reflected a belief on the part of some governmental officials that the Indian
New Deal had failed to solve the problems that the Dawes Act had created.
Therefore, a new direction and a new solution were needed to solve the
problems of Indian poverty and reliance upon the federal government for
subsistence and at the same time to gain access to Indian lands and natural
resources. For some supporters, termination would "liberate" the Indians
from the "bonds" of reservation life by forcing them to rely upon them-

selves, rather than upon the federal government, for their livelihood. By so doing, it was nothing less than the Emancipation Proclamation for the American Indians.[10]

In 1949 the movement for termination gained support when the Commission on the Organization of the Executive Branch of the Government, commonly known as the Hoover Commission, reported that "assimilation must be the dominant goal of public policy." Both Commissioners of Indian Affairs John R. Nichols and Dillon S. Myer, the latter of whom assumed office on 5 May 1950, supported assimilationist policy. In fact, Myer became so supportive of assimilation and termination that Harold L. Ickes, an old New Dealer, referred to him as a "Hitler and Mussolini rolled into one." By the time Myer left office on 20 May 1953, he had substantially furthered termination by identifying tribes that could be severed from federal support and by drafting legislation for the termination of those tribes— the largest of which were the Klamath in Oregon and the Menominee in Wisconsin.[11]

Many whites believed that termination would enable them to quickly acquire Indian lands once those holdings were no longer under federal protection, but termination was more complex than merely a land-grabbing scheme. Many who supported this policy believed that the Indians should live unhindered by special federal regulations. Although westerners controlled the committees and subcommittees that dealt with Indian affairs, philosophical conservatism, rather than geographical representation, provided their motives for supporting termination. These congressmen assumed that many Indian groups had a standard of living comparable to whites and that "cultural" differences between the more advanced Indians and whites had been virtually eliminated. Economic opportunity, not tradition, they reasoned, should now determine the relationship between red men and white men. Moreover, they also believed that state governments could handle the services that the bureau was providing and that the Indians were capable of protecting their interests under state and federal law. Some also argued that many Indians favored the termination of federal trusteeship.[12]

On 1 August 1953, Congress gave formal support to the policy of termination with the adoption of House Concurrent Resolution 108. This resolution declared that "it is the policy of Congress, as rapidly as possible, to make the Indians within the territorial limits of the United States subject to the same laws and entitled to the same privileges and responsibilities as are applicable to other citizens of the United States, to end their status as wards of the United States and to grant them all of the rights and prerogatives pertaining to American citizenship." In response, the bureau intensified its efforts to draft legislation that would free the Indians from federal supervision, regulation, and support. During the next year, more than twenty termination bills were introduced in Congress.[13]

For Indian agriculturists, termination had ominous consequences. It meant that the federal government no longer would provide access to the revolving credit fund and to the cattle pool; that trust status would end; that Indian farmers no longer would have their lands protected from taxation, liens, and foreclosure; that extension agents no longer would be specifically charged to the reservations, because those political divisions would no longer exist; and that Indian landowners would lose their property to whites if they did not have the financial, political, and legal ability to prevent the alienation of their lands. In contrast to the intent of the policy makers, termination would accomplish nothing less than the ultimate disintegration of the Indian land base and the destruction of any lingering hopes that agriculture would provide the ultimate means by which Indians could become assimilated and acculturated within white society.[14]

The drive for termination, however, began to slow down in 1956, when the Democratic party increased its congressional membership, thereby changing the committee structure. Indian and white opponents of termination also became more critical of a program that ensured disorganization and poverty, rather than freedom and liberation, for the tribesmen. They argued that the Bureau of Indian Affairs should provide massive economic aid for the reservations, rather than sever all relationships with the Indians. In short, the bureau needed to expand its involvement with the Indians, rather than abdicate its responsibility.[15]

In response to these critics, both Congress and the Department of the Interior began to withdraw from termination policy. By the autumn of 1958, administration officials were trying to deemphasize termination as much as possible. In September, Fred A. Seaton, secretary of the interior, declared that termination was merely an "objective," rather than an "immediate goal," and that no tribe should have its relationship with the federal government ended unless the tribe was supportive of that venture and had developed a program to ensure its success. For Seaton, forced termination was nothing less than a "criminal" action. With this change of attitude within the government, termination essentially had come to an end by the early 1960s. During the decade in which it posed as a viable Indian policy, only 13,263 Indians, or approximately 3 percent, had had their federal support withdrawn, and only about 3 percent, or nearly 1.4 million acres had been withdrawn from federal protection. The tribes that lost the most land from termination were the Klamath in Oregon, the Menominee in Wisconsin, and the Mixed-blood Ute in Utah.[16]

While the tribes either fought or drifted along with the rise and fall of termination policy, their farmers continued as best they could. In the Great Plains, however, Indian farmers often were unable to continue their farming operations on even the meager level of the past. On North Dakota's

Fort Totten Reservation, for example, few Indians had the resources to operate more than 40 acres, yet more than 160 acres were needed to ensure commercial viability. Heirship lands so divided the reservation that it was nearly impossible to find tracts of sufficient size to support the raising of livestock. This problem, together with an absence of adequate credit for machinery and livestock, prevented even subsistence-level farming. As a result, the superintendent leased much of the reservation land, some-times without the heirs knowing about it or receiving their portion of the rental fees, because they could not be identified or located. In the absence of heirs or with the inability of the heirs to agree to lease their lands, white stockmen grazed cattle on Indian lands without making pay-ments. With problems such as these, the tribesmen did not have the incen-tive to improve their agricultural operations. As a result, Indian cattlemen grew lethargic and neglected the proper care of their livestock. In addition, in adequate welfare services, low income, and hunger sometimes prompted these Fort Totten stockmen to butcher breeding cattle for food, thereby further limiting their ability to increase the size and profitability of their herds.[17]

Between 1950 and 1960, two-thirds, or 600,000 acres, of South Dakota's Rosebud Reservation was leased to whites. Although these Sioux did not alienate much land during that decade, the number of Sioux cattle-men shrank from 203 to 129, with 90 percent of the decline occurring among full bloods. With the bureau's funding reduced because of the termination policy, Rosebud ranchers were forced to seek needed credit from private sources, but the lack of collateral and established credit rat-ings, combined with discrimination, prevented them from receiving ade-quate funding. Lacking the necessary capital to buy cattle in order to begin or to expand production, many Sioux cattlemen chose to sell their land to the tribe. Ultimately, much of this land was leased to non-Indian cattle-men instead of being used to support Indian livestock operations. Even if the Sioux had utilized all of their trust lands for grazing purposes, cattle ranching would have supported only one-third of the reservation's popula-tion. In many respects, then, an inadequate land base was a more serious problem than the lack of adequate credit, which was needed in order to make existing operations more productive.[18]

As credit sources dried up for the Sioux, the number of extension agents also declined by half in 1952. Three years later, however, the South Dakota State College Extension Service provided approximately seventeen extension agents to aid all Indian farmers within the state. Three of those agents were assigned to the Rosebud Reservation, but they did not know how much time they should give to Indian farmers. In addition, conflicting directives from the bureau and the state extension service hindered per-formance. The Sioux at the Pine Ridge Reservation experienced similar problems; by 1956, only 21.4 percent of the tribesmen were engaged in agricultural work. As a result, the most productive reservation lands in North and South Dakota were being leased to white operators by the end of the decade.[19]

To the west, approximately 80 percent of the Crow Reservation in Montana was leased to whites. In 1953 the Crow were using only about 12 percent of their grazing lands, 7 percent of their irrigable cropland (less than half of which actually had water), and 5 percent of their dry cropland. Most Crow farmers had too few cattle and too little machinery to maintain viable agricultural operations. The most successful cattlemen shared or rented lands with family members or partners, although not all of the land might be contiguous. At least seven tribal grazing associations used Indian lands and raised "good quality" Hereford cattle. The Crow sold most of their cattle in Billings, but they also made shipments to Omaha, Nebraska, and Sheridan, Wyoming. Some of the larger ranchers also raised wheat as a commercial crop, while the smaller operators produced oats and barley for feed. Even so, cattle provided 81 percent of the cash income for Indian ranchers in 1953, while crops provided only 11 percent and livestock products and livestock other than cattle contributed an additional 5 percent.[20]

Crow cattlemen who grazed fewer than fifty head generally did not earn enough income to meet operating expenses. Although small-scale as well as large-scale operators could borrow from the Crow Tribal Council and the Bureau of Indian Affairs, under a jointly administered program designed to promote livestock production, as well as from the Farmers Home Administration, the Production Credit Association, and commercial institutions, few were able to meet credit requirements. Trust lands could not be used for collateral, and the livestock experience of many potential borrowers did not make creditors feel confident. These problems, however, were not unique to the Crow; they applied to all Indians who were still trying to maintain farming operations.[21]

In 1953 the Crow agreed to pay in cash rather than kind for all loans from the cattle pool, so as to simplify administration. Although previous borrowers had the option of repaying their loans in yearling heifers, new applicants were required to pay cash. This procedure, however, placed an undue burden on new stockmen, who lacked the business skills necessary to market their cattle at the right time and to accumulate the cash required to meet their loan obligations. These loans henceforth carried an interest rate of 4 percent; and to make matters worse, the price of cattle dropped from $161 per head in 1950 to $71 per head in 1953, which further discouraged Crow cattlemen.[22]

The agricultural situation was similar on Montana's Blackfeet Reservation, approximately 50 percent of which was leased to white operators. While most Blackfoot cattlemen were knowledgeable about ranching and held well-bred herds of Herefords, about 40 percent of Blackfoot families owned fewer than twenty head, and most cattle received only a "moderate" degree of care. Indeed, negligence caused greater losses among the Indians' calves than white cattlemen incurred. Most Blackfoot cattlemen shared land with relatives on a rent-free basis. The Blackfeet sold 57 percent of their cattle through a local marketing pool; the remainder either went to buyers who visited the reservation, or they were shipped to Great Falls or Harve, Montana, to be auctioned. Most were marketed as feeders; the Blackfeet did not emphasize either raising forage crops for winter feed or finishing their animals for sale. Even so, 77 percent of their income came from nonagricultural employment. Although Montana experiment-station officials believed that the Blackfeet had made "considerable progress" in the transformation from a nomadic to a sedentary agricultural people, most of the Blackfeet lived in poverty.[23]

To the south in Oklahoma, not more than a dozen Kiowa, Comanche, Apache, and Wichita remained as full-time farmers by the early 1950s. They, like many small-scale white farmers, simply could not compete in the postwar world, where the extensive use of both land and capital often meant the difference between success and failure. By the end of the decade, the majority of Quapaw, Cherokee, Chickasaw, Choctaw, Creek, Cheyenne, and Arapaho lands were being leased to white farmers.[24]

The problems of the past continued to plague Indian agriculturists in the Southwest during the 1950s. Although sheep units had declined from 620,000 in 1940 to 460,000 units by 1950, drought had stunted the grazing lands and was contributing to soil erosion and diminished productivity. Indeed, the drought was so severe that some officials estimated that five years of normal precipitation would be needed to restore the range to full carrying capacity. And in the autumn of 1951 the Navaho Tribal Council urged stockmen to transport their sheep to winter pasturelands in the wheat belt and offered to pay half of the shipping costs to anyone who would do so. This "revolutionary" plan, however, was culturally unacceptable for most Navaho flock masters, who could not and would not separate their sheep from family control, although they did send about eight hundred cattle to winter ranges in Arizona and California. The political costs for regulating the Navaho range also became too high, and after 1956, when the tribe became responsible for range management, the council stopped enforcing grazing regulations. Thereafter, compliance became voluntary; so overstocking again became a major problem, and the number of livestock on Navaho ranges soon approached that of the prereduction days—well above the "official" carrying capacity of the grazing lands. In 1943, for example, the carrying capacity of the Navaho range had been set at 512,922 sheep units; but in 1957 the range was carrying 527,989 sheep units.[25]

Even so, by 1957, more than half of all Navaho families did not own any livestock, and about 36 percent of families that did had fewer than 100 sheep units. At this time, subsistence still required at least 250 sheep units, but only 2.3 percent Navaho stockmen had herds of this size. In addition, the Navaho continued to graze too many horses and "poor quality" goats. Cultural tradition, therefore, still was more important than the proper use of their agricultural resources. Horse ownership remained a mark of prestige, and the Navaho believed that living things should not be needlessly destroyed. That belief, in addition to no help with marketing, caused the goat population to increase by 14 percent between 1950 and 1958. Drought continued to compound the problem of overstocking, and during the winter of 1956/57, $5.6 million in feed grains had to be shipped to the reservation to maintain herds. Although this Emergency Feed Grain Program helped Navaho stockmen to cope with feeding problems, it did not aid in the restoration of the grazing lands.[26]

The Navaho did, however, make some agricultural progress with the addition of 4,334 acres of new cropland between 1950 and 1958. Eighteen extension workers helped to meet tribal needs by making home visits and by organizing off-the-reservation farm tours, which were designed to "broaden the perspective" of Navaho agriculturists. Some Navaho also found agricultural employment harvesting sugar beets, cotton, vegetables, citrus fruits, and broom corn in Arizona, New Mexico, Colorado, Utah, and Idaho. This employment lasted only six to twelve weeks and paid an average of $6 per day for a total of 2.5 percent of Navaho income. Navaho income, however, did not improve substantially from the sales of agricultural products. In 1958 the Navaho marketed $2.2 million in wool, mohair, and livestock, and they consumed an additional $1.7 million in agricultural products at home. This amount, however, when divided among the nine thousand reservation families, averaged a mere $439 per family. As a result, by 1958, only an estimated 10 percent of Navaho income was derived from agriculture.[27]

In contrast, the San Carlos Apache maintained a strong livestock economy. By 1955, eleven cattlemen's associations marketed about twelve thousand head annually, for a return of more than $1 million. Each association had an area of the tribal range fenced for grazing purposes. The San Carlos also provided separate ranges for their two tribal herds—the IDT, or tribal herd sponsored by the Department of the Interior, and the registered herd. The local extension agent, however, made the major decisions for the associations; he controlled the budget and the accounts. This transferal of power to the extension agent indicated the continued paternalistic domination of San Carlos agriculture. In comparison to the past, when the superintendent had governed their daily lives, the extension agent had, by the mid 1950s, assumed that power for the associations. Only two associations, however, had a "stockman," that is, an overseer in charge of daily activities, such as branding, dehorning, and vaccinating: the stockman freed the cattle owners from daily concerns and enabled them to hold jobs elsewhere.[28]

Semiannual roundups provided yearling steers and heifers for sale to the feeder industry. Before an association cattleman received his check, fees for branding and tribal grazing were deducted, as were charges for feed at the sale pens, inspection and auctioneer charges, and a debit to maintain a "compensation fund" for employees who were injured on the job. In addition, the receipts from cattle sales also were applied to a stockman's account at the tribal stores; before a new credit line could be established, all previous indebtedness had to be cleared up. Once these requirements had been met, the cattleman received a check for the balance. This procedure caused some stockmen to complain that they should not be expected to work during the association's roundup, because their cattle belonged to the tribal store.[29]

By the mid 1950s, San Carlos Apaches who wanted to become stockmen had greater difficulty entering the business than ever before. In order to join an association, an individual had to be either twenty-one years of age or eighteen and the head of a household. Then he could obtain an application from the extension agent and file it with the association's board of directors. Upon formal approval, the application passed to the chairman of the Tribal Council, the bureau's forester, and the superintendent of the agency. This application procedure took time, and while it was designed to help maintain the carrying capacity of the range, it fostered bitterness among cattlemen who wanted their sons to enter the livestock business. Upon final approval, the new association member could acquire twenty or more head, depending upon council policy, from the tribal or registered herds or from the mavericks bearing the association's brand. The new member could pay cash for the cattle, get a loan for payment, or make repayment in kind within eight years by furnishing eleven head for each ten borrowed. Cash payment was based on the "average market value" at the time of issue. No individual or family, however, was to graze more than seventy head of breeding stock on association or tribal ranges. This restriction was designed to prevent overgrazing. In 1954, officials estimated that seventy head of cattle would provide an annual income of $3,400, which they considered adequate for family support, although they did not state the size of the family. Most San Carlos Apache did not own that many breeding cows and considered even that number insufficient to provide an adequate income; but few San Carlos Apache knew how many cattle they had between roundups, and the restriction was not enforced anyway.[30]

In October 1956, in order to make the cattle business more efficient on the reservation, the San Carlos Tribal Council consolidated the eleven associations into five, thereby improving the economy of scale and the financial condition of the smaller associations. The new associations were incorporated under Arizona law and were thus able to borrow money from commercial institutions. Each association adopted cost-accounting procedures and began to hold monthly meetings of its board of directors and annual meetings for the members—all of which placed the livestock associations under the total control of the Indians. Cattle owners paid $3 per head annually to the Tribal Council for a grazing fee, half of which was returned to the associations to hire the workers necessary to tend the herds. The associations no longer assessed penalty fees. These changes in operating procedure proved beneficial, and although total tribal income cannot be determined for that period, by 1960, cattle sales contributed more than $1.5 million in gross income to the San Carlos Apache.[31]

In contrast to the San Carlos Apache, the Jicarilla Apache in New Mexico lost interest in livestock raising after World War II. The new affluence that the war had stimulated and a drop in the demand for farm commodities damaged the Jicarilla economy. Small-scale stockmen experienced declining profits and increased operating costs. Although the number of cattle on the reservations held relatively steady between 1945 and 1959, the number of sheep declined from 36,698 to 15,768. During the late 1940s and the early 1950s, drought caused shortages of feed and water, both of which encouraged many small- to medium-scale livestock operators to leave the business. During the decade of the 1950s the Jicarilla never received more than 30 percent of their reservation income from the livestock industry, and many of them moved to Dulce, New Mexico, in search of relief. The bureau accepted this situation and did not attempt to aid the Jicarilla livestock industry by providing management training, new technology, or marketing aid, even though the tribe had the necessary capital, land, and experience to weather the drought and the downturn in the economy.[32]

The Papago in Arizona also suffered from drought during the early 1950s. The Coolidge Dam, built to provide irrigation water from the Gila River, was not enough to meet their agricultural needs. Engineers overestimated its capacity, and water allotments in 1951 were only 1 acre-foot, when the Indians needed 4 acre-feet to maintain agricultural subsistence. The bureau attempted to deepen wells and to dig others, but this action lowered the water table still further and necessitated the digging of still deeper wells. Well water also contained minerals that, over time, ruined or severely damaged the land for agricultural purposes. The Papago organized three cattle associations during the 1950s, but none was particularly successful, primarily because their cultural restraints prevented them from engaging in the collective action needed to make the associations function efficiently. About 50 percent of the Papago owned livestock, but fewer than 5 percent of these owned approximately 80 percent of the reservation's cattle.[33]

During the 1950s, too many cattle were grazing Papago lands: the carrying capacity of their range land was 11,000 cattle by the end of the decade, but 18,000 head were grazing on it. And because only a few families owned most of the cattle, the income gained from cattle sales was not dispersed widely, nor did it contribute significantly to the gross income of the tribe. Although cattle sales reached an estimated $634,000 in 1955 and jumped to approximately $750,000 in 1959, they still accounted for less than 30 percent of the Papago income. And though cattle sales were bringing increased receipts by the end of the decade, overgrazing had seriously depleted the ground cover, and only about 40 percent of the Papago range lands had adequate stock water. Like Indian farmers elsewhere, the Papago also continued to lease their lands to white operators who had the financial ability to cultivate large acreages and to provide irrigation with a level of technical expertise that the Indians did not possess. The Papago farmers could not match the relatively consistent and reliable income that this non-Indian use of the land brought to the reservation.[34]

Leasing continued to be a serious problem, and unless it was resolved, Indian agriculture had little chance for success. In the mid 1950s, most of the reservation land in the Pacific Northwest was checkerboarded with non-Indian holdings. The Coeur d'Alene Reservation in Idaho leased 90 percent of its land; the Nez Perce, 80 percent; and the Fort Hall, 50 percent. In Oregon the Klamath and the Umatilla reservations leased 66 percent of the land to non-Indian operators. The lands that remained in Indian possession were marginal for agriculture and insufficient even for subsistence farming. By 1956 the Indians nationwide were losing 500,000 acres annually to white operators. To make matters worse, Indian farmers still did not have access to the credit resources that would have enabled them to acquire more land and thereby strengthen their agricultural operations, and the Indians lost almost 2.6 million acres of trust land between 1947 and 1957. Although tribal groups purchased some of this alienated land, much of it passed from Indian control.[35]

In the late 1950s it was still relatively easy for the Indians to alienate their lands. Federal policy required an Indian to apply to the Bureau of Indian Affairs in order to sell trust land. If the bureau determined that the sale would be in the best interest of the owner, approval was granted provided that bureau officials accepted the Indian's reasons for wanting to sell and his plans for using the money that he would receive. In contrast to the past, Indian lands now could be fee patented without regard for the effect that such an action would have on the tribe or on land management. On 16 May 1955, Commissioner Glenn L. Emmons instructed bureau officials that "an individual Indian's right to the ownership of his land

in fee simple need not be subordinated to the interests of his tribe nor
to the management of the land as a part of a timber or grazing unit.'' If
an Indian was competent, he or she could receive title to land. The rationale behind this policy was an old one—give the Indians ownership of
enough land to provide subsistence and thereby enable the federal government to end its expenditures for their care. This policy was injurious to
Indians who had the misfortune of being allotted lands without adequate
water. Once an Indian received a patent and sold his land, the new owner
could deny Indian neighbors the use of the water that they had depended
upon in the past. In addition, the bureau no longer required owners to
maintain the right of access to nonpatented lands which might be surrounded by newly alienated land. Among tribes that had received allotments, this policy further damaged the potential for Indian agriculturists
to expand crop and livestock production.[36]

Thus, by 1960, Indian agriculture was continuing to decline, though
the policy of termination was in the process of being reversed. The Ken-
nedy administration and successive ones chose to emphasize a policy that
enabled self-determination without the severance of all tribal ties to the
federal government. Still, while the federal government was taking a
renewed interest in the economic development of the Indians, most tribesmen were living in poverty. Although they were the most rural of the
minority ethnic groups in the nation, the Indians were basically a nonagricultural people. By 1960, fewer than 10 percent were farmers, down from
nearly 45 percent in 1940. In 1960 the Indians made up less than 1 percent
of the population, but approximately 70 percent lived in rural areas. The
insufficient land base continued to decline, although at a slower rate than
in the past; therefore, commercial crop production and livestock raising
were beyond the realm of possibility for most Indians. Even subsistence
agriculture did not meet the basic economic and nutritional needs of the
Indians. They simply could not compete with white operators who had
greater capital resources, technical skill, and managerial ability in an age
when farm operating costs exceeded income even for many white agriculturists. As a result, the median income for reservation Indians in 1964
was $1,800, compared to $5,710 for non-Indians. In Mississippi, many
Choctaw sharecroppers earned less than $300 annually, while the $900
median income for Indians in South Dakota fostered a further decline in
agricultural endeavors.[37]

At a time when cattlemen needed at least one hundred head to maintain viable operations, by providing a family income of between $3,000
and $3,500 annually, few Indian cattlemen had that many animals. On
Montana's Fort Peck Reservation, more than 80 percent of the Indians had
fewer than one hundred head, and only 14.2 percent of the reservation
Indians in the Dakotas owned cattle at all. Only 445 of the 7,499 Navaho

families that owned grazing permits earned more than $1,500 annually from the combined efforts of livestock raising and crop production. At best, Indian stockmen produced only approximately half as much per acre or per unit livestock as their white neighbors did.[38]

The Bureau of Indian Affairs had an economic-development program for the Indians by the early 1960s, but it did not provide substantive aid for Indian agriculture. In fact, Indian farmers were virtually forgotten, and with so few Indians engaged in agriculture, that neglect was not apparent. At that time, for example, the bureau did not have plans to speed up or increase the development of irrigable lands. Nor did it have plans for solving the persistent problems of land tenure in relation to the matter of heirships, and it was withdrawing from the areas of extension and credit services. Instead, the bureau emphasized industrial development on the reservations and relocation of Indians to the cities, and as a result, the Indians did not have the opportunity to establish viable farming operations. With few tribesmen able to meet the requirements of "character, collateral and capacity" necessary to obtain loans from traditional lending institutions, Indian farmers were doomed. As a result, few Indian boys exhibited interest in agricultural education, because that training would not enable them to establish self-sufficient or commercially viable farms. The financial and land-tenure problems on the reservations were too great for them to overcome.[39]

Indian farmers were therefore caught in a vicious cycle. They could not obtain the necessary capital to expand and develop viable economic enterprises. Their inefficient operations and limited managerial abilities, compared to those of many university-trained white farmers, prevented lenders from having the confidence needed for granting loans to Indian farmers, and without land reform, even the best agricultural managers could not make Indian lands economically viable. Under these conditions, Indian landowners, such as those among the Cheyenne, the Arapaho, the Kiowa, the Comanche and the Wichita in Oklahoma, usually chose to lease their holdings; but this, too, prevented them from gaining managerial experience, and the heirship problem continued to limit the land base which would, in part, make them credit worthy. Nor could Indian farmers apply the expertise gained from the agricultural-extension agents, because they could not afford to do so. Even when they had the information necessary to improve their economic situation, financial and land problems blocked any progress.[40]

By the mid 1960s, however, the Indian cattlemen and livestock associations on the Fort Hall Reservation in Idaho were showing a "gradual but steady" improvement. The breeding herd of 3,800 cows had nearly doubled between 1955 and 1965. In addition, the Indians helped the extension agent develop water holes, build fences, and spray sagebrush and weeds. The Fort Hall Indian-Stockmen's Association and the Bannock Creek Stockmen's Association sold cattle to the feeder industry, averaging nearly $27 per hundredweight. This economic success prompted one official to predict a "promising future for progressive operators."[41]

To the east, Montana's Northern Cheyenne were, by the mid 1960s, as technically proficient and knowledgeable about the livestock industry as white cattlemen, but the heirship problem continued to foil the expansion of the livestock industry. All heirs had to agree in order to permit the sale of land, and the approval of court-appointed guardians was required for the sale of lands owned by minors, both of which were difficult to achieve. Without adequate acreage for successful ranching, the Northern Cheyenne had little choice but to abdicate control of their lands and to accept bureau administration and leasing policies.[42]

In the Southwest, drought continued to plague Indian agriculturists into the 1960s. At the beginning of the decade, the Navaho range was in "bad shape." Although the bureau pressed for stock reduction through the Tribal Council to relieve the "critical condition" of the range, Navaho livestock owners resisted. Cultural tradition still was more important than technical flexibility, which would have enabled them to reduce their livestock and thereby conserve and restore the parched grazing lands. With 5.5 million acres severely overgrazed, the Navaho range land supported 539,323 sheep units, when the carrying capacity was estimated at no more than 387,000, for an overstocking rate of about 40 percent. By 1960, an estimated 676,000 acres of range land was unusable because of drought and overgrazing, and the emergency feed program had cost $8.4 million since 1957—more than the returns from the sale of livestock during that period. Even had rainfall been adequate and the range lands lush with grass, reservation agriculture could not have supported the Navaho people. In 1960, 50 percent of the families in the southern Navaho jurisdiction did not have grazing permits, and income from crop production and livestock provided only an estimated 10 percent of their income—down from 58.4 percent in 1940. More than half of those who owned sheep had fifty head or fewer.[43]

In early November 1960 the Tribal Council appropriated $597,250, of which it spent $355,145 for an emergency livestock-purchase program. This program was designed to create a market for Navaho livestock and to provide desperately needed income. Ultimately, 44,731 sheep units were removed from the grazing lands under this program. Although this aid was important for those Navaho who received it, they still were subsistence operators; large-scale commercial production was beyond their reach. By 1960, for example, the Navaho at the Crownpoint Subagency in New Mexico earned only $450,000 from livestock and crop production. At that time, they earned almost $28.8 million from wages. An estimated $2.4 million of that income came from migratory agricultural work, that is, from unskilled "stoop" labor. In that agency, an estimated 54 percent of the households did not own livestock, and among those who did, the average sheep units per capita was only 58, when 250 to 300 sheep units per person were necessary for support.[44]

By the late 1960s, sheep raising largely remained a subsistence operation for the Navaho. Flocks were kept, not for profit, but for insurance in time of need or for ceremonial purposes and cultural identity. Because many Navaho were engaged in part-time wage work, cattle raising became more attractive. Cattle required less care than sheep and returned greater profits. For this transition to occur, however, the Navaho needed more land, wells, and stock-watering ponds. By the end of the decade, the Navaho range was capable of supporting only about 40,000 people, yet the population exceeded 120,000. Thus, the economic future of most Navaho lay elsewhere, but tradition and the lack of education prevented them from seeking it. Indeed, many Navaho supported efforts to expand water resources by digging deep wells so that they might remain on the reservation and maintain subsistence livestock operations.[45]

On a broader scale, the heirship problem continued to plague the tribesmen during the late 1960s. By 1968, more than 6 million acres, or about 11 percent of Indian lands, were in heirship status. Half of that acreage was claimed by five or more heirs. While 10 percent of heirship land lay idle, Indian agriculturists used only 65 percent of it, and non-Indians used 25 percent of it. Tribal lands, such as those among the Pima, were so "fractionated" that the lease fees returned less than one dollar per acre annually.[46]

Indian agricultural productivity also lagged, because the tribesmen preferred to lease their lands. By 1968, Indians were using less than 30 percent of their irrigated lands, 25 percent of their dry lands, and 80 percent of their range lands. Leasing enabled the tribesmen to earn an income from their lands while they held nonagricultural jobs elsewhere. The tribesmen's decision to lease their lands primarily resulted because they could not gain a sufficient return by farming it themselves. Low productivity also stemmed from cultural limitations as much as from inadequate access to capital, land, and credit. Indeed, culture often prevented the Indians from farming as productively as whites. Even by the late 1960s, the western tribesmen still did not have an agricultural tradition. With the exception of the Pueblos and a few other tribes, most Indian males did not have fathers who could teach them that vocation. Without experience in farming, Indian males had to learn agriculture, in part, from governmental agents or from schools, most of which provided inadequate training. In contrast, white farmers, who had a long agricultural tradition and who had adequate access to land, capital, and credit, usually succeeded where Indian farmers failed. Moreover, as among the Papago and the San Carlos Apache, Indian cultural values, which stressed voluntary cooperation and the sharing of surpluses within the extended family, hindered agricultural advancement in a market economy where whites emphasized initiative, responsibility, aggressiveness, and individualism.[47]

Indian farmers did not have the necessary technical assistance to become successful agriculturists, partly because agricultural extension remained underfunded. Although agricultural-extension services largely had been transferred to the United States Department of Agriculture in 1956, the Bureau of Indian Affairs still provided funds for that work on a contract basis. In 1969, that funding amounted to only about $1 million, which provided two hundred full- or part-time extension agents, or about one agent for each reservation that needed those services. At that time, Indian farmers on the reservations needed three times that number, according to one estimate.[48]

By the late 1960s the revolving loan fund was also inadequate. Without access to credit or capital for the purchase of land, machinery, fertilizer, seed, and livestock, Indian farmers had little chance of improving their agricultural positions. Although Indian operators could apply to the Farmers Home Administration, that agency was not able to meet the capital needs of Indian farmers. With problems such as these, few Indians desired to become full-time farmers. Although cooperative agricultural ventures on a tribal basis provided some economic return, the number of Indian families that engaged full-time in agriculture continued to decline during the 1960s. Between 1965 and 1968, for example, that number declined from 9,253 to 5,080, a reduction of nearly 45 percent. Although this trend was similar to that of white farmers who also were leaving the land in increasing numbers, the loss of Indian farmers was relatively greater because fewer Indian agriculturists existed to begin with. As a result, only about one-third of the gross agricultural income earned on the reservation lands went to Indian farmers; non-Indian operators earned the rest.[49]

By 1970, Indian agriculture had reached a low ebb. Although an estimated 55.5 percent of the Indian population lived in rural areas, only 11.2 percent were engaged in agriculture. The problems of heirship, capital, credit, inadequate agricultural education, insufficient access to science and technology, and limited managerial abilities remained unsolved. The Indians also continued to fear a reemergence of termination policy that would deprive them of their tribal estate. While Indian agriculturists and others grappled with these problems, an even greater difficulty had emerged by the 1970s, which, in many respects, held the key to the future development of Indian agriculture. This was the matter of Indian water rights. By the 1970s, the agricultural development of many reservations depended upon access to water. For most Indians, water was their preeminent natural resource, without which much tribal and allotted land would be unusable for agriculture. During the 1970s and 1980s the fight to gain or to ensure Indian water rights became the paramount issue affecting the lives of Indian farmers.[50]

13

Quagmires

Water is the lifeblood of the West, and by 1970, water rights had become a major concern of the Indians. With 75 percent of the reservation population and 55 percent of Indian lands located in a region that receives less than twenty inches of precipitation annually, water holds the key to economic development, including agriculture. Rapid population growth of non-Indians and the subsequent increased demands by city governments as well as the pressing needs, if not covetous desires, of industrialists intent on developing the West's natural resources have placed the matter of water rights in the forefront of Indian-white relations. The legal maneuvering over Indian water rights has been complex and confusing; it has also been bitter. Tribesmen fear they may lose their water rights as easily as they did their lands during the late nineteenth and early twentieth centuries. Indeed, Indian water rights are at the heart of the "water grab" in the West.[1]

Until the early nineteenth century, whites claimed western waters by "riparian right" as well as by the right of "prior appropriation." Riparian right is a common-law concept which holds that a landowner whose property abuts a stream has the right to use that water. To claim a riparian right, a landowner does not need to put the water to beneficial use; therefore, riparian rights cannot be lost. Furthermore, riparian rights are not limited by quantity of use; instead, reasonable use of those waters restricts riparian rights, so that other landowners have equal rights along the stream. In addition, the date of initial use does not determine priority of use. Consequently, landowners share water shortages equally.[2]

In the arid West, however, the concept of riparian rights did not serve the white immigrants as well as it had users in the East, where access to water was generally not a problem. Therefore, settlers developed a new concept of water rights to meet the needs of those who were located far from stream sides. Specifically, it is the right of prior appropriation, which was first developed in the California gold fields, where miners needed to divert water far from the stream beds where riparian rights governed water use. The doctrine of prior appropriation is based on the concept "first in time, first in right," or more colloquially, "first come, first served." Under this system, one's priority depends upon the date when the water was first put to beneficial use, or, more technically, when the user applied for a permit to divert water for a specific purpose. In contrast to riparian rights, whenever water shortages occur, the latest or junior appropriators can lose a portion or all of the water that they claim, while the earlier or senior appropriators are the last to bear the burden of decreased quantities. The right of prior appropriation also depends upon actual use. If the water claimed is not put to beneficial use, the right to it can be lost. Moreover, the user cannot increase the quantity appropriated for a specific use under the terms of the original application.[3]

The doctrine of prior appropriation was well suited to the needs of westerners who had the capital and the skill to use water efficiently, but it totally neglected the needs of the Indians, because they did not have the funds to divert and to use the western waters in an extensive irrigation system. In addition, the Indians generally lacked the cultural competitiveness to appropriate water before the settlers claimed the land that surrounded the reservations and diverted most of the water for their own use.[4]

In 1908, western water law was fundamentally changed when the United States Supreme Court ruled in the case of *Winters* v. *United States*, which is commonly known as the *Winters* decision. Since that time, Indian water rights essentially have been secure, but the drive to modify those rights to meet newly perceived needs of non-Indian users is gaining momentum. The substance of the *Winters* case dates to 1 May 1888, when Congress created the Fort Belknap Reservation in Montana for the Gros Ventre and Assiniboin Indians. The Milk River, which bordered the reservation on the north, drained those lands. White settlers soon claimed surrounding lands, which the tribe had surrendered as part of the reservation agreement. Upstream from the reservation the settlers began to divert the waters of the Milk River, thus leaving insufficient water for a government-sponsored irrigation project on reservation lands. Although the settlers claimed the right to use those waters under the doctrine of prior appropriation, according to the laws of the territory and, later, the state of Montana, the federal government sued to prevent them from diverting the upstream waters away from the reservation.[5]

When the Ninth Circuit Court of Appeals in Montana ruled on the case, it held in favor of the federal government, on the grounds that the reservation had been created "to encourage farming among the Indians." Therefore, access to water for irrigation was indispensable to the agreement between the Indians and the federal government, even though the matter of water rights was not explicitly stated in the treaty. The case ultimately reached the United States Supreme Court, which affirmed the trial court's opinion. The Supreme Court held that because the water necessary for Indian use had been reserved when the federal government created the Fort Belknap Reservation, it was exempt from appropriation under state law. Thus, Indian water rights were superior to the claims of the settlers, or at least were superior to all appropriators who claimed water rights after the founding date of the reservation.[6]

By so ruling, the Supreme Court struck down the concept of prior appropriation based on beneficial use, which had been the foundation of water law in the western states. Fundamentally, the court contended that the Indians had a water right far different from the concept of riparian rights or the doctrine of prior appropriation. Indian water rights were special "reserved rights" which superseded state law. Those rights dated from the creation of the reservation and did not depend upon actual diversion or continued use. This ruling gave a "priority date" to the water rights of the Indians which were senior to all appropriators except those who had established their rights before the creation of the reservation. Thereafter, Indian water rights were supreme. The Court did not fix the amount of water reserved for the Indians, but also in 1908 the Ninth Federal Circuit Court ruled, in *Conrad Investment Co.* v. *United States*, that the Blackfeet had a "paramount right" to use water "to the extent reasonably necessary for the purpose of irrigation and stock raising, and domestic and other useful purposes." Thus, the reserved Indian water rights were unquantified and open ended.[7]

The *Winters* decision seemed relatively clear at the time, although the white settlers bitterly resented it. More than a half-century later, however, the *Winters* decision has left the matter of Indian rights in murky waters. The heart of the problem involves determining who reserved the water; the Court did not clarify the matter. Was it the federal government or the Indians? In addition, how much water was reserved and for what purposes could the water be used? The answers to these crucial questions can determine how those waters will be used, as well as how much water will be used and by whom. If the Indians have the right to specific western waters based on the right of occupancy from "time immemorial," their right is "prior and paramount" to the water rights of all other non-Indian appropriators. In this case, the water right essentially is a property right protected by the Constitution, although under certain circumstances the federal government could change it. Most important, however, if the Indians reserved the water, they could claim all of it for any purpose.[8]

This interpretation, however, threatens western cities and states that might eventually have to pay for access to water that they are currently using. Thus, water rights based on occupancy of the land essentially would enable the tribes to preempt water that is currently being used by non-Indians. In contrast, if the federal government reserved the water as the sovereign or trustee of the reservation, then the reserved right extends back only to the creation of the reservation. Therefore, the water rights of the settlers who arrived prior to the establishment of the reservation would be superior to those of the Indians. This quandary has perplexed legal experts since the *Winters* decision, because the Supreme Court did not resolve the issue; rather, it seemed to support both interpretations. As a result, the proponents of both interpretations have marshaled mountains of evidence to support their respective views. While each side has argued forcefully and skillfully, the issue has not yet been decided, and the prospects for resolution in the immediate future are dim.[9]

Indian water rights also are muddled, because the courts did little to clarify the *Winters* decision for more than half a century. In 1963, however, the Supreme Court again ruled on Indian water rights in relation to five lower-Colorado Indian reservations. At that time, many non-Indian agriculturists and developers hoped that the High Court would resolve whether Indian water rights dated from "time immemorial" or from the creation of a reservation. They were disappointed in that desire, because the Court held, in the case of *Arizona* v. *California,* that the water rights in question dated only from the creation of these reservations. The problem was that those reservations had been created either by executive order or by an act of Congress. None had been established by an agreement or a treaty with the Indians. While this ruling clarified the matter of Indian water rights on reservations established by the president or by Congress, it did not resolve the issue of whether the Indians or the federal government reserved those rights when treaties had created those reservations.[10]

The case of *Arizona* v. *California* also complicated Indian water rights even further, because the Court held that the amount of water reserved was based on the number of "practicably irrigable acres of the reservations." By so contending, the Court quantified those reserved rights based on the amount of water needed to irrigate all of the irrigable acreage within those five reservations. Although the Court did not intend to quantify the maximum amount of water reserved for the Indians, it created great uncertainty over the amount of water to which the reservations were entitled. Thus, while Indian water rights were clarified for reservations created by two methods, the quantity of those rights still remained in question, and this uncertainty drove the western states, agriculturists, and developers to press for some specific system of quantification during the 1970s and 1980s.[11]

Although the measure of water in terms of practicably irrigable acreage on the reservations seemed a reasonable way to quantify Indian water rights and although Indian water rights are currently based on the size and irrigable acreage of the reservations, rather than on the population or the intent of the tribesmen to develop the reservations, the courts have not yet resolved whether the Indians can use water for purposes other than agriculture or whether a special treaty water right exists. These uncertainties have caused frustration and anger on the part of the Indians and the whites alike. In relation to the matter of irrigation, for example, "irrigable acreage" can be defined fairly easily, but "practicably irrigable acres" means vastly different things to the Indians and the non-Indians who are competing for the limited water reserves in the West.[12]

In addition to the problem of surface-water rights, the Indians and other westerners are contesting subsurface, or ground water, rights. Today, four systems govern ground-water rights in the United States. The first is that of "absolute ownership" under the rule of capture. In other words, the first person to tap an aquifer has the right to use as much of it as he or she pleases. The second rule is that of "reasonable use," whereby landowners can make beneficial use of the ground-water supply, provided they respect the rights and needs of others who also withdraw from the same source. The third governance is that of "correlative rights," which allocates water proportionally, based on landownership rather than on reasonable use. The last rule is that of "prior appropriation," that is, first in time, first in right, provided the landowner does not infringe upon the rights of other appropriators or waste the water. These four rules, however, are state regulations which give reasonable certainty and rationality to the use of ground water.[13]

In 1976 the United States Supreme Court ruled, in *Capaert* v. *United States,* that ground water under the reservations is reserved for the Indians. Thus, ground-water rights are merely the extension of reserved rights to surface waters. This doctrine creates a preemptive right to ground water, which enables Indian users to satisfy their needs prior to those of others, particularly in times of drought or shortage. This right is not lost if it is not used. Where the Indians do not currently exercise that right, they may do so in the future. White agriculturists fear either that they eventually will lose all of their water rights or that the Indians will seriously deplete the ground-water supply. Both results would seriously jeopardize, if not ruin, non-Indian agricultural, mining, and other investments.[14]

This preemptive right does not restrict non-Indians from mining ground water as long as the Indians are not using it. Therefore, this aboriginal water right encourages water mining as non-Indians extract as much water as possible to meet their immediate and speculative needs. Consequently, preemptive Indian water rights have the potential of damaging future Indian agricultural development. It also encourages excessive or unwise use by the Indians, because they are entitled only to the water that they can use. Thus, they have no incentive to conserve it for the future. The preemptive Indian water right also blocks state regulation of water mining. Still, ground water is a relatively new matter for the courts, and no one can yet say with certainty that the reserved or preemptive Indian rights to ground water will stand the test of time.[15]

Indian water rights became even more complex in relation to the Pueblos. When the federal government gained control over the Southwest after the Mexican War, it recognized and agreed to preserve the property rights of the inhabitants, which both the Spanish and the Mexican governments had granted. A decade later, Congress confirmed the land claims of most Pueblos. Therefore, the Pueblos' water rights depended upon both Spanish and Mexican law. Also, Pueblo lands were never part of the public domain—an added problem, which requires a resolution of the date when the Pueblos gained control over their reservation lands. For example, is that date 1933, when a congressional land act authorized compensation to these Indians for lands that non-Indians were using? Or does it date from an "immemorial" time, because the Pueblos have used their lands continuously for approximately nine centuries. If those rights date only from the twentieth century, then a host of non-Indians can legally claim water rights on land that they have acquired. But if those rights date from time immemorial, the Pueblos have a paramount right to those waters which no one can infringe. Future claims by the Pueblos could seriously damage non-Indian economic development or interrupt urban water supplies. If that happens, the Indians, who claim water rights from time immemorial, could become water brokers in the Southwest, unless Congress or the federal government, acting through the Department of the Interior, should interfere.[16]

Although this issue is far from being resolved, compromise probably is inevitable. Indeed, the Navaho already have compromised their reserved rights under the *Winters* decision in order to secure funds for the Navaho Indian Irrigation Project, which began in 1962. They did so by putting their *Winters* rights on a parity with those associated with the Bureau of Reclamation's San Juan–Chama Diversion Project, because sufficient water did not exist to meet entirely the needs of both groups. The western states are also demanding the quantification of Indian water rights, to make rational state planning possible. The Indians, however, continue to claim, under the *Winters* decision, the right to use all the water necessary to meet every purpose or need currently, as well as in the future. Only the exhaustion of the water supply will limit their claim without compromise. Yet, "quantification," "negotiation," and "compromise" threaten the loss of their water rights and water. In reality, of course, a great deal of difference exists between "paper" water and "wet" water, and theoretical rights can easily, though perhaps illegally, be superseded if the general welfare of the inhabitants of the Southwest is endangered by Indian claims to water.[17]

Indian water rights also remain uncertain on allotted lands that have been alienated. In 1939 the United States Supreme Court ruled, in *United States* v. *Powers,* that on Montana's Crow Reservation, some portion of the water essential for cultivation passed to the new owners upon sale of the land. The Court did not, however, specify the amount of water or the extent of the water right that was alienated at the time of the land sale, and it did not indicate whether water rights could be sold independently or whether those rights were inextricably attached to the land. In addition, the Court did not resolve whether the water rights that the tribe held in common could be alienated or leased. The Supreme Court also has not ruled as to whether Indian water rights that transfer to non-Indians have become subject to state law. Although a lower court has ruled that once sold, Indian water rights pass into state jurisdiction, legal scholars have argued both sides of the issue.[18]

The difficulty in resolving problems concerning Indian water rights stems from two causes. First, since the *Winters* decision in 1908, the courts have determined Indian water rights. Judicial or case law is an uncertain process, because no one can tell how far-reaching or how confusing a decision may be. Moreover, the courts are slow to act, and the timely resolution of a problem seldom can be made. Second, Congress and the executive branch have failed to act. Congress, as the legislative branch of government, has the authority and the ability to resolve the confusing problems that have developed concerning Indian water rights, and the increasing clamor of non-Indians for the quantification of Indian water rights may pressure Congress into addressing this issue. In the meantime, few Indians can afford to use their reserved water rights to irrigate croplands, because irrigation is expensive and the opponents of Indian water rights have successfully blocked congressional appropriations for that purpose. At the same time, the Indians fear that any attempt to clarify the doctrine of reserved rights will weaken their claim to the western waters. This fear seemed well founded when, in 1976, the Supreme Court ruled in *Colorado River Water Conservation District* v. *United States* that state as well as federal courts could adjudicate federal rights, and state courts have not traditionally treated the Indians sympathetically. Yet even in this case, the issue has not been entirely clarified; no one yet knows, for example, whether state courts have jurisdiction over Indian water rights when the state constitution disclaims any jurisdiction over Indian property.[19]

While the western tribesmen argued for the protection of their water rights, Indian farmers muddled along as best they could. By 1970 the Indians held about 50 million acres, but the old problems of inadequate land, insufficient credit, and limited technology continued to restrict their efforts. Most of the 400,000 reservation Indians, for example, were living in grim poverty and nearly hopeless conditions. With earnings of only one-third to one-fourth those of other citizens, with an unemployment rate of 40 percent, and with only half the education and two-thirds the life expectancy of non-Indians, the tribesmen faced a bleak future. The heirship problem had not yet been resolved, and they had no reason to believe that needed land reform would be forthcoming soon. With about 12 million acres of allotted trust land in heirship status, about half of which was owned by six or more heirs, the tribesmen had little choice but to acquiesce as the bureau leased more than half of it to white operators.[20]

Indeed, the acreages required for successful agriculture in the West also were far greater than the Indians could command. On the Blackfoot Reservation, for example, agriculturists thought that 2,400 acres were essential for cattle ranching and 640 acres for wheat cultivation, although farming was possible with 320 acres of irrigated land. Yet the allotment system provided only 40 to 160 acres, which heirships further divided. Consequently, the Blackfeet virtually had no chance to practice successful commercial agriculture on an individual basis. With land values low, heirs had little incentive to incur the expenses required to join others to use, lease, or sell their lands; instead, they preferred to let the bureau lease their lands to whites. At that same time, the tribes and individual Indians generally did not have adequate funds to purchase allotments when those lands became available. Moreover, the revolving loan fund of the bureau was not large enough to permit the consolidation of tribal land, and the Indians continued to lack the credit and collateral that private institutions required for loans.[21]

By 1970 the land problem also was acute for the Navaho, whose population had tripled since 1930. Little unclaimed land existed, and agricultural expansion depended upon chemical fertilizers, machinery, irrigation, and improved farming practices based on scientific knowledge. Without access to additional lands and unable to afford new technology, the Navaho were caught in a "Malthusian pincher" as the population increased while agricultural production remained static. Indeed, the Navaho raised crops on only 45,869 acres, while grazing livestock on 9 million acres. According to a 1971 bureau report on land use, the Navaho were cultivating 9,039 acres of dry land and 36,830 acres of irrigated land. At best the Navaho could only conduct subsistence agriculture. With few paved roads and no railroad service, commercial agriculture seemed far from possible by the early 1970s.[22]

By 1973, although about one-third of the Navaho families either raised some livestock or worked small farms, their agricultural earnings provided only a meager subsistence. Moreover, because of the absence of an adequate irrigation system, the Navaho had not been able to make full use

of their water rights. Therefore, with 55 percent of their reservation classified as desert and with an additional 37 percent classified as semiarid,
the Navaho had little choice but to emphasize livestock production, though
drought and overgrazing continued to plague them. Among the Navaho
near Shonto, Arizona, for example, the range was in poor condition by
the early 1970s. Although only 5,845 sheep units were permitted on the
nearby range, the Shonto Navaho were grazing 12,517 sheep units. Overgrazing and drought caused lamb weights to decline from a 67-pound
average in 1955 to 48 pounds in 1971. Although the average fleece weights
increased from 5.0 to 7.2 pounds during that same period, a sharp drop
in prices, from forty-two cents per pound in 1955 to nineteen cents per
pound in 1971, offset that production gain. Thus, where the Shonto
Navaho averaged $8.77 of income per sheep in 1955, in 1971 they earned
only $5.68 per head. During that same time, prices skyrocketed for almost
every item of daily need.[23]

The Shonto Navaho began to place more emphasis on cattle raising
than ever before, and in 1971, cattle contributed approximately 19 percent
of their income. Cattle brought higher prices per pound than sheep, cattle
did not graze the grass as closely as sheep, and cattle did not require close
supervision, as sheep did. As a result, the Shonto Navaho could earn a
higher income from cattle production and, at the same time, have more
time available for wage work. At this same time, however, the Shonto Navaho were cultivating only 1 percent of their land. The acreage in crops
was steadily declining, in part because of the availability of food programs
through the United States Department of Agriculture and because of the
shift to livestock grazing and wage work.[24]

For the Navaho as a whole, by January 1972, a glutted wool market
and depressed prices were threatening to drive an estimated twelve to fourteen thousand families onto the federal welfare rolls. Better grading and
handling, however, which improved sales and prices, soon alleviated much
of the problem. Other difficulties were less easily solved. In 1972 a Navaho
ten-year plan estimated that $520 million would be needed in order to
transform reservation agriculture and elevate it to a viable economic endeavor: $250 million for irrigation development; $150 million for reseeding grasslands and building fences; $20 million for digging wells and con-
structing stock tanks; and $100 million for maintaining income during per-
iods of soil restoration. But the Navaho had little hope of gaining this financial support from Congress.[25]

Instead of a major agricultural redevelopment program, the Navaho
were locked in the trap of overgrazing. Although they used more than
90 percent of their reservation for grazing livestock, they could not practice that agrarian craft wisely. A 1973 estimate indicated that overgrazing

reached 737 percent on the Navaho-Hopi Joint Land Use Area. There, 120,000 sheep grazed lands suited for only 16,278 head. The herds needed to be reduced by almost 90 percent to alleviate the problem. In 1974 the carrying capacity of the Navaho range was 1.5 million animal units, but an estimated 2.5 million units were actually grazing reservation lands, and more than 526,800 of those units were sheep, which were particularly destructive to the grazing lands. As a result, 63 percent of the reservation suffered from overgrazing, and profits fell. By 1977, 50 sheep could provide an income of $2,450 annually on good grazing lands, but 100 sheep returned only an estimated $732 annually from overgrazed lands. Because of the poor distribution of water and the lack of resources to develop stock watering ponds, the Navaho were unable to redistribute their livestock to make best use of the grazing lands. This depressing agricultural situation had not changed by the late 1970s. Moreover, the Tribal Council has not enforced grazing regulations since 1956, because that action remains unpopular among the Navaho and because council members believe the political cost of doing so is too great.[26]

While the Navaho continue their livestock tradition, their population growth has stimulated the development of more small herds to provide for family subsistence. By the late 1970s the average Navaho family, however, grazed fewer than 50 sheep, while a herd of at least 600 frequently is required to return an income of $3,000 annually. The inheritance of grazing permits and the continued division of those permits has increased the number of sheep while keeping herd size small. All suggestions for the consolidation of herds or for the enforcement of grazing permits has fallen on deaf ears. The Navaho remain culturally independent and prefer their traditional life style. Federal emergency feed-grain programs, which supplied 82 million pounds of milo in 1972 and 115 million pounds in 1978, served to subsidize overgrazing on the Navaho range. Thus, federal aid in time of dire need, together with an unenforced ceiling of 350 sheep units that the Tribal Council imposed on individual owners does nothing to alleviate the problems of overgrazing and inadequate land diversification on the Navaho reservation. As a result, the range has decreased in quality, and income from sheep raising has declined in quantity.[27]

On the northern Great Plains, the few remaining Indian agriculturists continued to struggle as well. In 1970, Congress passed legislation that provided $300,000 for the tribes on the Fort Berthold Reservation in North Dakota, through the Farmers Home Administration, to enable them to purchase and consolidate divided holdings. As a result, reservation acreage increased from 41,181 acres in 1971 to 57,954 acres in 1975. In addition, these tribesmen gained control over 34,542 acres that non-Indians had controlled. Even so, whites were using more than twice as much of the Indian cropland as were the tribesmen. Between 1973 and 1975, whites cultivated 141,329 acres and grazed livestock on 83,002 acres of Indian land. In the mid 1970s, Indian agriculturists earned only $1.9 million, compared to white operators on Indian lands, who earned $2.8 million annually.[28]

By 1970, the agricultural problems on the North Dakota reservations of Fort Berthold, Fort Totten, Turtle Mountain, and Standing Rock, the latter of which extended into South Dakota, reflected the dismal state of Indian agriculture everywhere. On those four reservations, only 9.1 percent of the Indians were farmers or farm managers, compared to 26.3 percent of the white population in the state. Although this average was three times higher than the percentage of non-Indians engaged in agriculture nationwide, these Indian farmers were far from equal on the economic scale. In fact, the economic disparity was great: while 69 percent of the Indians had annual incomes less than $4,000, only 43 percent of the white population were in that category. The median income for the Indians was $2,287, compared to $4,930 for whites; and 17 percent of the white population earned more than $10,000 annually, while only 3 percent of Indian men did so.[29]

Too many allotments and too many heirs prevented efficient agricultural operations on the reservations. Tribal lands were so divided that the sale and rental of Indian property still held in trust absorbed a great deal of the bureau's time; the cost of time required to calculate the various shares of sale and rental income often exceeded the value of those shares. The tribesmen continued to lease much of their land to white operators; and therefore, until the heirship problem could be resolved and until Indian lands could be consolidated, little agricultural progress could be made. The Indian agriculturists on these North Dakota reservations needed more technology and irrigation as well as improved management skills, but with an average of only one extension agent per reservation, this aid was wholly inadequate to meet the needs of Indian farmers.[30]

A similar situation existed among the Sisseton and Wahpeton Sioux on the Lake Traverse Reservation in South Dakota, where poor lands, inadequate capital, and the heirship problem combined to prevent the agriculturists among them from reaching their productive potential. By the late 1970s, agriculture employed less than 10 percent of the 1,100-man labor force. On average, each tribal member had an interest in eight tracts of land, and each tract had thirteen owners, so most of these Sioux preferred to lease their lands to whites, rather than to deal with this complicated problem. In 1976, however, the tribesmen, with the aid of a $4.25 million loan from the Farmers Home Administration, established two farms on tribal lands, one of which emphasized crop production, while the other stressed livestock raising. Personnel from South Dakota State University provided technical expertise to establish these farms. Although this development was encouraging, it offered too little too late to create a basis for a viable commercial agriculture.[31]

To the south, on the Pine Ridge Reservation, by 1970 the Sioux were losing 30,000 acres annually through land sales to non-Indians. Although the Tribal Council attempted to consolidate and increase the landholdings, the tribe owned only about 13 percent of the land. Without adequate funds, the Tribal Council could not purchase all of the reservation lands that periodically were offered for sale. Although the Indians held 39.1 percent of the land in allotments, whites and state and county governments owned 45.6 percent of the original reservation. Thus, the Sioux owned only 1.7 million acres of the reservation's approximately 2.8 million acres by 1971. They used 1.4 million acres for grazing land and 63,568 acres for dry-land agriculture; only 1,100 acres had irrigation. The Pine Ridge farmers used most of the dry-land acreage to raise feed grains.[32]

On the 92 percent of the Indian-owned land in range land at Pine Ridge, livestock grazing was the primary agricultural endeavor; but with 51 percent of the reservation area leased to whites, the Sioux were unable to expand their livestock industry to its fullest extent. In 1971 they organized the Oglala Sioux Farm and Ranch Enterprise, a corporate agricultural project that farmed 2,580 acres, 1,405 of which were irrigated. They used the remaining 1,175 acres for pasture and hay lands. This enterprise grazed three hundred cattle and raised a few hogs, for a total capital investment of $110,491. All farm profits were designated to fund the training of family heads for agricultural employment.[33]

While the Oglala Sioux Farm and Ranch Enterprise offered some hope for the development of a sound agricultural base on the reservation, the Pine Ridge Sioux had a long way to go before they could become commercial farmers. In 1971, only 270 families on the reservation earned their living through agriculture. Moreover, the heirship problem left small holdings of widely scattered lands across the reservation. Cultivation and grazing were not practical on these lands; the only profitable way to use them was to allow the bureau to lease the heirships to white cattlemen, who had funds to attain the economy of scale needed to make the reservation grazing lands profitable. Because in this area, one cow required thirty acres or more for subsistence and because a family needed at least 250 head of cattle to make a minimum income, scant prospect existed for the development of a large-scale Indian livestock industry. Moreover, even if every acre of the Pine Ridge Reservation had been used for grazing and cultivation, less than half the population would have been able to earn a living entirely from agriculture. The traditional cultural independence of the Oglala Sioux also continued to prevent the successful development of cattle cooperatives which might have alleviated some of the problems caused by inheritance and insufficient capital and credit.[34]

Indian farmers struggled in Oklahoma as well, where agriculture among the Five Civilized Tribes was nearly a thing of the past. By October 1977, for example, Seminole individuals owned only 29,544 of the original 369,854 acres of the reservation; the tribe owned a mere 380.10 acres. Although it is unclear how much of the allotted land remained in private Indian use or control, most of the fee-patented land had apparently been sold to whites by the end of the decade.[35]

By 1980 the Indians controlled 52 million acres, of which 42 million were under tribal ownership and approximately 10 million were under individual management. Despite the problems of Indian farmers, the federal government provided several aid programs to the few who were still engaged in agriculture. The first, known as the Indian Acute Distress Donation Program, was administered by the Agricultural Stabilization and Conservation Service. This program provided feed grains from the Commodity Credit Corporation to tribal stockmen who were suffering "chronic acute distress" from "severe drought, flood or other catastrophes." A second program, known as the Indian Land Acquisition Loan Program, provided funds to enable the tribesmen to purchase land within their reservations. The Farmers Home Administration also provided a more general program by which it assisted Indians to become owner-operators of their own farms. Farmers Home Administration loans could be used to purchase or expand a family farm or to make improvements. These loans were available to Indians who could not obtain credit through other institutions. The borrower's land secured each loan. The federal government also provided loans to Indians who needed capital to organize grazing associations, and it granted loans for irrigation and drainage projects. The programs of the Soil Conservation Service also were available to the Indians, although none were designed specifically for them. In addition, all other programs of the United States Department of Agriculture were available to Indian farmers, although none were exclusively developed to aid them.[36]

During the early 1980s, however, non-Indian agriculturists and developers in the West began to press the federal government to open its lands to private ownership and development. This "Sagebrush Rebellion" currently threatens the agricultural programs of the federal government that aid Indian farmers, as non-Indians seek to diminish the land base of the reservations and perhaps to limit the Indians' reserved water rights through federal or state quantification.[37]

Confronted with the frustrating and debilitating problems of insufficient land and inadequate capital and access to science and technology, only 7,211 Indian farmers remained by 1982. Although this number represented an increase of 322 since 1978, it clearly reflected the dismal state of Indian agriculture. Although the Indians devoted almost 46.2 million acres to agriculture, only 705,378 acres of cropland were harvested by

4,727 farmers. The majority of Indian farmers were small-scale operators; 4,252 had less than 140 acres; only 1,270 farmers had more than 500 acres. Thus, the vast majority of Indian farms in the semiarid and arid West were far too small to be considered viable economic operations. At best, these Indian farms provided some subsistence, but commercial agriculture was far from the grasp of these small-scale operators.[38]

Most of the 7,211 Indian farmers who remained were full- or part-owners. The 4,701 full owners controlled about 39.5 million acres, the 1,761 part-owners managed 6.2 million acres, and the 749 tenant farmers worked only 481,248 acres. Indian families held 6,301 of these farms; partnerships, 477; incorporated families, 43; other than family-held corporations, 11; and cooperative, estate or trust, and institutional farms, 379. In 1982, Indian farmers sold over $236.9 million of agricultural products, of which $138.5 million came from livestock and poultry and $98.4 million from crops. However, 4,750 farmers sold less than $10,000 annually, which kept them, in the absence of outside income, below the poverty line. Only 792 farmers earned between $10,000 and $20,000 while 1,476 sold more than $20,000 of agricultural products. Because of the low returns for individual farms, agriculture was the principal occupation for only 3,267 operators, and only 2,225 Indians spent all of their time in agriculture. Thus, more than half of the Indian farmers held jobs elsewhere. Agriculture clearly was not attractive to young Indians: in 1982, the average age of Indian farmers was 50.4 years, and only 946 Indian farmers were below thirty-five years of age. Farming particularly was not an attractive or viable occupation for Indian women; only 775 of them operated farms.[39]

By the early 1980s, then, most of the tribes, with the exception of the Navaho and a few others, did not have the land base or the financial means to exist independently in an agricultural economy, and even these tribes struggled under less than optimal circumstances. Farming was insignificant. It offered little hope for a brighter future, and like many of their white counterparts, the young were leaving the land to seek a better life in urban America. Thus, assimilation based on agricultural development remained an unrealized, if not an impossible, goal. Indian agricultural policy was a failure. New concerns over the development of natural resources and flood control in the West increasingly took the time of the tribesmen and the bureau. Little prospect existed for the economic and technical aid and the land reform needed to stimulate Indian agriculture and to make farming a viable economic endeavor for the tribesmen. Indeed, by the early 1980s, the achievements of Indian agriculturists were lost in the distant past, and the future for Indian farmers remained as bleak as it was uncertain.[40]

14

Epilogue

The history of Indian agriculture is the story of supreme achievement and dismal failure. From the prehistoric period to European contact, Indian farmers were responsible for considerable accomplishments; thereafter, governmental policy caused the ultimate collapse of that economic endeavor. Over centuries, Mesoamerican farmers domesticated and cultivated a host of plants which enabled them to gain greater control over their environment and made their food supply more dependable and secure. These developments fostered a sedentary life and further experimentation to control and use the environment in the best fashion for agricultural purposes. Mesoamerican farmers not only became skilled at plant breeding, at clearing and terracing their lands, and at irrigation; they also developed an accurate understanding of planting time. At first, they followed the progress of the seasons by observing changes of weather and vegetation and by watching the migration of birds. Without a calendar, however, they could not measure time, but they were remarkably capable of observing natural signs, such as the position of the sun and the size of leaves and blossoms for indications about when to plant, cultivate, and harvest crops. Once they understood those principles, they passed that knowledge from generation to generation, and along with other agricultural techniques, it became part of their cultural tradition. In time, the Maya and the Aztec developed accurate calendars to track the seasons and to minimize the risks of planting too early or too late.[1]

Beyond the development of the agricultural process, the domestication of corn was the greatest gift that Mesoamerican farmers gave to civilization. Indeed, corn became the most important crop for Mesoamerican and other Indian farmers. It had a high nutritional value, and each seed could produce three to four hundred kernels. Consequently, corn was the most productive plant that Indian farmers raised. In addition, corn could be easily planted, cultivated, harvested, and stored under a wide variety of climatic conditions. Indian farmers could raise it, for example, in the hot, wet lowlands or in the cool, dry highlands. Once Indian agriculturists had adopted corn, they became increasingly dependent upon it.[2]

Mesoamerican agricultural techniques and crops slowly spread northward and, in time, significantly influenced aboriginal life styles in what is now the contiguous continental United States. Corn gradually spread northward sometime after 3400 B.C. and became the most important food plant for the agricultural tribes. It possibly moved northward along the eastern coast of Mexico into Texas and into the Mississippi and Ohio river valleys, or Indian farmers may have introduced it from the American Southwest.[3]

Indian farmers in the region of the contiguous United States also were skilled at adapting agriculture to their environment. They not only bred certain plants, such as corn and cotton, so that they would produce under varied and extreme conditions, but they also developed numerous technical skills that enabled them increasingly to meet their food needs through agriculture. During the prehistoric era, for example, Mississippian farmers developed a fallowing cycle that permitted at least half of their cropland to rest each year. In the Southwest the Hohokam increasingly controlled their environment for agricultural purposes by building an extensive irrigation network.[4]

While Indian farmers improved their agricultural methods, they also developed rational systems of land tenure. By placing ultimate control of the land with the tribe and by allocating specific lands to individuals, the Indians were able to meet their communal and personal needs, both culturally and physically, in a most efficient manner. Although cultural differences prevented a unified system of land tenure, each group met its needs in a logical fashion. Generally, however, the women, who primarily were the farmers, controlled the land east of the Mississippi River. In contrast, tribal culture in the desert Southwest permitted the more active participation of Indian men in the agricultural process, and they usually controlled the use of the land. No matter whether Indian farmers were located east or west of the Mississippi River, their particular land-tenure systems provided for the acquisition, use, and transfer of agricultural lands. In contrast to white civilization, group rights superseded individual claims to the land, while the property rights of women were far more extensive then those under the English common law.

Thus, by the time of European contact, Indian farmers in many areas of the contiguous United States were tilling fertile fields along river flood plains. They cultivated a wide variety of plants, but corn was their most important crop. They had developed efficient harvesting and storage techniques, but the absence of iron kept them in the stone age, and they did not have domesticated animals. Even so, at the time of European contact, many Indians were engaged in varying degrees of agriculture which they integrated into the hunting-and-gathering processes. Although some nomadic tribes, such as the Sioux, traded for agricultural goods with the farming tribes on the Great Plains, others, such as the Pawnee, tried to maintain a balance between agriculture and hunting, until the introduction of the horse thoroughly changed their cultural traditions. Among the agricultural tribes, however, farming provided greater security of food supply and reduced the danger of famine or want during the winter.

Contact with European agricultural tradition was not the cause of the most significant and disruptive change in Indian farming. Rather, governmental policy, first in relation to the acquisition of Indian lands and later in relation to the development of subsistence and commercial agricultural policies among the tribesmen, altered Indian farming for all time. The first policy emanated from the chancelleries of the European powers. Soon thereafter, the colonial legislatures and, later, the central government, under the Articles of Confederation and the Constitution, devised moral, Biblical, and political rationales for the alienation of Indian lands. Ultimately, the acquisition of Indian lands through the cession-treaty process created an embarrassing paradox in the relationship between Indian and white civilizations. In order to encourage acculturation and assimilation into white society through the agricultural process, policy makers first sought the removal of the Indians east of the Mississippi River and later urged that they be concentrated on reservations, by arguing that removal and segregation would place the Indians far from the corrupting influences of white civilization. Once removed, the Indians could learn to farm in the self-sufficient manner of white men. At the same time, the idea spread that only the allotment of lands in severalty would solve the Indian problem. Before that concept reached fruition, however, governmental officials, missionaries, and friends of the Indians encouraged Indian agriculture by providing aid and education to the various tribesmen in the trans-Mississippi West.

The attempt to make the Indians into white farmers was slow and unsuccessful. Governmental economic, technical, and educational aid was never enough for the needs at hand. Consequently, only the most agricultural tribes, such as the Cherokee, showed much aptitude and success in farming once they were removed west of the Mississippi River. The tribes that had no agricultural traditions, such as the Sioux, found the sedentary farming life culturally unacceptable. At the same time, farmers, such as the Navaho, also believed that federal agricultural policy was destructive, because it did not meet their economic needs and it was not compatible with their traditions.

After the passage of the Dawes Act in 1887, the agricultural affairs of the Indians rapidly deteriorated. The allotment of individual landholding was a travesty of justice, because the vast majority of Indian lands were located in the semiarid Great Plains and the arid West. There, allotments of 160 acres or fewer were totally inadequate to enable the Indians to become self-sufficient farmers and thereby to end their burden on the public coffers. Remarkably, white farmers during the late nineteenth century were learning this lesson as well. Thousands of white homesteaders went bankrupt and gave up their lands. Those who remained eked out a tenuous existence until they were able to apply the science, technology, and capital needed to cope with the harsh western environment and to develop profitable agricultural operations.

Not only did the federal government neglect to furnish enough educational and financial support to aid the development and expansion of Indian agriculture in the trans-Mississippi West; it also failed to provide training and support to Indian women. Among the eastern tribes, the women were the primary agriculturists. In the trans-Mississippi West, however, the government was confronted for the most part with the difficult task of transforming nomadic male hunters into sedentary farmers who would practice women's work. By forcing an agricultural life upon the Indian men in that region, federal officials were doomed to failure. Had the federal government encouraged Indian women to take a more active role in the farming process, it probably could have improved the transition of the tribal groups from a nomadic to a sedentary life based on agriculture. Federal officials might also have been more successful if they had emphasized livestock raising among the men. Still, these are moot points. Federal officials chose to do neither. In 1879, for example, Commissioner E. A. Hayt succinctly explained the reluctance on the part of governmental officials to encourage livestock raising when he wrote: "The number of persons who can be employed in stockraising is small, since comparatively little labor is required, and a few men can herd and take care of a thousand head of cattle; but the cultivation of the soil will give employment to the whole race." Unfortunately, an adequate amount of fertile and well-watered lands was not available for the Indians. Usually, the tribes that were moved onto the reservations occupied the most infertile lands or lands that, in the absence of irrigation, could not be profitably farmed either by red men or by white men. Reservation lands, however, frequently were well suited for grazing livestock, and white cattlemen were quick to press the government and the Indians for the acquisition and lease of those grasslands. Without technical or financial support, the Indians had little choice but to lease their land, because they could not use it for agriculture. Quickly, millions of acres passed into the control and ownership of whites.[5]

Inheritance policies that divided estates among the heirs further fragmented Indian landholdings and made leasing not only a necessity but a way of life. As a result, the dreams and promises for an agricultural life that would enable the Indians to acculturate and assimilate moved further and further from reality. At the same time, governmental policy made the alienation of Indian lands increasingly easy, and the land base of the tribesmen was rapidly disappearing by the early twentieth century. With every acre of land lost to white control or ownership, the creation of a viable agricultural economy that would provide subsistence and, eventually, commercial profitability became increasingly harder to achieve.

Although environmental and cultural restraints prevented the Indians from becoming self-sufficient farmers within white civilization, they faced another handicap that was equally negative. This was white prejudice. From the time of European contact until the late twentieth century, most whites believed that the Indians were too lazy and too ignorant to become self-sufficient farmers. Collectively, whites treated the Indians like second- or third-class citizens. The Indians, though they were invited to compete with whites on an equal basis, were not provided with adequate lands and assistance. Moreover, most whites expected the Indian farmers to fail. Thus, when the Indians did fail, it became a self-fulfilling prophecy. Ultimately, the federal government essentially abdicated its responsibility and reneged on its promises to the Indians. Federal policy did not substantively encourage Indian agriculture. In fact, it hindered Indian farming with its policies of allotment, leasing, heirship, and land alienation. When John Collier tried to stem the tide of failure, he faced a nearly impossible task.

When Collier did attempt to reverse the damages of past federal policy by ending the allotment system and by encouraging communal agricultural activities, not all Indians greeted him with open arms. The most successful Indian agriculturists believed the new governmental policy would damage their farming operations, and they largely refused to cooperate with this new direction in tribal agricultural policy. For all of his accomplishments, such as halting the depletion of the Indian land base and providing funds for soil conservation and the improvement of livestock raising, Collier faced a difficult assignment. By the Great Depression, Indian farmers were separated by too great a gap—technologically, financially, and politically—ever to catch up with white farmers. Not only did the western environment, governmental policy, and their own cultural heritage prevent many Indians from becoming successful agriculturists, even on a subsistence level; the white agricultural tradition also overpowered them.

Indeed, white agriculture was far different from the cultural practices of the Indians, including the farming tribes. Most of the colonizing Europeans knew something about agriculture, or soon learned because of stern necessity. Over time, European cultural tradition adapted to the environmental conditions of a new continent and changed the use of the land. A large immigration rate and a high birth rate created the demand for more farmland by a people who saw landownership and agriculture as a golden opportunity for self-sufficiency and prosperity. It mattered little whether they acquired lands from the public domain, from speculators, or from the Indians. In the last case, most did not care whether they made such acquisitions legally or illegally. Unused land begged for the taking, or so they argued.

As white agricultural methods improved with advances in science, technology, financing, marketing, and transportation, the Indians fell farther and farther behind. Without access to adequate land, science, technology, finance, markets, and transportation, the Indians did not have a viable chance to become self-sufficient farmers by the twentieth century. Indeed, self-sufficiency required more than canning an adequate amount of vegetables and smoking enough meat to satisfy basic needs. By the twentieth century, a host of consumer goods became necessities. Innovations, such as the internal-combustion engine, for example, eventually required farmers to purchase tractors and trucks if they were to operate on a scale large enough to survive.

After World War II, termination policy prevented agricultural progress among the tribesmen, for whom termination harkened back to the days of the past, when whites could acquire tribal lands with relative ease. Although the federal government quickly discarded that policy, it left an enduring scar upon Indian-white relations, because it clearly indicated that the federal government no longer had any aspirations of supporting Indian agriculture in order to promote the acculturation and assimilation of the tribesmen into white society.

Today, most Indians are residents of rural America, but they are not agricultural, and they do not have much hope of becoming farmers. In fact, if the Indians were to do so, they probably would not improve their fortunes at all. In an age when white farmers increasingly are leaving agriculture because they cannot maintain profitability and in an age when success depends upon the expansion of landholdings, efficient management, and the application of science and technology in order to reduce unit costs, Indian farmers have little chance to succeed. Moreover, few young Indians, like their white counterparts, see much future in agriculture. It merely promises hard work and little remuneration. The extremely small size of individual landholdings or even the communal reservation lands, for example, prevents the development of the economy of scale necessary for successful agriculture, given current governmental policy.

In the final analysis, governmental policy designed to fashion the Indians into subsistence farmers required too great a cultural change for most of them, even under the best of circumstances. The lands upon which the government placed them, however, would have challenged even the best white farmers. The development of commercial agriculture would have required an even more substantive and radical change in the lives of most Indians west of the Mississippi River. Because whites and Indians used the environment differently, inevitably the most technically advanced and politically aggressive civilization would become dominant, particularly when that culture was far more agriculturally oriented than was the other one.

Ultimately, assimilation and acculturation depended upon the development of commercial agriculture for the tribesmen. Subsistence agriculture meant continued segregation and poverty. Only by developing their farming practices to such an extent that they could participate in a market economy could the Indians have acquired the wealth, education, consumer goods, and status that would have enabled them to move freely and equally among whites. Indeed, by the turn of the twentieth century, subsistence agriculture guaranteed poverty in an industrialized nation, because industrialization created new wants and needs that subsistence farming could not meet. Moreover, the trans-Mississippi West largely was suitable only for large-scale, extensive one-crop agriculture, not diverse, small-scale subsistence farming.

In the end, white civilization ruined, rather than promoted, Indian agriculture. After the creation of the United States, Indian agriculture expanded and intensified, but it did so, not as a result of economic forces, but because of governmental policy. Yet in most cases, federal officials ignored tribal wishes and cultural traditions and treated all Indians as a homogeneous group. Federal agricultural policy in the case of the Navaho, for example, destroyed their livestock-raising practices by violating sacrosanct cultural beliefs and by failing to provide the financial and technical support necessary to enable them to adopt white farming techniques. In contrast, federal policy prevented the development of farming among the nonagricultural tribes, such as the Sioux, because it placed them on inadequate lands, failed to provide necessary support, and ignored cultural tradition. In the end, the failure of federal agricultural policy for the Indians helped to create the foundation for the social and economic problems that plague the reservations during the late twentieth century.

Ironically, by the late twentieth century, agricultural conditions had changed dramatically for white farmers. With the flight of white agriculturists from the land, rural Anglo-Americans became an economic, if not an ethnic or cultural, minority. Rapidly changing agricultural conditions during the late twentieth century also relegated Indian farmers to minority status within the tribal community—a position far removed from the time when their ancestors were among the most skilled agriculturists on the North American continent.

Notes

1. Kent V. Flannery, ed., *Guilá Naquitz: Archaic Foraging and Early Agriculture in Oaxaca, Mexico* (Orlando, Fla.: Academic Press, 1986), pp. 272–74; Richard I. Ford, "The Processes of Plant Food Production in Prehistoric North America," in *Prehistoric Food Production in North America*, ed. Richard I. Ford, Anthropological Papers no. 75, Museum of Anthropology, University of Michigan (Ann Arbor, 1985), pp. 1–18; R. S. MacNeish, "The Origins of American Agriculture," *Antiquity* 39 (June 1965): 88; Richard Stockton MacNeish, "Food Production and Village Life Developed in the Tehuacan Valley, Mexico," *Archaeology* 24 (Oct. 1971): 310.

2. MacNeish, "Food Production and Village Life," p. 312; Richard S. MacNeish, "Ancient Mesoamerican Civilization," *Science* 143 (7 Feb. 1964): 535; idem, "Origins of New World Civilization," *Scientific American* 211 (Nov. 1964): 36.

3. Flannery, *Guilá Naquitz*, pp. 3–4, 6, 9–10; Donald W. Lathrap, "Our Father the Cayman, Our Mother the Gourd: Spinden Revisited, or a Unitary Model for the Emergence of Agriculture in the New World," in *Origins of Agriculture*, ed. Charles A. Reed (The Hague: Mouton Publishers, 1977), p. 721; Richard S. MacNeish, "A Summary of Subsistence," in *Environment and Subsistence*, vol. 1 of *The Prehistory of the Tehuacan Valley*, ed. Douglas S. Byers (Austin: University of Texas Press, 1967), p. 295; Paul C. Mangelsdorf, Richard S. MacNeish, and Gordon R. Willey, "Origins of Agriculture in Middle America," in *Prehistoric Agriculture*, ed. Stuart Struever (Garden City, N.Y.: Natural History Press, 1971), p. 291; MacNeish, "Ancient Mesoamerican Civilization," p. 535; E. Earle Smith, Jr., "Agriculture, Tehuacan Valley," *Fieldiana: Botany* 31 (22 Jan. 1965): 84, 95.

4. Mangelsdorf, MacNeish and Willey, "Origins of Agriculture in Middle America," pp. 489, 491; MacNeish, "Origins of New World Civilization," p. 35.

5. Smith, "Agriculture, Tehuacan Valley," pp. 76–77.

6. Flannery, *Guilá Naquitz,* pp. 3, 8–9, 14, 504.

7. MacNeish, "Origins of New World Civilization," p. 36; idem, "Food Production and Village Life," pp. 310–12; idem, "Ancient Mesoamerican Civilization," pp. 534–35; C. Earle Smith, Jr., and Richard S. MacNeish, "Antiquity of American

Polyploid Cotton," *Science* 143 (14 Feb. 1964): 676; C. Earle Smith, Jr., "The Archaeological Record of Cultivated Crops of New World Origins," *Economic Botany* 19 (Oct.–Dec. 1965): 330; idem, "Plant Remains," in *Environment and Subsistence*, p. 254.

8. MacNeish, "Food Production and Village Life," p. 312; idem, "Ancient Mesoamerican Civilization," pp. 535–37; idem, "Origins of American Agriculture," p. 88.

9. Smith, "Plant Remains," p. 254; MacNeish, "Origins of New World Civilization," pp. 36–37; idem, "Food Production and Village Life," p. 315; idem, "Ancient Mesoamerican Civilization," pp. 536–37; idem, "Origins of American Agriculture," p. 90.

10. Mangelsdorf, MacNeish, and Willey, "Origins of Agriculture in Middle America," pp. 491–96; MacNeish, "Origins of American Agriculture," p. 90; idem, "Summary of Subsistence," p. 294; Smith, "Agriculture, Tehuacan Valley," p. 79.

11. Ursala M. Cowgill, "An Agricultural Study of the Southern Maya Lowlands," in *Ancient Mesoamerica*, ed. John A. Graham (Palo Alto, Calif.: Peek Publications, 1966), pp. 105, 108, 115; B. L. Turner II, "Implications from Agriculture for Maya Pre-history," in *Pre-Hispanic Agriculture*, ed. Peter D. Harrison and B. L. Turner II (Albuquerque: University of New Mexico Press, 1978), pp. 337–38; John S. Henderson, *The Ancient World of the Maya* (Ithaca, N.Y.: Cornell University Press, 1981), pp. 98, 116, 148; Edward Higbee, "Agriculture in the Maya Homeland," *Geographical Review* 38 (July 1948): 459; Robert F. Heizer, "Agriculture and the Theocratic State in Lowland Southeastern Mexico," *American Antiquity* 26 (Oct. 1960): 215; Paul Weatherwax, *Indian Corn in Old America* (New York: Macmillan Co, 1954), p. 122.

12. Cowgill, "Agricultural Study of the Southern Maya Lowlands," pp. 108, 112; Higbee, "Agriculture in the Maya Homeland," pp. 459–61.

13. Norman Hammond, "The Myth of the Milpa: Agricultural Expansion in the Maya Lowlands," in *Pre-Hispanic Agriculture*, p. 34; Turner, "Implications from Agriculture," pp. 338, 346; B. L. Turner II, "Ancient Agricultural Land Use in the Central Maya Lowlands," in *Pre-Hispanic Agriculture*, pp. 171, 173; Henderson, *Ancient World of the Maya*, pp. 149–50; B. L. Turner II and Peter D. Harrison, *Pulltrouser Swamp: Ancient Maya Habitat, Agriculture and Settlement in Northern Belize* (Austin: University of Texas Press, 1983), p. 247.

14. Turner, "Ancient Agricultural Land Use," pp. 169–71, 181; idem, "Implications from Agriculture," pp. 347, 349–50.

15. Pedro Armillas, "Gardens on Swamps," *Science* 174 (12 Nov. 1971): 653–54.

16. Ibid., pp. 656–58, 660.

17. Flannery, *Guilá Naquitz*, pp. 6–7; George F. Carter, "A Hypothesis Suggesting a Single Origin of Agriculture," in *Origins of Agriculture*, p. 125; Kent V. Flannery, "The Origins of Agriculture," in *Annual Review of Anthropology*, ed. Bernard J. Siegel et al. (Palo Alto, Calif.: Annual Review, Inc., 1973), pp. 288–89.

18. Lawrence Kaplan, "Archaeology and Domestication of American Phaseolus (Beans)," *Economic Botany* 19 (Oct.–Dec. 1965): 363–64; Barbara Pickersgill and Charles B. Heiser, Jr., "Origins and Distribution of Plants Domesticated in the New World Tropics," in *Origins of Agriculture*, pp. 810–11.

19. Kaplan, "Archaeology and Domestication of American Phaseolus," pp. 519–20, 532; C. Earle Smith, "The New World Centers of Origin of Cultivated Plants and the Archaeological Evidence," *Economic Botany* 22 (July–Sept. 1968): 258.

20. Paul C. Mangelsdorf, "The Mystery of Corn: New Perspectives," *Proceedings of the American Philosophical Society* 127, no. 4 (1983): 215; George W.

Beadle, "The Origin of Zea Maize," in *Origins of Agriculture*, p. 633; idem, "The Ancestry of Corn," *Scientific American* 242 (Jan. 1980): 112–19; Hugh H. Iltis, "From Teosinte to Maize: The Catastrophic Sexual Transmutation," *Science* 222 (25 Nov. 1983): 886–94; Walton C. Galinat, "Domestication and Diffusion of Maize," in *Prehistoric Food Production in North America*, pp. 245–78.

21. Flannery, "Origins of Agriculture," pp. 290–96; idem, *Guilá Naquitz*, p. 26; Paul C. Mangelsdorf, Richard S. MacNeish, and Walton C. Galinat, "Domestication of Corn," *Science* 143 (7 Feb. 1964): 544.

22. Paul Weatherwax, "The Indian as a Corn Breeder," *Proceedings of the Indiana Academy of Sciences* 51 (1941): 18–19; idem, "The Origin of the Maize Plant and Maize Agriculture in Ancient America," in *Symposium on Prehistoric Agriculture*, ed. Donald D. Brand, University of New Mexico Bulletin 296, no. 5 (1936): 17; idem, *Indian Corn in Old America*, p. 183; Mangelsdorf, MacNeish, and Galinat, "Domestication of Corn," p. 545.

23. Smith, "Plant Remains," pp. 224, 251, 254; Richard B. Woodbury and James A. Neely, "Water Control Systems of the Tehuacan Valley," in *Chronology and Irrigation*, ed. Richard MacNeish and Frederick Johnson, vol. 4 of *The Prehistory of the Tehuacan Valley*, ed. Douglas Byers (Austin: University of Texas Press, 1972), p. 150; MacNeish, "Origins of New World Civilization," p. 36; idem, "Summary of Subsistence," p. 308.

24. Kent V. Flannery, V. T. Kirby, Michael J. Kirby, and Aubrey W. Williams, Jr., "Farming Systems and Political Growth in Ancient Oaxaca," *Science* 158 (27 Oct. 1967): 450.

25. William E. Doolittle, "Aboriginal Agricultural Development in the Valley of Sonora, Mexico," *Geographical Review* 70 (July 1980): 332–36.

26. Barbara Bender, *Farming in Prehistory* (London: John Baker, 1975), p. 160; Flannery et al., "Farming Systems," p. 450; Doolittle, "Aboriginal Agricultural Development," p. 328; Pickersgill and Heiser, "Origins and Distribution of Plants," p. 825.

27. Ralph Linton, "Crops, Soils and Cultures in America," in *Ancient Mesoamerica*, p. 57; Gordon R. Willey, Gordon F. Ekholm, and Rene F. Millon, "The Patterns of Farming Life and Civilization," in *Natural Environments and Early Cultures*, ed. Robert C. West, vol. 1 of *Handbook of Middle American Indians*, ed. Robert Wauchope (Austin: University of Texas Press, 1964), pp. 446–48.

28. MacNeish, "Origins of American Agriculture," p. 93; Michael D. Coe and Kent V. Flannery, "Microenvironments and Mesoamerican Prehistory," in *Prehistoric Agriculture*, p. 142; Heiser, "Agriculture and the Theocratic State," p. 215.

CHAPTER 2. PREHISTORIC AGRICULTURE

1. Frances B. King, "Early Cultivated Cucurbits in Eastern North America," in *Prehistoric Food Production in North America,* ed. Richard I. Ford, *Anthropological Papers* no. 75, Museum of Anthropology, University of Michigan (Ann Arbor, 1985), pp. 73–97; David L. Asch and Nancy B. Asch, "Prehistoric Plant Cultivation in West-Central Illinois," in *Prehistoric Food Production in North America,* pp. 149–203; Stephen A. Chomko and Gary W. Crawford, "Plant Husbandry in Prehistoric Eastern North America: New Evidence for Its Development," *American Antiquity* 43 (July 1978): 405; Richard A. Yarnell, "Early Plant Husbandry in Eastern North America," in *Cultural Change and Continuity,* ed. Charles E. Cleland (New York: Academic Press, 1976), p. 66. The most recent archaeological evidence sug-

gests that cucurbits were present in eastern North America by about 5000 B.C. and that they were widely distributed throughout the central Mississippi River Valley by about 1000 B.C.

2. Richard I. Ford, "Patterns of Prehistoric Food Production in North America," in *Prehistoric Food Production in North America,* pp. 341–49; Charles B. Heiser, Jr., "Some Botanical Considerations of the Early Domesticated Plants North of Mexico," in *Prehistoric Food Production in North America,* pp. 57–72; Stuart Struever and Kent D. Vickery, "The Beginnings of Agriculture in the Midwest-Riverine Area of the United States," *American Anthropologist* 75 (Oct. 1973): 1213–15; Stuart Struever, "The Hopewell Interaction Sphere in Riverine–Western Great Lakes Culture History," in *Hopewellian Studies,* ed. Joseph R. Caldwell and Robert L. Hall, Illinois State Museum Scientific Papers, vol. 12 (Springfield: n.p., 1964), p. 100; Melvin L. Fowler, "The Origin of Plant Cultivation in the Central Mississippi Valley: A Hypothesis," in *Prehistoric Agriculture,* ed. Stuart Struever (Garden City, N.Y.: Natural History Press, 1971), pp. 123–25.

3. Joseph R. Caldwell, "Eastern North America," in *Prehistoric Agriculture,* p. 380; Struever and Vickery, "Beginnings of Agriculture in the Midwest-Riverine Area," p. 1197; Yarnell, "Early Plant Husbandry," p. 271. A variety of stone hoes has been excavated. The larger ones were, perhaps, used for breaking the ground, while the medium-sized hoes may have been used for tillage. If this is true, it suggests a primitive form of technological specialization. Corn planters also have been found that were fashioned from roughly pointed stones. These stones were lashed or hafted onto stick handles. Stone spades, with blades fashioned into pointed, straight, and convex shapes, also have been discovered.

4. James L. Murphy, *An Archaeological History of the Hocking Valley* (Athens: University of Ohio Press, 1975), p. 122; Melvin L. Fowler, "Agriculture and Village Settlement in the North American East: The Central Mississippi Valley," in *Prehistoric Agriculture,* pp. 393–94, 402; Caldwell, "Eastern North America," p. 379; Francis B. King, "Plant Remains from Phillips Spring, a Multicomponent Site in the Western Ozark Highland of Missouri," *Plains Anthropologist* 25 (Aug. 1980): 224–26; George Quimby, "Human Origins: An Introductory General Course in Anthropology," Selected Readings, ser. 31 (Chicago: University of Chicago Press, 1946), pp. 207–9; Richard A. Yarnell, "Iva annua var macropa: Extinct American Cultigen?" *American Anthropologist* 74 (June 1972): 335, 339.

5. William S. Webb, "The Archaic Cultures and the Adena People," *Ohio Archaeological and Historical Quarterly* 61 (Apr. 1952): 179; Patty Jo Watson, "Prehistoric Horticulturists," in *Archaeology of the Mammoth Cave Area,* ed. Patty Jo Watson (New York: Academic Press, 1970), p. 234; Robert M. Goslin, "Food of the Adena People," in *The Adena People No. 2,* ed. William S. Webb and Raymond S. Baby (Columbus: Ohio Historical Society, 1957), pp. 41–45; David L. Asch, Kenneth B. Farnsworth, and Nancy B. Asch, "Woodland Subsistence and Settlement in West-Central Illinois," in *Hopewell Archaeology,* ed. David S. Brose and N'omi Greber (Kent, Ohio: Kent State University Press, 1979), p. 81; Patrick H. Munson, "The Origins and Antiquity of Maize-Beans-Squash Agriculture in Eastern North America: Some Linguistic Implications," in *Variations in Anthropology: Essays in Honor of John C. McGregor,* ed. Donald Lathrap and Jody Douglas (Urbana: Illinois Archaeological Survey, 1973), pp. 108, 111. The Adena cultivated two varieties of pumpkin—*Cucurbita pepo* and *Curcurbita moschata.*

6. Raymond S. Baby, "The Hopewell Culture," *Ohio Archaeological and Historical Quarterly* 61 (Apr. 1952): 182; Martha A. Potter, *Ohio's Prehistoric Peoples*

(Columbus: Ohio Historical Society, 1968), p. 40; Caldwell, "Eastern North America," p. 370; George Irving Quimby, *Indian Life in the Upper Great Lakes: 11000 B.C. to A.D. 1800* (Chicago: University of Chicago Press, 1960), pp. 72–73, 80; Richard I. Ford, "Gathering and Gardening: Trends and Consequences of Hopewell Subsistence Strategies," in *Hopewell Archaeology,* p. 234; Olaf H. Prufer, "The Hopewell Complex in Ohio," in *Hopewellian Studies,* p. 77; Watson, "Prehistoric Horticulturists," p. 234; Paul C. Mangelsdorf, Richard S. MacNeish, and Walton C. Galinat, "Domestication of Corn," in *Prehistoric Agriculture,* p. 482; Asch, Farnsworth, and Asch, "Woodland Subsistence and Settlement in West-Central Illinois," pp. 83–85; Richard A. Yarnell, *Aboriginal Relationships between Culture and Plant Life in the Upper Great Lakes Region,* Anthropological Papers no. 23, Museum of Anthropology, University of Michigan (Ann Arbor, 1964), p. 119; Struever and Vickery, "Beginnings of Agriculture in the Midwest-Riverine Area," p. 1198.

7. Ford, "Gathering and Gardening," pp. 237–38; Caldwell, "Eastern North America," p. 361, 372–73, 379; Fowler, "Agriculture and Village Settlement," p. 399.

8. Yarnell, "Early Plant Husbandry," p. 272.

9. Gregory Waselkov, "Prehistoric Agriculture in the Central Mississippi Valley," *Agricultural History* 51 (July 1977): 513, 515; Yarnell, *Aboriginal Relationships between Culture and Plant Life,* p. 128.

10. Waselkov, "Prehistoric Agriculture in the Central Mississippi Valley," pp. 513, 515; William L. Brown and Edgar Anderson, "The Northern Flint Corns," *Annals of the Missouri Botanical Garden* 34 (Feb. 1947): 10.

11. Joffre L. Coe, "The Indian in North Carolina," *North Carolina Historical Review* 64 (Spring 1979): 159; Roy S. Dickens, Jr., *Cherokee Prehistory: The Pisgah Phase in the Appalachian Summit Region* (Knoxville: University of Tennessee Press, 1976), pp. 14, 19, 204, 223; Robert L. Neuman, "Domesticated Corn from the Fort Walton Mound Site in Houston County, Alabama," *Florida Anthropologist* 14 (Sept.–Dec. 1961): 77, 80; Gordon R. Willey, *Archeology of the Florida Gulf Coast,* Smithsonian Institution, Miscellaneous Contributions, vol. 113 (Washington, D.C., 1949), pp. 455, 539.

12. Lewis H. Larson, *Aboriginal Subsistence on the Southeastern Coastal Plain during the Late Prehistoric Period* (Gainesville: University Presses of Florida, 1980), pp. 184, 206.

13. Dickens, *Cherokee Prehistory,* pp. 211, 222; Lewis H. Larson, "Functional Considerations of Warfare in the Southeast during the Mississippian Period," *American Antiquity* 37 (July 1972): 389; Bennett H. Young, *The Prehistoric Men of Kentucky,* Filson Club, Publications no. 25 (Louisville, Ky., 1910), pp. 179–80.

14. Paul E. Minnis, "Domesticating People and Plants in the Greater Southwest," in *Prehistoric Food Production in North America,* pp. 309–39; Herbert W. Dick, "The Bat Cave Pod Corn Complex: A Note on Its Distribution and Archaeological Significance," *El Palacio* 61 (May 1954): 141, 143; Herbert W. Dick, *Bat Cave,* Monograph no. 27 (Santa Fe, N.Mex.: School of American Research, 1965), pp. 97, 107, 110; Michael S. Berry, "The Age of Maize in the Greater Southwest: A Critical Review," in *Prehistoric Food Production in North America,* pp. 279–307; Richard B. Woodbury and Ezra B. W. Zubrow, "Agricultural Beginnings, 1000 B.C.–A.D. 500," in *Southwest,* ed. Alfonso Ortiz, vol. 9 of *Handbook of North American Indians,* ed. William C. Sturtevant (Washington, D.C.: Smithsonian Institution, 1979), p. 43; Joseph Winter, "The Distribution and Development of Fremont Maize Agriculture: Some Preliminary Interpretations," *American Antiquity* 38 (Oct. 1973): 442; Albert H. Schroeder, "Unregulated Diffusion from Mexico into the

Southwest Prior to A.D. 700," *American Antiquity* 30 (Jan. 1965): 299–300, 302; Walton G. Galinat and James H. Gunnerson, "Spread of Eight-Rowed Maize from the Prehistoric Southwest," Harvard University Botanical Museum Leaflets, no. 20 (1 May 1963): 118; G. N. Collins, "A Drought-Resisting Adaptation in Seedlings of Hopi Maize," *Journal of Agricultural Research* 1 (10 Jan. 1914): 293, 298–300.

15. Dick, *Bat Cave*, p. 98; A. T. Erwin and E. P. Lana, "The Seminole Pumpkin," *Economic Botany* 10 (Jan.–Mar. 1956): 34; Thomas W. Whitaker and H. C. Cutler, "Cucurbits and Cultures in the Americas," *Economic Botany* 19 (Oct.–Dec. 1965): 348; Joseph Charles Winter, "Aboriginal Agriculture in the Southwest and Great Basin" (Ph.D. diss., University of Utah, 1974), p. 97.

16. Dick, *Bat Cave*, pp. 98–99; Vorsila Bohrer, "Ethnobotanical Aspects of Snaketown," *American Antiquity* 35 (Oct. 1970): 413; Lawrence Kaplan, "The Cultivated Beans of the Prehistoric Southwest," *Annals of the Missouri Botanical Garden* 43 (May 1956): 219, 223–24; Charles Steen and Volney H. Jones, "Prehistoric Lima Beans in the Southwest," *El Palacio* 47 (Sept. 1947): 198.

17. Kaplan, "Cultivated Beans of the Prehistoric Southwest," pp. 194, 218–19, 222; Emile W. Haury, *The Hohokam: Desert Farmers and Craftsmen* (Tucson: University of Arizona Press, 1976), p. 118.

18. George F. Carter, *Plant Geography and Culture History in the American Southwest,* Viking Fund Publications in Anthropology no. 5 (New York, 1945), pp. 78–79; Kaplan, "Cultivated Beans of the Prehistoric Southwest," p. 197, 222–23; Steen and Jones, "Prehistoric Lima Beans," p. 200; Vorsila L. Bohrer, Hugh C. Cutler, and Jonathan D. Sauer, "Carbonized Plant Remains from Two Hohokam Sites, Arizona BB: 13:14 and Arizona BB: 13:50," *Kiva* 35 (Oct. 1969): 1.

19. Kaplan, "Cultivated Beans of the Prehistoric Southwest," p. 197.

20. Kate Peck Kent, "The Cultivation and Weaving of Cotton in the Prehistoric Southwestern United States," *Transactions of the American Philosophical Society* 47 (1957): 467; Haury, *Hohokam,* p. 118; Winter, "Aboriginal Agriculture in the Southwest and Great Basin," pp. 103–4.

21. Kent, "Cultivation and Weaving of Cotton," p. 467.

22. Ibid., pp. 465, 469–70.

23. Ibid., p. 467; Volney H. Jones and Elizabeth Ann Morris, "A Seventh-Century Record of Tobacco Utilization in Arizona," *El Palacio* 67 (Aug. 1960): 115, 117; Yarnell, "Early Plant Husbandry," p. 266.

24. Carter, *Plant Geography*, pp. 95–96.

25. Haury, *Hohokam,* p. 119; Albert B. Reagan, "Some Ancient Indian Granaries," *Proceedings of the Utah Academy of Sciences, Arts and Letters* 11 (1934): 39–40; Woodbury and Zubrow, "Agricultural Beginnings," p. 52; Samuel James Guernsey and Alfred Vincent Kidder, "Basketmaker Caves of Northeastern Arizona," Harvard University, Papers of the Peabody Museum of American Archaeology and Ethnology, vol. 18, no. 2 (1921): 110.

26. Kirk Bryan, "Pre-Columbian Agriculture in the Southwest, as Conditioned by Periods of Alluviation," *Annals of the Association of American Geographers* 31 (Dec. 1941): 224–25.

27. Henry C. Shetrone, "A Unique Prehistoric Irrigation Project," *Annual Report of the Smithsonian Institution, 1945* (Washington, D.C.: Government Printing Office, 1946), p. 381; Odd S. Halseth, "Prehistoric Irrigation in Central Arizona," *Masterkey* 5 (Jan. 1932): 165; idem, "Prehistoric Irrigation in the Salt River Valley," in *Symposium on Prehistoric Agriculture,* ed. Donald D. Brand, University of New Mexico Bulletin, vol. 296, no. 5 (1936), p. 42.

28. Albert H. Schroeder, "Prehistoric Canals in the Salt River Valley, Arizona," *American Antiquity* 8 (Apr. 1943): 381–82; Halseth, "Prehistoric Irrigation in Central Arizona," pp. 165–66; Haury, *Hohokam,* p. 150; O. A. Turney, "Prehistoric Irrigation," *Arizona Historical Review* 2 (Apr. 1929): 11, 42; Richard B. Woodbury, "A Reappraisal of Hohokam Irrigation," *American Anthropologist* 63 (Apr. 1961): 556.

29. Haury, *Hohokam,* pp. 137, 141, 148–49.

30. Frank Midvale, "Prehistoric Irrigation of the Casa Grande Ruins Area," *Kiva* 30 (Feb. 1965): 83; F. W. Hodge, "Prehistoric Irrigation in Arizona," *American Anthropologist,* o.s., 6 (July 1893): 325; Richard B. Woodbury, "The Hohokam Canals at Pueblo Grande Arizona," *American Antiquity* 26 (Oct. 1960): 269.

31. Haury, *Hohokam,* pp. 141, 144; Emil Haury, "The Snaketown Canal," in *Symposium on Prehistoric Agriculture,* p. 50.

32. Schroeder, "Prehistoric Canals in the Salt River Valley," p. 385; Shetrone, "Unique Prehistoric Irrigation Project," p. 385; Haury, *Hohokam,* p. 151.

33. Guy R. Stewart and Maurice Donnelly, "Soil and Water Economy in the Pueblo Southwest II: Evaluation of Primitive Methods of Conservation," *Scientific Monthly* 56 (Feb. 1943): 134–37, 144; Arthur H. Rohn, "Prehistoric Soil and Water Conservation on Chapin Mesa, Southwestern Colorado," *American Antiquity* 28 (Apr. 1963): 441–42, 446, 454; Guy R. Stewart, "Conservation in Pueblo Agriculture I: Primitive Practices," *Scientific Monthly* 51 (Sept. 1940): 209–10.

34. Richard B. Woodbury, "Prehistoric Agriculture at Point of Pines Arizona," *Memoirs of the Society of American Archaeology,* no. 17 (1961), pp. 36–38; Alexander J. Lindsay, Jr., "The Beaver Creek Agricultural Community on the San Juan River, Utah," *American Antiquity* 27 (Oct. 1961): 175, 182–84.

35. Woodbury, "Prehistoric Agriculture at Point of Pines," p. 37.

36. A. B. Stout, "Vegetable Foods of the American Indians," *Journal of the New York Botanical Garden* 15 (Mar. 1914): 53; Yarnell, *Aboriginal Relationships between Culture and Plant Life,* pp. 91–92; Erminie Wheeler Voegelin, "The Place of Agriculture in the Subsistence Economy of the Shawnee," *Papers of the Michigan Academy of Science and Letters* 26 (1940): 513.

37. Caldwell, "Eastern North America," p. 373; William Albert Setchell, "Aboriginal Tobaccos," *American Anthropologist* 23 (Oct.–Dec. 1921): 401; Charles C. Willoughby, *Antiquities of the New England Indians* (Cambridge, Mass.: Peabody Museum of American Archaeology and Ethnology, 1935), p. 296.

38. Carl L. Johnnessen, Michael R. Wilson, and William A. Davenport, "The Domestication of Maize: Process or Event?" *Geographical Review* 60 (July 1970): 405, 409; Richard B. Woodbury, "Climatic Changes and Prehistoric Agriculture in the Southwestern United States," *Annals of the New York Academy of Science* 95 (Oct. 1961): 708; Woodbury and Zubrow, "Agricultural Beginnings," p. 59; George F. Carter, "A Preliminary Survey of Maize in the Southwestern United States," *Annals of the Missouri Botanical Garden* 32 (Sept. 1945): 313.

39. Halseth, "Prehistoric Irrigation in Central Arizona," p. 165; Turney, "Prehistoric Irrigation," p. 21; Haury, *Hohokam,* p. 121; Shetrone, "Unique Prehistoric Irrigation Project," p. 384; Frank Midvale, "Prehistoric Irrigation in the Salt River Valley, Arizona," *Kiva* 34 (Oct. 1968): 28.

40. Woodbury and Zubrow, "Agricultural Beginnings," pp. 44, 51; Paul S. Martin, "Prehistory: Mogollon," in *Southwest,* pp. 64–65; Winter, "Aboriginal Agriculture in the Southwest and Great Basin," p. 12.

CHAPTER 3. EASTERN FARMING
AT THE TIME OF EUROPEAN CONTACT

1. Lewis H. Larson, *Aboriginal Subsistence on the Southeastern Coastal Plain during the Late Prehistoric Period* (Gainesville: University Presses of Florida, 1980), pp. 212, 214, 217, 219; Charles W. Spellman, "The Agriculture of the Early North Florida Indians," *Florida Anthropologist* 1 (Nov. 1948): 37, 41–42; John R. Swanton, *The Indians of the Southeastern United States*, Smithsonian Institution, Bureau of American Ethnology, Bulletin 137 (Washington, D. C.: Government Printing Office, 1946), p. 308; Carl Ortwin Sauer, *Sixteenth Century North America* (Berkeley: University of California Press, 1971), p. 180.

2. Larson, *Aboriginal Subsistence*, pp. 212, 214, 217, 219; Spellman, "Agriculture of the Early North Florida Indians," p. 43; U.S. Congress, House, *Final Report of the United States DeSoto Expedition Commission*, 76th Cong., 1st sess., 1939, Document 71, p. 143; James W. Covington, "Agriculture as Practiced by the Eastern Timucuans of Florida, 1564–1590," *Associates of the NAL Today* 3 (Jan.–June, 1978): 9–11; Swanton, *Indians of the Southeastern United States*, pp. 308–9.

3. Spellman, "Agriculture of the Early North Florida Indians," pp. 44–46.

4. Ibid., p. 41; *Final Report of the DeSoto Commission*, p. 143; Sauer, *Sixteenth Century North America*, pp. 167, 181, 295.

5. Spellman, "Agriculture of the Early North Florida Indians," p. 48; Sauer, *Sixteenth Century North America*, pp. 294–95.

6. Sauer, *Sixteenth Century North America*, p. 252; Swanton, *Indians of the Southeastern United States*, p. 306.

7. Christian F. Feest, "Virginia Algonquians," in *Northeast*, ed. Bruce G. Trigger, vol. 15 of *Handbook of North American Indians*, ed. William C. Sturtevant (Washington, D.C.: Smithsonian Institution, 1978), p. 258.

8. Swanton, *Indians of the Southeastern United States*, pp. 268–70, 306; Sauer, *Sixteenth Century North America*, p. 260; Charles C. Willoughby, "The Virginia Indians in the Seventeenth Century," *American Anthropologist*, n.s., 9 (Jan.–Mar. 1907): 82.

9. Willoughby, "Virginia Indians in the Seventeenth Century," pp. 83–84; Sauer, *Sixteenth Century North America*, p. 287; G. Melvin Herndon, "Indian Agriculture in the Southern Colonies," *North Carolina Historical Review* 44 (Summer 1977): 284.

10. Edward Arber, ed., *Travels and Works of Captain John Smith* (Edinburgh: John Grant, 1910), pp. cxi–cxii, 62; Herndon, "Indian Agriculture in the Southern Colonies," p. 287.

11. Robert Beverly, *The History and Present State of Virginia*, bk. 2 (London, 1705), pp. 12, 14–15, 21–22, 26–28, bk. 3, p. 56; Swanton, *Indians of the Southeastern United States*, pp. 276, 279.

12. William N. Fenton and John Gulick, *Symposium on Cherokee and Iroquois Culture*, Smithsonian Institution, Bureau of American Ethnology, Bulletin 180 (Washington, D.C.: Government Printing Office, 1961), pp. 94–95, 111; John R. Swanton, *Early History of the Creek Indians and Their Neighbors*, Smithsonian Institution, Bureau of American Ethnology, Bulletin 73 (Washington, D.C.: Government Printing Office, 1922), p. 359; John R. Swanton, "Social Organization and Social Usages of the Indians of the Creek Confederacy," *Forty-Second Annual Report of the Bureau of American Ethnology to the Secretary of the Smithsonian Institution* (Washington, D.C.: Government Printing Office, 1928), pp. 443–44; Richard White, *The Roots of Dependency* (Lincoln: University of Nebraska Press,

1983), pp. 16, 19–20; Swanton, *Indians of the Southeastern United States*, pp. 288–89, 309–10; John R. Swanton, "Aboriginal Culture of the Southeast," *Forty-Second Annual Report of the Bureau of American Ethnology to the Secretary of the Smithsonian Institution* (Washington, D.C.: Government Printing Office, 1928), p. 691.

13. Fenton and Gulick, *Symposium on Cherokee and Iroquois Culture*, pp. 95–97; Swanton, *Early History of the Creek Indians*, p. 359; Swanton, "Social Organization," pp. 443–44.

14. Fenton and Gulick, *Symposium on Cherokee and Iroquois Culture*, pp. 96–97; Spellman, "Agriculture of the Early North Florida Indians," p. 47; Charles Hudson, *The Southeastern Indians* (Knoxville: University of Tennessee Press, 1976), p. 313; Swanton, *Early History of the Creek Indians*, p. 359; Swanton, "Social Organization," pp. 443–44; White, *Roots of Dependency*, pp. 20–24; Swanton, *Indians of the Southeastern United States*, pp. 289, 309–10; Fenton, *Symposium on Cherokee and Iroquois Culture*, p. 111.

15. Fenton, *Symposium on Cherokee and Iroquois Culture*, p. 111; Swanton, *Indians of the Southeastern United States*, pp. 285–87.

16. Reuben Gold Thwaites, ed., *The Jesuit Relations and Allied Documents*, vol. 31 (Cleveland, Ohio: Burrows Brothers, 1898), p. 209; Bruce G. Trigger, *The Children of Aataentsic: A History of the Huron People to 1660*, vol. 1 (Montreal: McGill-Queens University Press, 1976), p. 359.

17. Bruce G. Trigger, *The Huron Farmers of the North* (New York: Holt, Rinehart & Winston, 1969), pp. 26–27; George Irving Quimby, *Indian Life in the Upper Great Lakes: 11000 B.C. to A.D. 1800* (Chicago: University of Chicago Press, 1960), p. 114; Elizabeth Tooker, *An Ethnology of the Huron Indians, 1615–1649*, Smithsonian Institution, Bureau of Ethnology, Bulletin 190 (Washington, D.C.: Government Printing Office, 1964), p. 61; Trigger, *Children of Aataentsic*, p. 119. Many of the eastern Indians also raised muskmelons after the Europeans introduced them.

18. Reuben Gold Thwaites, ed., *Jesuit Relations and Allied Documents*, vol. 15 (Cleveland, Ohio: Burrows Brothers, 1898), p. 157; Trigger, *Children of Aataentsic*, pp. 34, 40; Quimby, *Indian Life in the Upper Great Lakes*, p. 114; W. Vernon Kinetz, *The Indians of the Western Great Lakes, 1615–1760* (Ann Arbor: University of Michigan Press, 1940), pp. 16–18; Christian Morissonneau, "Huron of Lorette," in *Northeast*, p. 390.

19. Conrad E. Heidenreich, "Huron," in *Northeast*, pp. 380–81.

20. Arthur D. Parker, *Iroquois Uses of Maize and Other Plant Foods*, New York State Museum, Bulletin 144 (Albany: University of the State of New York, 1910), pp. 21, 24, 30; Reuben Gold Thwaites, ed., *Jesuit Relations and Allied Documents*, vol. 67 (Cleveland, Ohio: Burrows Brothers, 1900), p. 25; Daniel K. Onion, "Corn Culture of the Mohawk Iroquois," *Economic Botany* 18 (Jan.–Mar. 1964): 61. In the Northeast the Narraganset utilized a two-field system, whereby they planted two continuous crops on one field while the other field lay fallow. At the end of the second year, they burned off the grass and underbrush from the fallowed land and planted the next corn crop on it. If they did not fallow the cropland, the soil soon became exhausted, thus necessitating the clearing of new fields (see Eva L. Butler, "Algonkian Culture and Use of Maize in Southern New England," *Bulletin of the Archaeological Society of Connecticut*, no. 22 [Dec. 1948]: 13, 15–16).

21. Onion, "Corn Culture of the Mohawk Iroquois," pp. 62–63, 66: Edmund B. Delabarre and Harris H. Wilder, "Indian Corn-Hills in Massachusetts," *American Anthropology* 22 (July–Sept. 1920): 205, 208, 225.

22. "Memoirs of DeGannes Concerning the Illinois Country," in *The French Foundations, 1680–1693*, ed. Theodore Calvin Pease and Raymond C. Werner, Collections of the Illinois State Historical Library, vol. 23 (Springfield, Illinois: Trustees of the Illinois State Historical Library, 1934), pp. 343–44; Milo Milton Quaife, ed., *The Western Country in the Seventeenth Century: The Memoirs of Lamothe Cadillac and Pierre Liette* (Chicago: Lakeside Press, 1947), pp. 125–26.

23. John Gilmary Shea, ed., *Early Voyages up and down the Mississippi River by Cavelier, St. Cosme, Le Sueur, Gravier and Guignas* (Albany, N.Y., 1861), p. 75; *Life of MA-KA-TAI-ME-SHE-KIA-KIAK or Black Hawk* (Boston, Mass., 1834), pp. 70, 73.

24. Emma Helen Blair, ed., *The Indian Tribes of the Upper Mississippi and Region of the Great Lakes*, vol. 1 (Cleveland, Ohio: Arthur H. Clark Co., 1911), pp. 102, 113; Gordon M. Day, "The Indians as an Ecological Factor in the Northeastern Forests," *Ecology* 34 (Apr. 1953): 333.

25. Wilbert B. Hinsdale, "Indian Corn Culture in Michigan," *Papers of the Michigan Academy of Science Arts and Letters* 8 (1927): 39; George W. Featherstonhaugh, *Canoe Voyage up the Minnay Sotor*, vol. 1 (London, 1847), pp. 350–51; Ulysses Printiss Hedrick, *A History of Agriculture in the State of New York* (n.p., 1933), p. 20. Champlain made a similar remark about the agricultural labors of the Iroquois women when he observed that "woman is the Indian's mule."

26. Bernard Coleman, "The Ojibwa and the Wild Rice Problem," *Anthropological Quarterly* 26 (July 1953): 80; Albert Ernest Jenks, "The Wild Rice Gatherers of the Upper Great Lakes," *Nineteenth Annual Report of the Bureau of American Ethnology*, 1897–1898, pt. 2 (Washington, D.C.: Government Printing Office, 1900), pp. 1038, 1057, 1061, 1064, 1066–68, 1070, 1072; H. Clyde Wilson, "A New Interpretation of the Wild Rice District of Wisconsin," *American Anthropologist* 58 (Dec. 1956): 1059; Wilma F. Aller, "Aboriginal Food Utilization of Vegetation by the Indians of the Great Lakes Region as Recorded in the Jesuit Relations," *Wisconsin Archaeologist* 35 (Sept. 1954): 66; Quimby, *Indian Life in the Upper Great Lakes*, p. 123; Francis Densmore, "Uses of Plants by the Chippewa Indians," *Forty-fourth Annual Report of the Bureau of American Ethnology*, 1926–1927 (Washington, D.C.: Government Printing Office, 1928), pp. 313–14; Henry R. Schoolcraft, *Personal Memoirs of a Thirty Years Residence with the Indian Tribes on the American Frontiers* (Philadelphia, 1851; reprint, New York: AMS Press, 1978), p. 385.

27. Gardner P. Stickney, "Indian Uses of Wild Rice," *American Anthropologist* 9 (Apr. 1896): 118–19; Jenks, "Wild Rice Gatherers," pp. 1080, 1083–84; Stickney, "Indian Uses of Wild Rice," pp. 119–20; Gerald C. Stowe, "Plant Uses among the Chippewa," *Wisconsin Archaeologist* 21 (Apr. 1940): 10–11.

28. Erhard Rostlund, "The Evidence for the Use of Fish as Fertilizer in Aboriginal North America," *Journal of Geography* 56 (May 1957): 223–28; Lynn Ceci, "Fish Fertilizer: A Native North American Practice?" *Science* 188 (4 Apr. 1975): 26–30. For a differing opinion see Howard Russell, "Indian Corn Cultivation," *Science* 189 (19 Sept. 1975): 944–45.

29. Arber, *Travels and Works of Captain John Smith*, p. 67; H. Maxwell, "The Use and Abuse of Forests by the Virginia Indians," *William and Mary Quarterly* 19 (1910): 80–81.

30. Earl L. Core, "Ethnobotany of the Southern Appalachian Aborigines," *Economic Botany* 21 (July–Sept. 1967): 199–200; Hudson, *Southeastern Indians*, pp. 290–91.

31. Herndon, "Indian Agriculture in the Southern Colonies," p. 294; Hudson, *Southeastern Indians*, p. 297.

32. Swanton, "Aboriginal Culture of the Southeast," pp. 691–92; John R. Swanton, *The Indian Tribes of the Lower Mississippi Valley and Adjacent Coast of the Gulf of Mexico,* Smithsonian Institution, Bureau of American Ethnology, Bulletin 43 (Washington, D.C.: Government Printing Office, 1911), p. 304; idem, *Indians of the Southeast,* pp. 296–97; Hudson, *Southeastern Indians,* p. 293.

33. Feest, "Virginia Algonquians," p. 259; Staley Pargellis, ed., "An Account of the Indians in Virginia," *William and Mary Quarterly* 16 (1959): 230; Daniel Coxe, *A Description of the English Province of Carolina* (St. Louis, Mo., 1840), p. 65.

34. Hudson, *Southeastern Indians,* pp. 498–99.

35. Sauer, *Sixteenth Century North America,* pp. 225, 289; Pargellis, "Account of the Indians in Virginia," p. 230.

CHAPTER 4. THE TRANS-MISSISSIPPI WEST

1. Guy R. Stewart, "Conservation in Pueblo Agriculture I: Primitive Practices," *Scientific Monthly* 51 (Sept. 1940): 203; Carl Ortwin Sauer, *Sixteenth Century North America* (Berkeley: University of California Press, 1971), pp. 293–94; Edward F. Castetter and Willis H. Bell, *Pima and Papago Indian Agriculture* (Albuquerque: University of New Mexico Press, 1942), p. 2.

2. Frederick L. Lewton, "The Cotton of the Hopi Indians: A New Species of Gossypium," Smithsonian Institution, Miscellaneous Collections no. 6 (Washington, D.C.: Smithsonian Institution, 1912), pp. 3–4; Leslie A. White, "The Cultivation of Cotton by Pueblo Indians of New Mexico," *Science* 94 (15 Aug. 1941): 162; Volney Jones, "A Summary of Data on Aboriginal Cotton of the Southwest," in *Symposium on Prehistoric Agriculture,* ed. Donald D. Brand, University of New Mexico Bulletin 296, no. 5 (1936), pp. 52–53, 58–59; Castetter and Bell, *Pima and Papago Indian Agriculture,* p. 199; Walter Hough, "Cotton," in *Handbook of American Indians North of Mexico,* pt. 1, ed. Frederick W. Hodge (Washington, D.C.: Government Printing Office, 1907), pp. 252–53. By the early 1940s, only the Pima raised cotton, but it was not their native variety. Because they emphasized cash crops and because cotton cloth and clothing were readily available in stores, their weaving industry had been abandoned (see George F. Carter, *Plant Geography and Culture History in the American Southwest,* Viking Fund Publications in Anthropology no. 5 [New York, 1945], p. 82).

3. George Parker Winship, "The Coronado Expedition, 1540–1542," *Fourteenth Annual Report of the Bureau of American Ethnology to the Secretary of the Smithsonian Institution,* 1892–93, pt. 1 (Washington, D.C.: Government Printing Office, 1896), p. 550; Wilfred William Robbins, John Peaby Harrington, and Barbara Frure-Marreco, *Ethnobotany of the Tewa Indians,* Smithsonian Institution, Bureau of American Ethnology, Bulletin no. 55 (Washington, D.C.: Government Printing Office, 1916), p. 76; Kirk Bryan, "Pre-Columbian Agriculture in the Southwest as Conditioned by Periods of Alluviation," *Annals of the Association of American Geographers* 31 (Dec. 1941): 233; Theodore E. Treutlein, trans., *Sonora: A Description of the Province* (Albuquerque: University of New Mexico Press, 1949), pp. 94, 103.

4. G. W. Hendry and Bellue Hendry, "An Approach to Southwestern Agricultural History through Adobe Brick Analysis," in *Symposium on Prehistoric Agriculture,* p. 67; Castetter and Bell, *Pima and Papago Indian Agriculture,* p. 115. As late as the 1930s, both Propo and Little Club were being raised to a limited extent in California.

5. Richard J. Morrisey, "Early Agriculture in Pimeria Alta," *Mid-America* 31 (Apr. 1939): 101, 104; Herbert Eugene Bolton, *Kino's Historical Memoir of Pimeria Alta*, vol. 1 (Cleveland, Ohio: Arthur H. Clark Co., 1919), pp. 50, 58. The Indian farmers in Pimeria Alta were the Pima, the Sobaipuri, the Papago, and the Yuma.
6. Castetter and Bell, *Pima and Papago Indian Agriculture, pp.* 134, 136–38.
7. Ibid., pp. 144–45, 147–48, 152–54.
8. Ibid., pp. 173–76.
9. Ibid., pp. 176–77.
10. Ibid., pp. 180–85. For a description of granary-making techniques see Mary L. Kissel, "Basketry of the Papago and Pima," *Anthropological Papers of the American Museum of Natural History*, vol. 17, pt. 4 (1916), pp. 173–90.
11. Castetter and Bell, *Pima and Papago Indian Agriculture*, pp. 188–91.
12. Ibid., pp. 191–94.
13. Morrisey, "Early Agriculture in Pimeria Alta," pp. 105–7; Paul H. Ezell, "The Hispanic Acculturation of the Gila River Pimas," *American Anthropologist, Memoir* 90, vol. 63, no. 5, pt. 2 (Oct. 1961), pp. 33–36; Castetter and Bell, *Pima and Papago Indian Agriculture*, pp. 49–50, 119. No one is certain when the Pima began raising such crops as chickpeas, onions, blackeyed peas, potatoes, and yams, which the Spanish introduced into the region.
14. Castetter and Bell, *Pima and Papago Indian Agriculture*, pp. 210–18; Edward F. Castetter and Willis H. Bell, *Yuman Indian Agriculture* (Albuquerque: University of New Mexico Press, 1951), p. 121. Possibly the southwestern farmers learned to cultivate tobacco from the Spaniards, who raised that crop in the region prior to 1626, because no evidence has been discovered that Yuman tribes on the lower Colorado and Gila rivers or the Pima and Papago cultivated it before the time of European contact. Two centuries later, Indian farmers raised tobacco in New Mexico; and in 1826, they cultivated it along the west bank of the Colorado River near the confluence of the Gila. The tobacco varieties that the Spanish introduced to Pueblo agriculture are not known, but one variety was called punche in the eighteenth century. That variety may or may not have been *Nicotiana rustica* or *Nicotiana attenuata*. In any event, the historical development of tobacco cultivation in the American Southwest is still very much a mystery (see Edward F. Castetter, "Early Tobacco Utilization and Cultivation in the American Southwest," *American Anthropologist*, n.s., 45 [Apr.–June 1943]: 321–22, 325; Leslie A. White, "Further Data on the Cultivation of Tobacco among the Pueblo Indians," *Science* 96 [17 July 1942]: 60; Leslie A. White, "Punche: Tobacco in New Mexico History," *New Mexico Historical Review* 18 [Oct. 1943]: 386, 388, 393).
15. Ezell, "Hispanic Acculturation of the Gila River Pimas," p. 46.
16. Ibid., pp. 32, 138–40.
17. Frank Russell, "The Pima Indians," *Twenty-Sixth Annual Report of the Bureau of American Ethnology to the Secretary of the Smithsonian Institution, 1904–1905* (Washington, D.C.: Government Printing Office, 1908), p. 87; Richard B. Woodbury, "A Reappraisal of Hohokam Irrigation," *American Anthropologist* 63 (Apr. 1961): 557. In California's Owen Valley the Paiute built simple irrigation canals which they used to flood tubers and berry bushes. Although they did not practice cultivation, they may have learned the principles of irrigation from the Indian farmers to the south, or they may have developed them independently (see R. W. Patch, "Irrigation in East-Central California," *American Antiquity* 17 [July 1951]: 50–52).
18. Castetter and Bell, *Pima and Papago Indian Agriculture*, pp. 159–62, 168–69.

19. Ibid., pp. 170–72.
20. Russell, "Pima Indians," pp. 87–89; Woodbury, "Reappraisal of Hohokam Irrigation," p. 557; Castetter and Bell, *Pima and Papago Indian Agriculture*, pp. 53–54. In 1854 the Pima in the Sacaton area cultivated approximately three thousand acres, or about three-quarters of an acre per person.
21. Russell, "Pima Indians," pp. 90–91, 94. By the mid nineteenth century, the Pima were also raising some barley, which was in great demand as a feed grain among the whites in Arizona.
22. Ibid., p. 98; Castetter and Bell, *Pima and Papago Indian Agriculture*, pp. 138–39.
23. Castetter and Bell, *Yuman Indian Agriculture*, pp. 38, 135, 140, 149, 248.
24. Ibid., 135; Bolton, *Kino's Historical Memoir*, vol. 1, p. 51; Ezell, "Hispanic Acculturation of the Gila River Pimas," p. 22.
25. Castetter and Bell, *Yuman Indian Agriculture*, pp. 90–91, 150–55, 158, 240–41, 248.
26. Ibid., pp. 116–18, 158–60.
27. William L. Brown, "Observations on Three Varieties of Hopi Maize," *American Journal of Botany* 39 (Oct. 1952): 597; Alfred F. Whiting, "Hopi Indian Agriculture II: Seed Sources and Distribution," Museum of Northern Arizona, *Museum Notes* 10 (Nov. 1937): 13; Alfred F. Whiting, "Ethnobotany of the Hopi," Museum of Northern Arizona, Bulletin no. 15 (June 1939), p. 14; Wilfred C. Bailey, "Tree Rings and Droughts," *American Antiquity* 14 (July 1948): 59.
28. Brown, "Observations on Three Varieties of Hopi Maize," p. 559; Whiting, "Ethnobotany of the Hopi," pp. 12–14; C. Daryll Forde, "Hopi Agriculture and Land Ownership," *Royal Anthropological Institute of Great Britain and Ireland* 61 (1931): 390. In the early twentieth century the Hopi preferred white and blue corn, because red and yellow required more water and produced smaller ears.
29. Brown, "Observations on Three Varieties of Hopi Maize," p. 599; Whiting, "Ethnobotany of the Hopi," pp. 14–15; Forde, "Hopi Agriculture and Land Ownership," p. 391; Paul R. Franke and Don Watson, "An Experimental Corn Field at Mesa Verde National Park," in *Symposium on Prehistoric Agriculture*, p. 40. The Hopi used a husking peg, but how it originated among them is not known.
30. Charles B. Heiser, Jr., "The Hopi Sunflower," *Bulletin of the Missouri Botanical Garden* 33 (Oct. 1945): 163–64.
31. Whiting, "Hopi Indian Agriculture," p. 15.
32. Whiting, "Ethnobotany of the Hopi," pp. 8–9.
33. W. W. Hill, *The Agricultural and Hunting Methods of the Navaho Indians*, Yale University Publications in Anthropology no. 18 (1938), pp. 20, 189; Joseph Charles Winter, "Aboriginal Agriculture in the Southwest and Great Basin" (Ph.D. diss., University of Utah, 1974), p. 59.
34. Hill, *Agricultural and Hunting Methods of the Navaho*, pp. 24–25, 183. The Navaho did not practice canal irrigation until about 1860.
35. Ibid., pp. 26–28. The Navaho men selected their corn seed for the next year at harvest time. They preferred seeds from ears that were about eight inches long and two and a half inches in diameter at the butt and which had straight rows. The ears of seed corn usually were husked. Squash, watermelon, and muskmelon seeds were selected from the largest fruits and were sorted in sacks or pits.
36. Ibid., pp. 28–32.
37. Ibid., pp. 34–35.
38. Ibid., pp. 35–38.
39. Ibid., pp. 39–42.

40. Ibid., pp. 42–45.

41. Ibid., pp. 48–49; Stephen C. Jett, "History of Fruit Tree Raising among the Navaho," *Agricultural History* 51 (Oct. 1977): 681, 684.

42. Hill, *Agricultural and Hunting Methods of the Navaho*, pp. 49–50; Jett, "History of Fruit Tree Raising among the Navaho," pp. 692, 695, 701.

43. Hill, *Agricultural and Hunting Methods of the Navaho*, pp. 48, 50. Winter wheat did not become important among the Navaho until the twentieth century.

44. Ruth M. Underhill, *The Navajos* (Norman: University of Oklahoma Press, 1983), pp. 33–34, 38, 60–61.

45. Waldo R. Wedel, *Environment and Native Subsistence Economies in the Central Great Plains*, Smithsonian Institution, Miscellaneous Collections, vol. 101, no. 3 (Washington, D.C.: Smithsonian Institution, 1941), p. 9; Melvin R. Gilmore, "Uses of Plants by the Indians of the Missouri River Region," *Thirty-Third Annual Report of the Bureau of American Ethnology,* 1911/12 (Washington, D.C.: Government Printing Office, 1919), pp. 67, 136; Richard White, *The Roots of Dependency (Lincoln: University of Nebraska Press,* 1983), pp. 159–60.

46. White, *Roots of Dependency*, pp. 160–61; Wedel, *Environment and Native Subsistence Economies*, pp. 9–10; Waldo R. Wedel, *An Introduction to Pawnee Archeology*, Smithsonian Institution, Bureau of American Ethnology, Bulletin 112 (Washington, D.C.: Government Printing Office, 1936) p. 57; Gilmore, "Uses of Plants by the Indians," p. 133; Waldo R. Wedel, *Prehistoric Man on the Great Plains* (Norman: University of Oklahoma Press, 1961), p. 109.

47. White, *Roots of Dependency*, pp. 162, 179, 180–81, 183–84.

48. Melvin R. Gilmore, "Aboriginal Tobaccos," *American Anthropologist*, n.s., 24 (Oct.–Dec. 1922): 481; Gilmore, "Uses of Plants by the Indians," p. 114.

49. Wedel, *Environment and Native Subsistence Economies*, pp. 8–9; Gilmore, "Uses of Plants by the Indians, pp. 67, 87, 97, 136.

50. George F. Will and George E. Hyde, *Corn among the Indians of the Upper Missouri* (Lincoln: University of Nebraska Press, 1964), p. 77; Gilbert Livingstone Wilson, *Agriculture of the Hidatsa Indians: An Indian Interpretation*, University of Minnesota, Studies in the Social Sciences, Bulletin no. 9 (1917), p. 13.

51. George F. Will and Herbert J. Spinden, *The Mandans,* Harvard University, Papers of the Peabody Museum of American Archeology and Ethnology, vol. 3, no. 4 (1906), p. 117; Will and Hyde, *Corn among the Indians*, pp. 79, 83, 88, 99–100, 102, 108; Roy W. Meyer, *The Village Indians of the Upper Missouri* (Lincoln: University of Nebraska Press, 1977), p. 63; Wilson, *Agriculture of the Hidatsa Indians*, pp. 22–23, 116–17. After the introduction of cattle, the Hidatsa also removed the cow dung from their fields.

52. Will and Hyde, *Corn among the Indians*, pp. 92–94, 113; Wedel, *Prehistoric Man on the Great Plains*, p. 173; Meyer, *Village Indians of the Upper Missouri*, p. 63.

53. Meyer, *Village Indians of the Upper Missouri*, p. 64; Will and Spinden, *Mandans*, p. 120; Will and Hyde, *Corn among the Indians,* pp. 115, 120, 127–28, 130, 133, 290.

54. Will and Hyde, *Corn among the Indians*, pp. 134, 137–38; Wilson, *Agriculture of the Hidatsa Indians*, pp. 42–57, 85–95, 116. The Hidatsa also saved some of the stalks of green corn for fodder; and after harvesting the mature corn in the autumn, they pastured their horses on the husks and stalks. By spring, most of the stalks had been eaten to the ground, thereby making the spring task of land clearing easier.

55. Ibid., pp. 68–69. The Arikara also sprouted corn in this manner (see Thaddeus A. Culbertson, *Journal of an Expedition to the Mauvaises Terres and the*

Upper Missouri in 1850, ed. John Francis McDermott, Smithsonian Institution, Bureau of American Ethnology, Bulletin 147 [Washington, D.C.: Government Printing Office, 1952], p. 98).

56. Wilson, *Agriculture of the Hidatsa Indians,* pp. 69–75, 78–80.

57. Ibid., pp. 82–85.

58. Ibid., pp. 17–18.

59. Ibid., pp. 21–22.

60. Ibid., pp. 123–24. The Hidatsa traded a large part of their tobacco crop to the Sioux.

61. Will and Hyde, *Corn among the Indians,* pp. 284–85, 290–91.

62. Meyer, *Village Indians of the Upper Missouri,* p. 15; Will and Hyde, *Corn among the Indians,* pp. 185–86, 189–91; Alfred Atkenson and M. L. Wilson, "Corn in Montana," Montana Agricultural Experiment Station, Bulletin no. 107 (1915), p. 29.

63. Wedel, *Prehistoric Man on the Great Plains,* p. 160; George F. Will, "Indian Agriculture at Its Northern Limits in the Great Plains Region of North America," *Annaes do XX Congresso Internacional de Americanistas,* vol. 1 (Rio de Janeiro: Imprenasa Nacionaly, 1924), p. 204.

64. Wedel, *Prehistoric Man on the Great Plains,* p. 208.

65. Will and Hyde, *Corn among the Indians,* pp. 65, 141–43, 204; Wayne Suttles, "The Early Diffusion of the Potato among the Coast Salish," *Southwestern Journal of Anthropology* 7 (Autumn 1951): 281–82; Wedel, *Environment and Native Subsistence Economies,* p. 23.

66. Wedel, *Prehistoric Man on the Great Plains,* p. 208.

67. Marc Simmons, "History of Pueblo-Spanish Relations to 1821," in *Southwest,* ed. Alfonso Ortiz, vol. 9 of *Handbook of North American Indians,* ed. William C. Sturtevant (Washington, D.C.: Smithsonian Institution, 1979), p. 190; Castetter and Bell, *Pima and Papago Indian Agriculture,* p. 39; Ezell, "Hispanic Acculturation of the Gila River Pimas," p. 39.

CHAPTER 5. LAND TENURE

1. A. L. Kroeber, "Nature of the Land-Holding Group," *Ethnohistory* 2 (Fall 1955): 303.

2. Ibid., p. 304.

3. Ralph M. Linton, "Land Tenure in Aboriginal America," in *The Changing Indian,* ed. Oliver La Farge (Norman: University of Oklahoma Press, 1942), pp. 42, 49–50; Anthony F. C. Wallace, "Political Organization and Land Tenure among the Northeastern Indians, 1600–1830," *Southwestern Journal of Anthropology* 13 (Winter 1957): 311.

4. Linton, "Land Tenure in Aboriginal America," p. 50.

5. Ibid., p. 51; John M. Cooper, "Land Tenure among the Indians of Eastern and Northern North America," *Pennsylvania Archaeologist* 8 (July 1938): 55–56; Elizabeth Tooker, *An Ethnology of the Huron Indians, 1615–1649,* Smithsonian Institution, Bureau of Ethnology, Bulletin 190 (Washington, D.C.: Government Printing Office, 1964), p. 60; George S. Snyderman, "Concepts of Land Ownership among the Iroquois and Their Neighbors," *Symposium on Local Diversity in Iroquois Culture,* ed. William N. Fenton, Smithsonian Institution, Bureau of American Ethnology, Bulletin 149 (Washington, D.C.: Government Printing Office, 1951), p. 16; Conrad E. Heidenreich, "Huron," in *Northeast,* ed. Bruce G. Trigger, vol.

15 of *Handbook of North American Indians*, ed. William C. Sturtevant (Washington, D.C.: Smithsonian Institution, 1978), p. 380. Occasionally, garden plots could be inherited, and Roger Williams reported that the Narragansett recognized the right of sale.

6. Snyderman, "Concepts of Land Ownership among the Iroquois," pp. 16–18, 28.

7. Tooker, *Ethnology of the Huron Indians*, p. 60; Arthur C. Parker, *Iroquois Uses of Maize and Other Plant Foods*, New York State Museum, Bulletin 144 (Albany: University of the State of New York, 1910), p. 92; Erminie Wheeler Voegelin, "The Place of Agriculture in the Subsistence Economy of the Shawnee," *Papers of the Michigan Academy of Science, Arts and Letters* 26 (1940): 515–19.

8. G. Melvin Herndon, "Indian Agriculture in the Southern Colonies," *North Carolina Historical Review* 44 (Summer 1977): 286.

9. Charles Hudson, *The Southeastern Indians* (Knoxville: University of Tennessee Press, 1976), p. 313; John Phillip Reid, *A Law of Blood: The Primitive Law of the Cherokee Nation* (New York: New York University Press, 1970), pp. 132–33, 140.

10. Reid, *Law of Blood,* pp. 132–33, 140.

11. Richard White, *The Roots of Dependency* (Lincoln: University of Nebraska Press, 1983), p. 159; Linton, "Land Tenure in Aboriginal America," p. 51.

12. Gilbert Livingstone Wilson, *Agriculture of the Hidatsa Indians: An Indian Interpretation*, University of Minnesota Studies in the Social Sciences no. 9 (1917), pp. 113–14.

13. W. W. Hill, "Notes on Pima Land Law and Tenure," *American Anthropologist* 38 (Oct.–Dec. 1936): 586; Edward F. Castetter and Willis H. Bell, *Pima and Papago Indian Agriculture* (Albuquerque: University of New Mexico Press, 1942), p. 127.

14. Hill, "Notes on Pima Land Law and Tenure," pp. 587–88; Castetter and Bell, *Pima and Papago Indian Agriculture*, p. 127.

15. Hill, "Notes on Pima Land Law and Tenure," p. 588; Castetter and Bell, *Pima and Papago Indian Agriculture*, pp. 127–28.

16. Hill, "Notes on Pima Land Law and Tenure," p. 588; Castetter and Bell, *Pima and Papago Indian Agriculture*, p. 127–28.

17. Kenneth M. Stewart, "Mohave Indian Agriculture," *Masterkey* 40 (Jan.–Mar. 1966): 10–11; Edward F. Castetter and Willis H. Bell, *Yuman Indian Agriculture* (Albuquerque: University of New Mexico Press, 1951), pp. 142–43.

18. Castetter and Bell, *Yuman Indian Agriculture*, pp. 143–44; Stewart, "Mohave Indian Agriculture," p. 11.

19. Castetter and Bell, *Yuman Indian Agriculture*, pp. 141–42; C. Daryll Forde, "Ethnography of the Yuma Indians," University of California, Publications in American Archaeology and Ethnology 38 (1930/31), pp. 114–15; Stewart, "Mohave Indian Agriculture," pp. 14–15.

20. C. Daryll Forde, "Hopi Agriculture and Land Ownership," *Royal Anthropological Institute of Great Britain and Ireland* 61 (1931): 367, 369, 378; Ernest Beaglehole, "Ownership and Inheritance in an American Indian Tribe," *Iowa Law Review* 20 (Jan. 1935): 311.

21. Forde, "Hopi Agriculture and Land Ownership," pp. 370–71; Bruce A. Cox, "Hopi Trouble Cases: Cultivation Rights and Homesteads," *Plateau* 39 (Spring 1967): 145.

22. Beaglehole, "Ownership and Inheritance," pp. 314–15; Cox, "Hopi Trouble Cases," p. 147. No one is certain when a man's son began to inherit land that his

father had cleared from waste. Because the principle of Hopi land inheritance was matrilineal, inheritance through the male line probably was not an ancient custom, but rather was an influence of white society.

23. Forde, "Hopi Agriculture and Land Ownership," p. 383.

24. John F. Martin, "A Reconsideration of Havasupai Land Tenure," *Ethnology* 8 (Oct. 1968): 451; Elman Service, "Recent Observations on Havasupai Land Tenure," *Southwestern Journal of Anthropology* 3 (Winter 1947): 365–66.

25. W. W. Hill, *The Agricultural and Hunting Methods of the Navaho Indians*, Yale University Publications in Anthropology no. 18 (1938), pp. 21–24. In the twentieth century, due to contact with white civilization, the Navaho began to pass land directly to a man's children at the time of his death.

26. Ibid., pp. 21–22, 183.

27. Ibid., p. 23.

28. John Province, "Tenure Problems of the American Indian," in *Land Tenure*, ed. Kenneth H. Parsons, Raymond J. Penn, and Philip M. Raup (Madison: University of Wisconsin Press, 1956), p. 421; Linton, "Land Tenure in Aboriginal America," p. 53.

29. Linton, "Land Tenure in Aboriginal America," p. 53; Marshall Harris, *Origin of the Land Tenure System in the United States* (Ames: Iowa State College Press, 1953), p. 67.

30. George B. Grinnell, "Tenure of Land among the Indians," *American Anthropologist* 9 (Jan.–Mar. 1907): 2–3.

31. Harris, *Origin of the Land Tenure System*, p. 68.

32. Anthony F. C. Wallace, "Woman, Land, and Society: Three Aspects of Aboriginal Delaware Life," *Pennsylvania Archaeologist* 17 (1947): 2.

CHAPTER 6. THE ALIENATION OF INDIAN LANDS

1. Charles C. Royce, "Indian Land Cessions in the United States," *Eighteenth Annual Report of the Bureau of American Ethnology to the Secretary of the Smithsonian Institution, 1896–'97*, pt. 2 (Washington, D.C.: Government Printing Office, 1899), pp. 527–28.

2. J. P. Kinney, *A Continent Lost—A Civilization Won: Indian Land Tenure in America* (Baltimore, Md.: Johns Hopkins Press, 1937) p. 2; Royce, "Indian Land Cessions," p. 544.

3. Richard Schifter, "Indian Title to Land," *American Indian* 7 (Spring 1954): 40–41; Felix S. Cohen, "Original Indian Title," *Minnesota Law Review* 32 (Dec. 1947): 44–45.

4. Harry P. Dart, "Louisiana Land Titles Derived from Indian Tribes," *Louisiana Historical Quarterly* 4 (Jan. 1921): 136–37, 143; R. E. Twitchell, "Pueblo Indian Land Tenures in New Mexico and Arizona," *El Palacio* 12 (1 Mar. 1922): 39, 43–45, 47–49, 53; Marshall Harris, *Origin of the Land Tenure System in the United States* (Ames: Iowa State College Press, 1953), p. 156.

5. Kinney, *A Continent Lost,* p. 2; Royce, "Indian Land Cessions," pp. 546, 548.

6. Harris, *Origin of the Land Tenure System*, p. 165; Cohen, "Original Indian Title," pp. 39–40.

7. Kinney, *A Continent Lost*, pp. 3–4; Royce, "Indian Land Cessions," p. 557; Paul W. Gates, *History of Public Land Law Development* (Washington, D.C.: Government Printing Office, 1968), pp. 33–34; Dwight L. Smith, "The Land Cession Treaty: A Valid Instrument of Transfer of Indian Title," in *This Land of Ours: The*

Acquisition of the Public Domain (Indianapolis: Indiana Historical Society, 1978), pp. 88, 90.

8. Peter A. Thomas, "Contrastive Subsistence Strategies and Land Use as Factors for Understanding Indian-White Relations in New England," *Ethnohistory* 23 (Winter 1976): 4–5, 15.

9. Harris, *Origin of the Land Tenure System*, p. 158.

10. Chester E. Eisinger, "The Puritans' Justification for Taking the Land," *Essex Institute Historical Collections* 84 (Apr. 1948): 131, 142; William T. Hagan, "Justifying Dispossession of the Indian: The Land Utilization Argument," in *American Indian Environments: Ecological Issues in Native American History*, ed. Christopher Vecsey and Robert W. Venables (Syracuse, N.Y.: Syracuse University Press, 1980), pp. 66–67; Wilcomb E. Washburn, "The Moral and Legal Justification for Dispossession of the Indians," in *Seventeenth Century America: Essays in Colonial History*, ed. James Morton Smith (Chapel Hill: University of North Carolina Press, for the Institute of Early American History and Culture, 1959), p. 22; Youngblood Henderson, "Unraveling the Riddle of Aboriginal Title," *American Indian Law Review* 5, no. 1 (1977): 101.

11. Alden T. Vaughan, *New England Frontier: Puritans and Indians, 1620–1675* (Boston, Mass.: Little, Brown & Co., 1965), pp. 107–15, 327–28; Howard Peckam, *The Colonial Wars, 1689–1762* (Chicago: University of Chicago Press, 1965).

12. Harris, *Origin of the Land Tenure System*, pp. 159–60.

13. Kinney, *A Continent Lost*, p. 13.

14. Ibid., pp. 18, 20.

15. Ibid., p. 23; Smith, "Land Cession Treaty," p. 91.

16. Royce, "Indian Land Cessions," pp. 533, 639–40; Smith, "Land Cession Treaty," pp. 91–92.

17. Royce, "Indian Land Cessions," p. 533.

18. Francis Paul Prucha, *American Indian Policy in the Formative Years: The Indian Trade and Intercourse Acts, 1790–1834* (Cambridge, Mass.: Harvard University Press, 1962), p. 34; Reginald Horsman, *Expansion and American Indian Policy, 1783–1812* (East Lansing: Michigan State University Press, 1967), p. 5.

19. Horsman, *Expansion and American Indian Policy*, p. 12.

20. Ibid., pp. 12, 15.

21. Prucha, *American Indian Policy*, p. 40; Horsman, *Expansion and American Indian Policy*, pp. 30–31.

22. Cohen, "Original Indian Title," p. 41; Horsman, *Expansion and American Indian Policy*, pp. 42, 44, 47, 56, 60.

23. John Joseph Davis, "Land Claims under the Indian Trade and Intercourse Acts: The White Settlement Exception Defense," *Boston Law Review* 60 (Nov. 1980): 913–14, 918–20.

24. Prucha, *American Indian Policy*, p. 144; Horsman, *Expansion and American Indian Policy*, p. 62.

25. Prucha, *American Indian Policy*, pp. 147, 186–87; Horsman, *Expansion and American Indian Policy*, pp. 62–63; Ester V. Hill, "The Iroquois Indians and Their Lands since 1783," *Quarterly Journal of the New York State Historical Association* 11 (Oct. 1930): 338.

26. Robert W. McCluggage, "The Senate and Indian Land Titles, 1800–1825," *Western Historical Quarterly* 1 (Oct. 1970): 416; Horsman, *Expansion and American Indian Policy*, p. 105.

27. Horsman, *Expansion and American Indian Policy*, pp. 110–11.

NOTES TO PAGES 86–91   253

28. Logan Esarey, ed., *Messages and Letters of William Henry Harrison*, vol. 1 of Indiana Historical Collections (Indianapolis: Indiana Historical Commission, 1922), p. 71.

29. Fred L. Israel, *The State of the Union Messages of the Presidents, 1790–1966*, vol. 1 (New York: Chelsea House, Robert Hector Publisher, 1966), p. 82; Horsman, *Expansion and American Indian Policy*, p. 112.

30. Horsman, *Expansion and American Indian Policy*, pp. 118, 122, 124.

31. Ibid., pp. 130–31, 145.

32. Ibid., pp. 159, 170.

33. Prucha, *American Indian Policy*, p. 226.

34. John Spencer Bassett, *Correspondence of Andrew Jackson*, vol. 2 (Washington, D.C.: Carnegie Institution, 1927), p. 331; Israel, *State of the Union Messages*, 1:152; Prucha, *American Indian Policy*, pp. 226–27; Albert K. Weinberg, *Manifest Destiny* (Chicago: Quadrangle Books, 1963), p. 81.

35. Weinberg, *Manifest Destiny*, pp. 81–82; *Johnson T. Graham's Lessee* v. *William McIntosh*, 21 U.S. (8 Wheaton) 543 (1823); Schifter, "Indian Title to Land," pp. 42–43. The unanimous opinion of the Supreme Court in the *McIntosh* case reflects a culmination of court decisions regarding Indian land title which stemmed from the mid seventeenth century. The attorney general of the United States voiced that same opinion in 1821, when he held: "So long as a tribe exists and remains in possession of his land its title and possession are sovereign and exclusive. . . . Although the Indian title continues only during their possession, yet that possession has been always held sacred, and can never be disturbed by their consent. They do not hold under the states, nor under the United States; their title is original, sovereign, and exclusive." Thus, tribal land tenure was a separate entity from federal and state land-tenure systems (see Henderson, "Unraveling the Riddle of Aboriginal Title," p. 96).

36. Kinney, *A Continent Lost*, pp. 39–40, 56–57, 61; Weinberg, *Manifest Destiny*, p. 79; Charles Francis Adams, *Memoirs of John Quincy Adams*, vol. 3 (Philadelphia, Pa.: J.B. Lippincott & Co., 1874), p. 28.

37. Prucha, *American Indian Policy*, p. 234; Francis Paul Prucha, "Andrew Jackson's Indian Policy: A Reassessment," *Journal of American History* 54 (Dec. 1969): 531.

38. Prucha, *American Indian Policy*, pp. 237–38; Mary Elizabeth Young, *Redskins, Ruffleshirts and Rednecks: Indian Land Allotments in Alabama and Mississippi, 1830–1860* (Norman: University of Oklahoma Press, 1961): p. 115; Hagan, "Justifying Dispossession of the Indian," pp. 68–69; Israel, *State of the Union Messages*, 1:308–10; Kinney, *A Continent Lost*, p. 78.

39. Prucha, "Andrew Jackson's Indian Policy," p. 534; U.S., Congress, House, *Messages of the President of the United States to the Two Houses of Congress, December 6, 1831*, Executive Document 2, ser. 216, 22d Cong., 1st sess., 1831, pp. 32, 34.

40. Kinney, *A Continent Lost*, p. 64; Ronald N. Satz, *American Indian Policy in the Jacksonian Era* (Lincoln: University of Nebraska Press, 1975), pp. 19–31; Prucha, *American Indian Policy*, p. 243; Weinberg, *Manifest Destiny*, p. 88.

41. *Cherokee Nation* v. *Georgia*, 5 Peters 1 (1831); *Worcester* v. *Georgia*, 6 Peters 515 (1832); Alfred H. Kelly and Winfred A. Harbison, *The American Constitution: Its Origin and Development* (New York: W. W. Norton & Co., 1963), pp. 302–3; Prucha, *American Indian Policy*, pp. 245–46.

42. Henderson, "Unraveling the Riddle of Aboriginal Title," p. 105; Young, *Redskins, Ruffleshirts and Rednecks*, p. 17.

43. Prucha, "Andrew Jackson's Indian Policy," p. 534–36; Mary E. Young, "Indian Removal and Land Allotment: The Civilized Tribes and Jacksonian Justice," *American Historical Review* 64 (Oct. 1958): 34.

44. Bernard W. Sheehan, *Seeds of Extinction: Jeffersonian Philanthrophy and the American Indian* (Chapel Hill: University of North Carolina Press, for the Institute of Early American History and Culture at Williamsburg, Virginia, 1973), pp. 4–5, 8–12, 167, 180, 241–42, 274–75.

45. Kinney, *A Continent Lost*, p. 82.

46. Ibid., pp. 83–84; Young, "Indian Removal and Land Allotment," p. 37; Young, *Redskins, Ruffleshirts and Rednecks*, p. 11.

47. Young, "Indian Removal and Land Allotment," pp. 37–38; Kinney, *A Continent Lost*, pp. 90–91; Royce, "Indian Land Cessions," pp. 686–88; Paul W. Gates, "Indian Allotments Preceding the Dawes Act," in *Frontier Challenge: Responses to the Trans-Mississippi West*, ed. John G. Clark (Lawrence: University Press of Kansas, 1971), pp. 142–43, 147. By spring 1824 the Cherokee and the Creek had ceded 15,744,000 acres in Georgia (see *American State Papers, Indian Affairs*, vol. 2, p. 463).

48. Mary E. Young, "The Creek Frauds: A Study in Conscience and Corruption," *Mississippi Valley Historical Review* 62 (Dec. 1955): 412, 414–16, 428; Young, *Redskins, Ruffleshirts and Rednecks*, pp. 44, 76–77; U.S., Congress, *Correspondence on the Subject of the Emigration of Indians,* Senate Document 512, 23rd Cong., 1st sess., 1835, ser. 246, p. 565; Satz, *American Indian Policy*, p. 105.

49. Young, "Indian Removal and Land Allotment," pp. 39–41, 43–44; Young, *Redskins, Ruffleshirts and Rednecks*, p. 175.

50. Howard W. Paulson, "The Allotment of Land to the Dakota Indians before the Dawes Act," *South Dakota History* 1 (Spring 1971): 134.

CHAPTER 7. AGRICULTURAL AID AND EDUCATION

1. Richard Farrell, "Promoting Agriculture among the Indian Tribes of the Old Northwest, 1789–1820," *Associates of the NAL Today*, n.s., 3 (Jan.–June, 1978): 13–18.

2. Douglas C. Wilms, "Cherokee Acculturation and Changing Land Use Practices," *Chronicles of Oklahoma* 56 (Fall 1978): 341; George Dewey Harmon, *Sixty Years of Indian Affairs: Political, Economic, and Diplomatic, 1789–1850* (Chapel Hill: University of North Carolina Press, 1941), p. 157.

3. Merle H. Deardorff and George S. Snyderman, "A Nineteenth-Century Journal of a Visit to the Indians of New York," *Proceedings of the American Philosophical Society* 100 (Dec. 1956): 587, 590; Robert F. Berkhofer, *Salvation and the Savage: An Analysis of Protestant Missions and American Indian Response, 1787–1862* (Lexington: University of Kentucky Press, 1965), pp. 75–77.

4. Berkhofer, *Salvation and the Savage*, pp. 75, 78.

5. Charles J. Kappler, ed., *Indian Treaties, 1778–1883* (New York: Interland Publishing, Inc., 1972), pp. 28, 31; William N. Fenton and John Gulick, *Symposium on Cherokee and Iroquois Culture*, Smithsonian Institution, Bureau of American Ethnology, Bulletin 180 (Washington, D.C.: Government Printing Office, 1961), p. 98; Joseph A. Parsons, "Civilizing the Indians of the Old Northwest, 1800–1810," *Indiana Magazine of History* 56 (Sept. 1960): 202.

6. Henry Warner Bowden, *American Indians and Christian Missions* (Chicago: University of Chicago Press, 1981), p. 165; Farrell, "Promoting Agriculture," p. 13.

7. Wilms, "Cherokee Acculturation," pp. 339–40; Abraham Eleazer Knepler, "Education in the Cherokee Nation," *Chronicles of Oklahoma* 21 (Dec. 1943): 389.

8. Farrell, "Promoting Agriculture," p. 13; Parsons, "Civilizing the Indians," pp. 199–200; Knepler, "Education in the Cherokee Nation," p. 390; Merritt B. Pound, *Benjamin Hawkins—Indian Agent* (Athens: University of Georgia Press, 1951), pp. 143–44.

9. Parsons, "Civilizing the Indians," p. 200; Berkhofer, *Salvation and the Savage*, pp. 70, 73–74.

10. Farrell, "Promoting Agriculture," p. 13; Albert Ellery Bergh, ed., *The Writings of Thomas Jefferson*, vol. 16 (Washington, D.C.: Thomas Jefferson Memorial Association, 1903), pp. 390–91.

11. Farrell, "Promoting Agriculture," p. 13; "Fur Trade on the Upper Lakes, 1778–1815," *Collections of the Wisconsin Historical Society* 19 (1910): 315.

12. Parsons, "Civilizing the Indians," p. 201; Herman J. Viola, *Thomas L. McKenney: Architect of America's Early Indian Policy, 1816–1830* (Chicago: Swallow Press, Inc., 1974), pp. 25–26.

13. *American State Papers, Indian Affairs*, vol. 2, p. 151; William T. Hagan, *American Indians* (Chicago: University of Chicago Press, 1979), pp. 87–88.

14. Berkhofer, *Salvation and the Savage*, p. 38; Wilms, "Cherokee Acculturation," p. 342; U.S., Congress, Senate, *Report from the Office of Indian Affairs*, Document 2, 18th Cong., 2d sess., 1824, ser. 108, p. 106; idem, *Report from the Office of Indians Affairs*, Document 1, 23d Cong., 1st sess., 1833, ser. 238, pp. 186, 198; Viola, *Thomas L. McKenney*, p. 195; Reginald Horsman, *Expansion and American Indian Policy, 1783–1812* (East Lansing: Michigan State University Press, 1967), pp. 354–55.

15. Ronald N. Satz, *American Indian Policy in the Jacksonian Era* (Lincoln: University of Nebraska Press, 1975), pp. 258, 260.

16. Francis Paul Prucha, "American Indian Policy in the 1840s: Visions of Reform," in *The Frontier Challenge: Responses to the Trans-Mississippi West*, ed. John G. Clark (Lawrence: University Press of Kansas, 1971), pp. 87–88; *Report of the Commissioner of Indian Affairs*, 1856, p. 116 (hereafter these reports will be cited simply as *Report*, followed by year, serial numbers, if appropriate, and page).

17. Satz, *American Indian Policy*, p. 271.

18. U.S., Congress, Senate, *Report from the Office of Indian Affairs*, Document 1, 24th Cong., 2d sess., 1836, ser. 297, p. 391; Norman Arthur Graebner, "Pioneer Indian Agriculture in Oklahoma," *Chronicles of Oklahoma* 33 (Autumn 1945): 233–34.

19. U.S., Congress, Senate, *Report of the Commissary General of Subsistence*, Document 1, 24th Cong., 1st sess., 1835, ser. 279, p. 289; *Report from the Office of Indian Affairs*, 1836, ser. 297, p. 391; *Report*, 1837, ser. 314, p. 540; Graebner, "Pioneer Indian Agriculture in Oklahoma," p. 234; Gilbert C. Fite, "Development of the Cotton Industry by the Five Civilized Tribes in Indian Territory," *Journal of Southern History* 15 (Aug. 1949): 345.

20. Graebner, "Pioneer Indian Agriculture in Oklahoma," p. 235.

21. Report from the Office of Indian Affairs, 1836, ser. 297, p. 391; *Report*, 1837, ser. 314, p. 541; Fite, "Development of the Cotton Industry," p. 346.

22. *Report*, 1837, ser. 314, p. 543; *Report*, 1838, ser. 338, p. 503.

23. Prucha, "American Indian Policy in the 1840s," p. 101; *Report*, 1842, ser. 413, p. 449; *Report*, 1845, ser. 470, p. 512.

24. *Report*, 1841, ser. 395, pp. 335, 337; *Report*, 1842, ser. 413, pp. 445, 449–50; *Report*, 1845, ser. 470, p. 519; *Report*, 1847, ser. 503, pp. 884, 887.

25. *Report,* 1842, ser. 413, p. 442; *Report,* 1843, ser. 431, p. 385; *Report,* 1847, ser. 503, p. 876; *Report of the Commissioner of Patents, 1849: Agriculture,* pp. 451–55; *Report,* 1849, ser. 570, p. 1131.

26. *Report,* 1842, ser. 413, p. 440; *Report,* 1843, ser. 431, pp. 387–88; *Report,* 1845, ser. 470, p. 514; *Report,* 1846, ser. 493, p. 268.

27. *Report,* 1848, ser. 537, p. 474; *Report,* 1849, ser. 570, p. 1069.

28. *Report,* 1841, ser. 431, p. 459; *Report,* 1844, ser. 449, p. 490.

29. Thomas P. Barr, "The Pottawatomie Baptist Manual Labor Training School," *Kansas Historical Quarterly* 43 (Winter 1977): 405, 410; *Report,* 1853, ser. 690, p. 325.

30. Frank A. Balyeat, "Review of Chickasaw Education before the Civil War," *Chronicles of Oklahoma* 34 (Winter 1956/57): 489.

31. *Report,* 1855, ser. 810, p. 507; *Report,* 1856, pp. 15–16, 237–38; *Report,* 1857, ser. 919, pp. 564, 678, 684; *Report,* 1859, ser. 1023, p. 718.

32. *Report,* 1854, ser. 746, pp. 306–7, 322; *Report,* 1855, ser. 810, p. 444; *Report,* 1856, pp. 129, 131, 170; *Report,* 1858, ser. 974, p. 647; *Report,* 1859, ser. 1023, p. 736. The Cherokee, who migrated to Indian Territory, had a long tradition of using slaves for agricultural labor. English traders introduced black slaves among the Cherokee during the colonial period, and by 1741, the Cherokee had purchased, sold, and worked slaves regularly in their fields of tobacco, indigo, wheat, oats, and potatoes. As the Cherokee became increasingly agricultural, the institution of slavery became more important to them, just as it did among white agriculturists in the South. During removal, many of the southeastern Indians took their slaves with them and even purchased additional slaves later in Mobile and New Orleans. By 1835, slaveholders in Indian Territory cultivated more acres and produced larger corn crops than those who did not own black servants. At that time, Cherokee slaveholders each cultivated an average of seventy-five acres and produced 1,040 bushels of corn. Nonslaveholders farmed only eleven acres and produced only 141 bushels of corn. Since the yield depended on cultivating the corn before the weeds could choke the crop, slave labor enabled Cherokee farmers to raise larger crops than could nonslaveholders. The larger acreage that could be cultivated convinced the slaveholders that bonded labor was profitable. By producing a surplus of agricultural products such as corn and cotton, some Cherokee slaveholders obtained capital for investment in milling and transportation. For a thorough study of slavery among the Indians see R. Halliburton, Jr., *Red over Black: Black Slavery among the Cherokee Indians* (Westport, Conn.: Greenwood Press, 1977); Theda Perdue, *Slavery and the Evolution of Cherokee Society, 1540–1866* (Knoxville: University of Tennessee Press, 1979); William G. McLaughlin, "Cherokee Slaveholders and Baptist Missionaries, 1845–1860," *Historian* 65 (Feb. 1983): 147–66.

33. *Report,* 1858, ser. 974, pp. 592, 603.

34. Waldo R. Wedel, *Environment and Native Subsistence Economies in the Central Great Plains,* Smithsonian Institution, Miscellaneous Collections, vol. 101, no. 3 (Washington, D.C.: Smithsonian Institution, 1941), pp. 21–22; *Report,* 1857, ser. 919, p. 678.

35. *Report,* 1860, p. 61.

36. *Report,* 1856, p. 171.

37. *Report,* 1857, ser. 919, p. 535; *Report,* 1859, ser. 1023, pp. 541, 771; *Report,* 1860, pp. 225–27.

38. *Report,* 1853, ser. 690, pp. 348–49; *Report,* 1857, ser. 919, p. 543; Graebner, "Pioneer Indian Agriculture in Oklahoma," p. 240.

39. Joe Jackson, "Schools among the Minor Tribes in Indian Territory," *Chronicles of Oklahoma* 32 (Spring 1934): 60; Donald Jackson, "William Ewing, Agricultural Agent to the Indians," *Agricultural History* 31 (Apr. 1957): 6.

40. *Report,* 1856, pp. 109–10.

41. *Report,* 1850, ser. 587, p. 36; Sven Liljeblod, "Epilogue: Indian Policy and the Fort Hall Reservation," *Idaho Yesterdays* 2 (Summer 1958): 15.

CHAPTER 8. FROM THE CIVIL WAR TO SEVERALTY

1. Edmund Jefferson Danziger, Jr., *Indians and Bureaucrats: Administering Reservation Policy during the Civil War* (Urbana: University of Illinois Press, 1974), pp. 153, 157–58; Lawrence F. Schmeckebier, *The Office of Indian Affairs: Its History, Activities and Organization* (Baltimore, Md.: Johns Hopkins University Press, 1927), pp. 49, 101; Theda Perdue, *Slavery and the Evolution of Cherokee Society, 1540–1866* (Knoxville: University of Tennessee Press, 1979), p. 125; Robert M. Utley, *The Indian Frontier of the American West, 1846–1890* (Albuquerque: University of New Mexico Press, 1984), pp. 67–70.

2. R. Halliburton, Jr., *Red over Black: Black Slavery among the Cherokee Indians* (Westport, Conn.: Greenwood Press, 1977), pp. 122–26, 132.

3. *Report,* 1863, ser. 1182, p. 144; *Report,* 1865, ser. 1248, p. 206.

4. *Report,* 1864, pp. 309–10, 346.

5. *Report,* 1865, ser. 1248, pp. 436–37, 446–50.

6. *Report,* 1864, pp. 320, 347.

7. Ibid., p. 306.

8. *Report,* 1862, pp. 11, 102–3; *Report,* 1863, ser. 1182, p. 356.

9. *Report,* 1862, p. 103; *Report,* 1864, p. 354.

10. *Report,* 1862, pp. 98, 115, 119, 121.

11. Ibid., pp. 184, 190, 239.

12. Ibid., pp. 278–79, 285; *Report,* 1863, ser. 1182, p. 190; *Report,* 1864, pp. 87–88, 91.

13. Schmeckebier, *Office of Indian Affairs,* pp. 101–2; Helga H. Harriman, "Economic Conditions in the Cherokee Nation, 1865–1871," *Chronicles of Oklahoma* 51 (Fall 1973): 531–32.

14. Schmeckebier, *Office of Indian Affairs,* p. 53; *Report,* 1866, pp. 9–13. Under these treaties the Choctaw and the Chickasaw lost 6,800,000 acres; the Creek, 3,250,560; and the Cherokee, 800,000.

15. *Report,* 1866, pp. 82, 109, 112, 136, 179; *Report,* 1867, pp. 68–69; *Report,* 1869, p. 209.

16. *Report,* 1867, pp. 63, 103, 162, 182; *Report,* 1866, p. 179.

17. Francis Paul Prucha, *American Indian Policy in the Formative Years: The Indian Trade and Intercourse Acts, 1790–1834* (Cambridge, Mass.: Harvard University Press, 1962), p. 36.

18. Ibid., pp. 46–49, 55; Schmeckebier, *Office of Indian Affairs,* p. 55. The Quakers were the first to receive assignments. They symbolized the government's desire for peace and honesty in dealing with the Indians. The Hicksite Friends received the northern superintendency, while the Orthodox Friends were given the central superintendency.

19. Prucha, *American Indian Policy,* pp. 63–70.

20. Ibid., pp. 103–7, 194, 270.

21. William D. Pennington, "Government Policy and Indian Farming on the Cheyenne and Arapaho Reservation, 1869–1880," *Chronicles of Oklahoma* 57

(Summer 1979): 172; Charles J. Kappler, ed., *Indian Treaties, 1778–1883* (New York: Interland Publishing, Inc., 1972), pp. 807–11.

22. Pennington, "Government Policy," pp. 174–75, 177, 180; *Report,* 1872, pp. 249–50; *Report,* 1874, p. 235; *Report,* 1875, p. 270; *Report,* 1876, p. 47.

23. Pennington, "Government Policy," pp. 183–89; *Report,* 1872, p. 250; *Report,* 1876, p. 48; *Report,* 1878, p. 56.

24. *Report,* 1870, p. 145; T. Ashley Zwink, "On the Whiteman's Road: Lawrie Tatum and the Formative Years of the Kiowa Agency, 1869–1873," *Chronicles of Oklahoma* 56 (Winter 1978/79): 434–35, 439; William D. Pennington, "Government Agricultural Policy on the Kiowa Reservation, 1869–1901," *Indian Historian* 11 (Winter 1978): 12; *Report,* 1878, pp. 184–85.

25. *Report,* 1872, p. 33; *Report,* 1873, p. 205; *Report,* 1874, p. 69.

26. *Report,* 1876, p. ii.

27. *Report,* 1879, pp. 75–76.

28. *Report,* 1872, pp. 34, 233–34.

29. *Report,* 1870, p. 203; *Report,* 1872, p. 266; *Report,* 1874, p. 28; Clyde A. Milner II, "Off the Whiteman's Road: Seven Nebraska Indian Societies in the 1870s—A Statistical Analysis of Assimilation, Population and Prosperity," *Western Historical Quarterly* 12 (Jan. 1981): 39, 48.

30. *Report,* 1870, p. 220; *Report,* 1872, p. 222; *Fifth Annual Report of the Board of Indian Commissioners,* 1873, pp. 40, 231; *Report,* 1873, p. 194.

31. *Sixth Annual Report of the Board of Indian Commissioners,* 1874, pp. 29, 34; *Report,* 1872, p. 259; *Report,* 1874, p. 209; *Report,* 1876, p. 32; *Report,* 1878, pp. 41, 47.

32. *Report,* 1870, pp. 116–17; *Report,* 1872, p. 167; *Report,* 1876, p. 7; Edward F. Castetter and Willis H. Bell, *Yuman Indian Agriculture* (Albuquerque: University of New Mexico Press, 1951), p. 78. In 1871 the Pima cultivated between ten and fifteen acres per farm and produced as much as 70 percent of their food supply.

33. *Report,* 1870, p. 149; *Report,* 1872, p. 53; *Report,* 1875, p. 7; *Report,* 1878, p. 109.

34. Schmeckebier, *Office of Indian Affairs,* p. 77; Paul Stuart, *The Indian Office: Growth and Development of an American Institution, 1865–1900* (Ann Arbor, Mich.: UMI Research Press, 1979), pp. 69–71; Robert L. Brunhouse, "The Founding of the Carlisle Indian School," *Pennsylvania History* 6 (Apr. 1939): 72.

35. *Report,* 1881, p. xxxii; Robert A. Trennert, " 'And the Sword Will Give Way to the Spelling Book': Establishing the Phoenix Indian School," *Journal of Arizona History* 23 (Spring 1982): 35–36; Brian W. Dippie, *The Vanishing American: White Attitudes and U.S. Indian Policy* (Middletown, Conn.: Wesleyan University Press, 1982), p. 117; Robert L. Brunhouse, "Apprenticeship for Civilization: The Outing System at the Carlisle Indian School," *Educational Outlook* 12 (May 1939): 30; Frank W. Blackmar, "Haskell Institute as Illustrating Indian Progress," *Review of Reviews* 5 (June 1892): 557–58, 560; *Report,* 1881, p. xxxiii; *Report,* 1882, p. xxxiii.

36. Brunhouse, "Apprenticeship for Civilization," pp. 33, 36; idem, "Founding of the Carlisle Indian School," pp. 78–79.

37. Trennert, " 'And the Sword Will Give Way to the Spelling Book,' " p. 38; *Sixteenth Annual Report of the Board of Indian Commissioners,* 1884, pp. 46–47; Blackmar, "Haskell Institute," p. 59; Charles R. Kutzleb, "Educating the Dakota Sioux, 1876–1890," *North Dakota History* 32 (Oct. 1965): 211. Not all whites supported the development of Indian boarding schools that would provide agricultural training. One individual expressed the fear that if the Pima were taught to

be better agriculturists, they would seriously compete with white farmers by increasing the demand for irrigation water which already was in short supply.

38. *Report,* 1880, p. xxii, 131, 144.

39. *Report,* 1881, p. xxxviii, lxi–lxii, 52.

40. *Report,* 1882, pp. 83–84; *Report,* 1883, p. 72; William T. Hagan, "Kiowas, Comanches, and Cattlemen, 1867–1906: A Case Study of the Failure of U.S. Reservation Policy," *Pacific Historical Review* 40 (1971): 336–39; Martha Buntin, "Beginning of the Leasing of Surplus Grazing Lands on the Kiowa and Comanche Reservations," *Chronicles of Oklahoma* 10 (Sept. 1932): 369; Prucha, *American Indian Policy,* p. 215; Edward Everett Dale, "Ranching on the Cheyenne-Arapaho Reservation, 1880–1885," *Chronicles of Oklahoma* 6 (Mar. 1928): 35–59; William D. Savage, Jr., *The Cherokee Strip Livestock Association* (Columbia: University of Missouri Press, 1979), pp. 7, 30; Robert M. Burrill, "The Establishment of Ranching on the Osage Reservation," *Geographical Review* 62 (Oct. 1972): 535.

41. H. M. Teller, secretary of the interior, to E. Feulin, 25 Apr. 1883, roll 31, M606, National Archives Record Group 48 (hereafter cited as NARG 48); Donald J. Berthrong, *The Cheyenne and Arapaho Ordeal: Reservation and Agency Life in the Indian Territory, 1875–1907* (Norman: University of Oklahoma Press, 1976), pp. 95, 98; Prucha, *American Indian Policy,* p. 215. While the Indians received two cents per acre as "grass money," the ranges were worth an estimated eight to twelve cents per acre.

42. *Report,* 1884, p. 96; Burrill, "Establishment of Ranching," p. 524.

43. Berthrong, *Cheyenne and Arapaho Ordeal,* p. 116; Burrill, "Establishment of Ranching," pp. 525, 531; Prucha, *American Indian Policy,* pp. 215–16; *Report,* 1885, pp. xvi–xviii; Savage, *Cherokee Strip Livestock Association,* p. 7; J. Orin Oliphant, "Encroachment of Cattlemen on Indian Reservations in the Pacific Northwest, 1870–1890," *Agricultural History* 24 (Jan. 1950): 42–58.

44. *Report,* 1881, pp. xxvi–xxix; *Report,* 1883, p. xviii; Burrill, "Establishment of Ranching," pp. 530–31.

45. *Report,* 1883, pp. xxiii–xxiv; Burrill, "Establishment of Ranching," pp. 530–31.

46. *Report,* 1880, p. 35; *Report,* 1881, pp. 38–40, 46–47, 52.

47. *Report,* 1883, pp. 28, 30, 35; *Report,* 1884, p. 31.

48. *Report,* 1884, p. 49; *Report,* 1885, p. 39.

49. *Report,* 1884, p. 153; *Report,* 1886, pp. vi, xxii–xxiii, 130.

50. *Report,* 1873, p. 44; *Report,* 1876, p. 69.

51. Berthrong, *Cheyenne and Arapaho Ordeal,* pp. 103–4; *Report,* 1873, p. 292.

52. *Report,* 1872, p. 46.

53. *Third Annual Report of the Board of Indian Commissioners,* 1871, pp. 156–57; *Report,* 1884, p. 85.

54. *Report,* 1884, p. 58.

CHAPTER 9. THE WHITE MAN'S ROAD

1. J. P. Kinney, *A Continent Lost—A Civilization Won, Indian Land Tenure in America* (Baltimore, Md.: Johns Hopkins Press, 1937), pp. 188–94; *Report,* 1881, p. xii; *Fourth Annual Report of the Indian Rights Association,* 1886, p. 9; Mary Antonio Johnston, *Federal Regulations with the Great Sioux Indians of South Dakota, 1887–1933* (Washington, D.C.: Catholic University of American Press,

1948), p. 8; David M. Holford, "The Subversion of the Indian Land Allotment System, 1887–1934," *Indian Historian* 8 (Spring 1975): 12; *Thirteenth Annual Report of the Board of Indian Commissioners*, 1881, pp. 7–8; William T. Hagan, "Private Property, the Indian's Door to Civilization," *Ethnohistory* 3 (Spring 1956): 126; *Report*, 1880, p. xviii.

2. Kinney, *A Continent Lost*, p. 82; Francis Paul Prucha, *American Indian Policy: The Indian Trade and Intercourse Acts, 1790–1834* (Cambridge, Mass.: Harvard University Press, 1962), pp. 228–33, 236–37, 241; D. S. Otis, *The Dawes Act, and the Allotment of Indian Lands*, ed. Francis Paul Prucha (Norman: University of Oklahoma Press, 1973), p. 5. For a detailed study of allotment policy before the Dawes Act see Kinney, *A Continent Lost*, pp. 103–213; Loring Benson Priest, *Uncle Sam's Stepchildren: The Reformation of United States Indian Policy, 1865–1887* (New Brunswick, N.J.: Rutgers University Press, 1942); Howard W. Paulson, "The Allotment of Land in Severalty to the Dakota Indians before the Dawes Act," *South Dakota History* 1 (Spring 1971): 132–53; and Paul W. Gates, "Indian Allotments Preceding the Dawes Act," in *The Frontier Challenge: Responses to the Trans-Mississippi West*, ed. John G. Clark (Lawrence: University Press of Kansas, 1971), pp. 141–70.

3. Prucha, *American Indian Policy*, pp. 242–43.

4. Ibid., pp. 243–47; Kinney, *A Continent Lost*, pp. 206–7. Founded on 29 Dec. 1882, the Indian Rights Association was an organization of friends of the Indians. It worked to ensure the political and civil rights of the tribesmen which had been granted by treaty and statute (see William T. Hagan, *The Indian Rights Association: The Herbert Welsh Years, 1882–1902* [Tucson: University of Arizona Press, 1985]).

5. Prucha, *American Indian Policy*, pp. 248, 252–55; Leonard A. Carlson, *Indians, Bureaucrats, and Land: The Dawes Act and the Decline of Indian Farming* (Westport, Conn.: Greenwood Press, 1981), p. 10; *Report*, 1887, pp. iv–v; Holford, "Subversion of the Indian Land Allotment System," p. 12; Otis, *Dawes Act*, pp. 3–7; Francis Paul Prucha, *Indian Policy in the United States: Historical Essays* (Lincoln: University of Nebraska Press, 1981), p. 237. The Dawes Act did not apply to the Five Civilized Tribes; to the Osage, the Miami, the Peoria, the Sac, the Fox, and the Seneca; or to some Sioux in Nebraska. It did, however, grant citizenship to the Indians who took allotments.

6. Otis, *Dawes Act*, pp. 17, 25, 27–28.

7. Kinney, *A Continent Lost*, p. 219; Otis, *Dawes Act*, pp. 41, 46, 50–53, 56; *Report*, 1886, pp. xx–xxi.

8. *Report*, 1887, pp. xxvii, 4, 12, 18, 138, 153; *Report*, 1888, pp. 5, 141–42.

9. *Report*, 1887, pp. 18, 37, 44, 48, 55–57, 151; *Report*, 1888, pp. 21, 49, 85; *Report*, 1889, pp. 140, 254; *Twenty-third Annual Report of the Board of Indian Commissioners*, 1890, p. 6.

10. *Report*, 1887, pp. 50–51, 81–82, 115, 123–24, 153; *Report*, 1888, p. 93; *Report*, 1889, p. 211.

11. *Report*, 1888, pp. 190–91; *Report*, 1889, pp. 116, 256.

12. *Report*, 1890, pp. xcic–xcv.

13. *Report*, 1889, pp. 30–32; *Report*, 1890, pp. lxxi–lxxiii.

14. Kinney, *A Continent Lost*, p. 221; Otis, *Dawes Act*, pp. 112–13, 116; Prucha, *American Indian Policy*, pp. 258–59; *Thirty-first Annual Report of the Board of Indian Commissioners*, 1899, pp. 19, 31; *Report*, 1892, p. 244; *Report*, 1893, p. 248.

15. *Report*, 1891, p. 438; *Report*, 1892, p. 221.

16. Kinney, *A Continent Lost*, p. 222; Prucha, *American Indian Policy*, pp. 261–62; *Twenty-sixth Annual Report of the Board of Indian Commissioners, 1894*, p. 7; Thomas Ryan, assistant secretary to the commissioner of Indian Affairs, 23 June 1897, roll 92, M606, NARG 48; Otis, *Dawes Act*, p. 120.

17. John W. Noble to James H. ?, 27 Apr. 1892, roll 75, M606, NARG 48; *Seventeenth Annual Report of the Indian Rights Association*, 1899, pp. 56–57; *Report*, 1895, ser. 3382, pp. 37–38, 199; *Tenth Annual Report of the Indian Rights Association*, 1892, pp. 33–34. In 1896 the secretary of the interior authorized leasing fees for the Wichita Reservation in Oklahoma to be set by sealed bids for that grazing season in order to increase the return for Indian lands (see William H. Sims, acting secretary to the commissioner of Indian Affairs, 4 May 1896, roll 88, M606, NARG 48).

18. Kinney, *A Continent Lost*, pp. 222–34.

19. *Report*, 1900, ser. 4101, p. 13; Otis, *Dawes Act*, pp. 122–23; *Eighteenth Annual Report of the Indian Rights Association*, 1900, p. 58; *Thirty-third Annual Report of the Board of Indian Commissioners*, 1901, p. 17.

20. "The Action of the Interior Department in Forcing the Standing Rock Indians to Lease Their Lands to Cattle Syndicates," Indian Rights Association, 12 May 1902, no. 61; Thomas R. Wessel, "Agriculture on the Reservations: The Case of the Blackfeet, 1885–1935," *Journal of the West* 18 (Oct. 1979): 20.

21. *Annual Report of the U.S. Indian Inspector for the Indian Territory, 1903*, pp. 39, 60; *Annual Report of the U.S. Inspector for the Indian Territory*, 1904, p. 55; *Thirty-fifth Annual Report of the Board of Indian Commissioners*, 1903, p. 16. In 1898 the Curtis Act provided for the allotment and leasing of lands of the Five Civilized Tribes. For a detailed study of leasing policy on the southern Great Plains see William T. Hagan, "Kiowas, Comanches and Cattlemen, 1867–1906: A Case Study of the Failure of U.S. Reservation Policy," *Pacific Historical Review* 40 (1971): 333–55; Donald J. Berthrong, *The Cheyenne and Arapaho Ordeal: Reservation and Agency Life in the Indian Territory, 1875–1907* (Norman: University of Oklahoma Press, 1976), pp. 197–205; Lloyd H. Cornell, "Leasing and Land Utilization of the Cheyenne and Arapaho Indians, 1891–1907" (Master's thesis, University of Oklahoma, 1954).

22. *Report*, 1891, pp. 360–61; *Report*, 1892, pp. 252–53; Hope Smith to the secretary of war, 21 Oct. 1894, roll 83, M606, NARG 48; Charles F. Meserve, "The Dawes Commission and the Five Civilized Tribes," Indian Rights Association, 1886, no. 33, pp. 32–33.

23. *Report*, 1890, p. 5; *Report*, 1891, p. 382; *Report*, 1892, p. 335; *Report*, 1893, pp. 1109–10, 1118; *Report*, 1895, ser. 3382, p. 25; *Report*, 1896, ser. 3641, pp. 112, 114, 190; *Report*, 1898, ser. 3757, p. 206.

24. *Report*, 1890, pp. 25, 37, 50, 59, 66, 87; *Report*, 1891, pp. 293–94; *Report*, 1892, p. 371; *Report*, 1894, pp. 224–25, 274, 303; *Report*, 1895, ser. 3382, p. 251; *Report*, 1896, ser. 3489, pp. 249, 253; *Report*, 1898, ser. 3757, pp. 227–28.

25. *Report*, 1890, p. 122; Wessel, "Agriculture on the Reservations," p. 18; *Report*, 1891, pp. 256, 343; Berthrong, *Cheyenne and Arapaho Ordeal*, p. 139; *Report*, 1893, p. 158; *Report*, 1894, pp. 276–77, 294; *Report*, 1895, ser. 3382, pp. 190, 203, 295.

26. *Twelfth Census of the United States*, vol. 5, 1900, pp. 721–22; *Report*, 1901, ser. 4290, pp. 271, 317.

27. *Twelfth Census*, 5: 722–23.

28. *Report*, 1897, ser. 3641, pp. 152–53; *Twelfth Census*, 5: 723–24.

29. *Report,* 1900, ser. 4101, pp. 326, 332; Berthrong, *Cheyenne and Arapaho Ordeal,* p. 300; *Twelfth Census,* 5:725–26.

30. *Twelfth Census,* 5:727, 730–32; *Thirty-third Annual Report of the Board of Indian Commissioners,* 1901, p. 15.

31. *Report,* 1903, ser. 4645, pp. 321, 261; *Annual Report of the U.S. Indian Inspector for the Indian Territory,* 1903, pp. 60–61; *Report,* 1904, ser. 4798, pp. 215, 321, 341; *Report,* 1905, ser. 4959, pp. 82, 286; *Report,* 1906, ser. 5118, pp. 242, 334; Donald J. Berthrong, "Federal Indian Policy and the Southern Cheyennes and Arapahoes, 1887–1907," *Ethnohistory* 3 (Spring 1956): 143, 151.

32. *Report,* 1902, ser. 4458, pp. 177, 222, 235; *Report,* 1904, ser. 4798, pp. 152, 321; *Report,* 1905, ser. 4959, pp. 167, 276, 286; *Report,* 1906, ser. 5118, pp. 182, 188.

33. Kinney, *A Continent Lost,* pp. 237–38; Holford, "Subversion of the Indian Land Allotment System," p. 14. For problems of heirship sales see Berthrong, *Cheyenne and Arapaho Ordeal,* pp. 280–84. Authorization for the sale of heirship lands was provided by the Indian Appropriation Act of 1902, which became known as the Dead Indian Land Act. Minors could sell lands only upon approval.

34. *Report,* 1906, ser. 5118, p. 28; *Twenty-fourth Annual Report of the Indian Rights Association,* 1906, pp. 45–48; Kinney, *A Continent Lost,* pp. 240–41.

35. Kinney, *A Continent Lost,* p. 242.

36. Otis, *Dawes Act,* pp. 69, 77–81; Herbert T. Hoover, "Arikara, Sioux, and Government Farmers: The Indian Agricultural Legacies," *South Dakota History* 13 (Spring/Summer 1983): 37–38; Thomas R. Wessel, "Agent of Acculturation: Farming on the Northern Plains Reservations, 1880–1910," *Agricultural History* 60 (Spring 1985): 239–45.

37. Thomas R. Wessel, "Agriculture, Indians and American History," *Agricultural History* 50 (Jan. 1976): 19; Hagan, "Kiowas, Comanches and Cattlemen," p. 341; William T. Hagan, *American Indians* (Chicago: University of Chicago Press, 1979), p. 146; Roy W. Meyers, *History of the Santee Sioux* (Lincoln: University of Nebraska Press, 1967), p. 191.

38. Untitled publication of the Indian Rights Association, 8 June 1898, no. 47; Prucha, *American Indian Policy,* pp. 256–57; Paul Stuart, *The Indian Office: Growth and Development of an American Institution, 1865–1900* (Ann Arbor, Mich.: UMI Research Press, 1979), p. 26.

CHAPTER 10. GOOD INTENTIONS

1. *Report,* 1909, ser. 5747, pp. 10, 12; *Report,* 1910, p. 53; *Report,* 1911, pp. 9–11; *Report,* 1912, p. 32; *Thirtieth Annual Report of the Indian Rights Association,* 1912, p. 27.

2. *Report,* 1911, pp. 11–12; *Report,* 1912, pp. 31–32; *Thirty-fourth Annual Report of the Indian Rights Association,* 1916, p. 13; *Report,* 1916, pp. 27–30.

3. *Report,* 1913, pp. 7–9.

4. *Report,* 1907, ser. 5296, p. 132; *Report,* 1908, ser. 5453, pp. 6–7, 138; *Report,* 1909, ser. 5747, p. 58; *Report,* 1912, p. 34; *Report,* 1914, pp. 10–14; *Report,* 1915, pp. 23–24, 26–27; *Report,* 1916, p. 33; *Thirty-fourth Annual Report of the Indian Rights Association,* 1916, p. 42; J. F. Wojta, "Indian Farm Institutes in Wisconsin," *Wisconsin Magazine of History* 29 (June 1946): 423–34. By 1910, demonstration farms had been established on twenty-two reservations. Sometimes the

Indians also received agricultural aid from the state board of agriculture, as they did in Oklahoma (see *Report*, 1911, pp. 9–11).

5. *Report*, 1908, ser. 5453, p. 28; *Report*, 1910, p. 5; *Thirtieth Annual Report of the Indian Rights Association*, 1912, p. 70; *Fortieth Annual Report of the Board of Indian Commissioners*, 1908, pp. 19–20; *Forty-second Annual Report of the Board of Indian Commissioners*, 1910–11, p. 9; *Forty-seventh Annual Report of the Board of Indian Commissioners*, 1916, pp. 15, 18.

6. *Thirtieth Annual Report of the Indian Rights Association*, 1912, pp. 26, 35; *Thirty-first Annual Report of the Indian Rights Association*, 1913, pp. 32, 57; *Thirty-fourth Annual Report of the Indian Rights Association*, 1916, p. 21; *Report*, 1913, p. 9; *Report*, 1914, pp. 17–20.

7. *Twenty-fifth Annual Report of the Indian Rights Association*, 1907, p. 48; *Report*, 1907, ser. 5296, p. 71; *Report*, 1908, ser. 5453, pp. 31, 74. The Indians were paid $4.50 per acre for harvesting, $6.00 per acre for thinning, $1.50 per acre for plowing, $4.00 per acre for irrigating and hoeing, $10.00 per acre for topping and loading, and $0.75 per ton for hauling.

8. *Report*, 1909, ser. 5747, p. 49; *Report*, 1910, p. 34; *Report*, 1911, pp. 26–28.

9. *Report*, 1914, pp. 16, 27; *Report*, 1915, p. 50; *Fifty-first Annual Report of the Board of Indian Commissioners*, 1920, p. 21; Janet Ann McDonnell, "The Disintegration of the Indian Estate: Indian Land Policy, 1913–1929" (Ph.D. diss., Marquette University, 1980), p. 59.

10. *Report*, 1915, p. 51.

11. *Report*, 1917, p. 25; David L. Wood, "American Indian Farmland and the Great War," *Agricultural History* 55 (July 1981): 249–52.

12. *Report*, 1917, p. 28; *Report*, 1918, p. 45; McDonnell, "Disintegration of the Indian Estate," p. 177; Wood, "American Indian Farmland," pp. 254–55; J. P. Kinney, *A Continent Lost—A Civilization Won: Indian Land Tenure in America* (Baltimore, Md.: Johns Hopkins Press, 1937), p. 296.

13. *Forty-first Annual Report of the Board of Indian Commissioners*, 1909, p. 7; *Twenty-seventh Annual Report of the Indian Rights Association*, 1909, p. 11; Janet A. McDonnell, "Land Policy on the Omaha Reservation: Competency Commissions and Forced Fee Patents," *Nebraska History* 63 (Fall 1982): 406; Donald J. Berthrong, "Legacies of the Dawes Act: Bureaucrats and Land Thieves at the Cheyenne-Arapaho Agencies of Oklahoma," *Arizona and the West* 21 (Winter 1979): 345–46.

14. McDonnell, "Land Policy on the Omaha Reservation," pp. 400–401, 404–5, 407; Janet McDonnell, "Competency Commissions and Indian Land Policy, 1913–1920," *South Dakota History* 11 (Winter 1980): 22–23, 29, 33–34; idem, "Disintegration of the Indian Estate," pp. 115, 170.

15. *Report*, 1917, pp. 3–5; *Report*, 1918, p. 19; Mary Antonio Johnston, *Federal Relations with the Great Sioux Indians of South Dakota, 1887–1933* (Washington, D.C.: Catholic University of America Press, 1948), pp. 20–22; Berthrong, "Legacies of the Dawes Act," p. 350; Francis Paul Prucha, *The Great Father*, vol. 2 (Lincoln: University of Nebraska Press, 1984), pp. 882–83.

16. *Report*, 1914, p. 45; McDonnell, "Disintegration of the Indian Estate," p. 204; *Thirty-ninth Annual Report of the Indian Rights Association*, 1921, p. 45.

17. *Report*, 1907, ser. 5296, p. 73; *Report*, 1908, ser. 5453, p. 71; *Report*, 1909, ser. 5747, p. 103.

18. David M. Holford, "The Subversion of the Indian Land Allotment System,

1887–1934," *Indian Historian* 8 (Spring 1975): 15–16; Kinney, *A Continent Lost,* p. 250; *Report,* 1910, pp. 40–41; *Report,* 1911, p. 26; *Report,* 1912, pp. 61–62.

19. *Report,* 1914, p. 59; *Thirty-fifth Annual Report of the Indian Rights Association,* 1917, p. 45; Kinney, *A Continent Lost,* p. 295; McDonnell, "Disintegration of the Indian Estate," p. 319.

20. *Thirty-sixth Annual Report of the Indian Rights Association,* 1918, pp. 14, 45; *Thirty-seventh Annual Report of the Indian Rights Association,* 1919, pp. 38, 67–68; Wood, "American Indian Farmland," pp. 264–65; Robert John Stahl, "Farming among the Kiowa, Comanche, Kiowa-Apache and Wichita" (Ph.D. diss., University of Oklahoma, 1978), p. 225.

21. Stahl, "Farming among the Kiowa," pp. 232, 241.

22. Ibid., p. 241.

23. *Thirty-ninth Annual Report of the Indian Rights Association,* 1921, p. 19; *Fifty-first Annual Report of the Board of Indian Commissioners,* 1920, pp. 33–34; Harry T. Getty, "Development of the San Carlos Cattle Industry," *Kiva* 23 (Feb. 1958): 2; *Minneapolis Sunday Tribune,* 22 June 1924; *Dakota Farmer,* 15 Feb. 1923.

24. Leonard A. Carlson, *Indians, Bureaucrats, and Land: The Dawes Act and the Decline of Indian Farming* (Westport, Conn.: Greenwood Press, 1981), p. 139; *Fifty-second Annual Report of the Board of Indian Commissioners,* 1921, p. 108; *Fifth-third Annual Report of the Board of Indian Commissioners,* 1922, pp. 19, 26; *Fifty-fourth Annual Report of the Board of Indian Commissioners,* 1923, p. 38.

25. *Fifty-fourth Annual Report of the Board of Indian Commissioners,* 1923, pp. 39–40, 42; Johnston, *Federal Relations with the Great Sioux Indians of South Dakota,* p. 122; *Fifty-third Annual Report of the Board of Indian Commissioners,* 1922, pp. 11–12, 29; Richard White, *The Roots of Dependency* (Lincoln: University of Nebraska Press, 1983), p. 233; *Fifty-fifth Annual Report of the Board of Indian Commissioners,* 1924, pp. 13, 31. Choctaw farmers in Mississippi were usually in "distress" by the end of each year, and the sharecropping system kept them poverty stricken. In North Carolina the Eastern Cherokee, located on the reservation known as the Qualla Boundary, struggled on small farms of twenty to thirty acres to raise chickens, hogs, vegetables, and small grains. The Board of Indian Commissioners hoped, however, that they could become commercial fruit growers.

26. *Report,* 1921, p. 14; *Report,* 1924, p. 9; *Fifth-sixth Annual Report of the Board of Indian Commissioners,* 1925, p. 20; *Forty-third Annual Report of the Indian Rights Association,* 1925, p. 23.

27. *Thirty-eighth Annual Report of the Indian Rights Association,* 1920, p. 47; *Report,* 1921, pp. 2–24; *Forty-third Annual Report of the Indian Rights Association,* 1925, p. 3. W. David Baird has argued that between 1906 and 1931 many of the members of the Five Civilized Tribes favored allotment and the alienation of their lands, because it was economically beneficial and politically inevitable, and that land-hungry whites were not solely responsible for that policy (see "The Reduction of the Allotted Indian Land Base in Oklahoma: Who Was Responsible?" paper presented at the annual meeting of the Organization of American Historians, 9 Apr. 1983, Cincinnati, Ohio).

28. Field notes, Mary Louise Mark Papers, box 1, folder 5–6, Ohio Historical Society.

29. "The Needs in Human Relationships," Mary Louise Mark Papers, box 1, folder 8, p. 5; Prucha, *Great Father,* 2:808; Lewis Meriam, *The Problem of Indian*

Administration (Baltimore, Md.: Johns Hopkins Press, 1928), p. vii. Given these problems, by 1925, only 32,234 Indians cultivated 644,873 acres, and production reached a mere $7,798,778. The livestock industry was in a similar state of decline; an estimated 44,847 Indians raised livestock on 29,098,459 acres.

30. Meriam, *Problem of Indian Administration,* pp. 14–15, 123–24, 135.
31. Ibid., pp. 384–85. Mary Louise Mark, one of the field observers, reported that agricultural agents needed to be "social engineers" (see "Needs in Human Relationships," p. 6).
32. Meriam, *Problem of Indian Administration,* pp. 460–61, 466–67.
33. Ibid., pp. 470–72, 475–76.
34. Ibid., pp. 477–79.
35. Ibid., pp. 488–91, 499; Flora Warren Seymour, "Our Indian Policy," *Journal of Land and Public Utility Economics* 2 (Jan. 1926): 109.
36. Meriam, *Problem of Indian Administration,* pp. 504–6.
37. Porter J. Preston and Charles A. Engle, "Report of Advisors on Irrigation on Indian Reservations," in "Survey of Conditions of the Indians of the United States," *Hearings before a Subcommittee of the Committee on Indian Affairs, United States Senate,* 71st Cong., 2d sess., 1930, pt. 6, pp. 2111, 2219, 2221, 2223, 2230–31, 2234–35; McDonnell, "Disintegration of the Indian Estate," p. 249; *Report,* 1932, pp. 18–19; Prucha, *Great Father,* 2:892–93; Meriam, *Problem of Indian Administration,* p. 510.
38. Preston and Engle, "Report of Advisors," pp. 2236, 2255.
39. *Forty-fourth Annual Report of the Board of Indian Commissioners, 1912–1913,* p. 7; Norris Hundley, Jr., "The Dark and Bloody Ground of Indian Water Rights: Confusion Elevated to Principle," *Western Historical Quarterly* 9 (Oct. 1978): 455–56; McDonnell, "Disintegration of the Indian Estate," p. 70; *Report,* 1915, p. 50; Prucha, *Great Father,* 2:893; Frederick E. Hoxie, *A Final Promise: The Campaign to Assimilate the Indians, 1880–1920* (Lincoln: University of Nebraska Press, 1984), pp. 170–72.
40. Preston and Engle, "Report of Advisors," p. 2217; Carlson, *Indians, Bureaucrats, and Land,* pp. 135, 140–41; Kinney, *A Continent Lost,* p. 282.
41. *Report,* 1928, pp. 30, 33; *Indian Truth,* July 1928, p. 4; *Forty-sixth Annual Report of the Indian Rights Association,* 1928, p. 9; Gary D. Libecap and Ronald N. Johnson, "Legislating Commons: The Navajo Tribal Council and the Navajo Range," *Economic Inquiry* 18 (Jan. 1980): 71.
42. *Forty-ninth Annual Report of the Indian Rights Association,* 1931, p. 4; *Report,* 1932, p. 3; *Sixty-third Annual Report of the Board of Indian Commissioners,* 1932, p. 2; R. Douglas Hurt, *The Dust Bowl: An Agricultural and Social History* (Chicago: Nelson-Hall, 1981), pp. 33–47.
43. William G. Robbins, "Herbert Hoover's Indian Reformers under Attack: The Failures of Reform," *Mid-America* 63 (Oct. 1981): 158, 164; *Indian Truth,* June 1930, p. 1, Aug.–Sept. 1931, p. 4, and Nov. 1932, p. 4; *Report,* 1932, p. 14; J. W. Hoover, "Tusayan: The Hopi Indian Country of Arizona," *Geographical Review* 20 (July 1930): 435, 438–39; William T. Hagan, *American Indians* (Chicago: University of Chicago Press, 1979), pp. 147, 150; Randolph C. Downes, "A Crusade for Indian Reform, 1922–1934," *Mississippi Valley Historical Review* 32 (Dec. 1945): 332; Prucha, *Great Father,* 2:896.
44. *Fifty-third Annual Report of the Board of Indian Commissioners,* 1922, p. 26; McDonnell, "Disintegration of the Indian Estate," pp. 88, 99; Meriam, *Problem of Indian Administration,* p. 471.

CHAPTER 11. THE INDIAN NEW DEAL

1. Francis Paul Prucha, *The Great Father,* vol. 2 (Lincoln: University of Nebraska Press, 1984), pp. 941, 944; Kenneth R. Philip, *John Collier's Crusade for Indian Reform, 1920–1954* (Tucson: University of Arizona Press, 1977), pp. xi, xiv, 97.

2. Kenneth R. Philip, "John Collier and the Controversy over the Wheeler-Howard Bill," in *Indian-White Relations: A Persistent Paradox,* ed. James F. Smith and Robert M. Kvasnicka (Washington, D.C.: Howard University Press, 1976), pp. 273–75; Philip, *John Collier's Crusade,* pp. xi, xiii, 46.

3. Philip, "John Collier," pp. 275–76.

4. *Annual Report of the Secretary of the Interior, 1933,* p. 69; Prucha, *Great Father,* 2:944–45; Donald L. Parman, "The Indian and the Civilian Conservation Corps," *Pacific Historical Review* 40 (Feb. 1971): 40–41; *Indians at Work,* 11 Nov. 1933, p. 8; Roger Bromert, "The Sioux and the Indian—CCC," *South Dakota History* 8 (Fall 1978): 345.

5. *Annual Report of the Secretary of the Interior, 1933,* pp. 69–70; Philip, *John Collier's Crusade,* pp. 120–21; Prucha, *Great Father,* 2:945; *Indians at Work,* 15 Sept. 1933, pp. 2, 33–34.

6. *Indians at Work,* Nov. 1933, p. 1, 1 Oct. 1933, pp. 18–21, and 15 May 1934, pp. 13–14, 16–18; Parman, "The Indian and the Civilian Conservation Corps," p. 45; Bromert, "Sioux and the Indian—CCC," pp. 347, 349–50.

7. Prucha, *Great Father,* 2:951.

8. *Indians at Work,* 15 Sept. 1933, p. 3, and 1 Jan. 1934, p. 6.

9. Ibid., 1 Oct. 1933, p. 7, and Nov. 1933, p. 21.

10. Philip, *John Collier's Crusade,* p. 142; Lawrence C. Kelly, *The Navajo Indians and Federal Indian Policy, 1900–1935* (Tucson: University of Arizona Press, 1968), p. 163; Lawrence C. Kelly, "The Indian Reorganization Act: The Dream and the Reality," *Pacific Historical Review* 44 (Aug. 1975): 293–96; Prucha, *Great Father,* 2:957–58.

11. *Annual Report of the Secretary of the Interior, 1934,* p. 79; Philip, *John Collier's Crusade,* p. 145; U.S., Congress, Subcommittee of the Senate Committee on Indian Affairs, *Survey of Conditions among the Indians of the United States,* Report 310, 78th Cong., 2d sess., 1944, pt. 2, ser. 10841, p. 1; Kelly, "Indian Reorganization Act," p. 296; Prucha, *Great Father,* 2:959–60.

12. Philip, *John Collier's Crusade,* p. 159; Prucha, *Great Father,* 2:961–63; Kelly, *Navajo Indians,* pp. 165–66. The Indian Reoganization Act did not apply to the Five Civilized Tribes, because they were among the most assimilated Indians in the nation. On 18 June 1936, however, Congress approved the Thomas-Rogers Indian Welfare Act, which extended the credit and land provisions of the IRA to the Oklahoma Indians. For a succinct overview of the Indian New Deal see Richard Lowitt, *The New Deal and the West* (Bloomington: Indiana University Press, 1984), pp. 122–37.

13. Prucha, *Great Father,* 2:964–65; *Indians at Work,* 1 July 1934, p. 1. Ultimately, ninety-three tribes wrote constitutions but only seventy-three created charters of incorporation (see Kenneth R. Philip, "John Collier, 1933–45," in *The Commissioners of Indian Affairs, 1824–1977,* ed. Robert M. Kvasnicka and Herman H. Viola [Lincoln: University of Nebraska Press, 1979], p. 278).

14. Prucha, *Great Father,* 2:969; E. H. Spicer, "Sheepmen and Technicians: A Program of Soil Conservation on the Navaho Indian Reservation," in *Human Problems in Technological Change: A Casebook,* ed. Edward H. Spicer (New York:

John Wiley & Sons, 1967), p. 185; Gary D. Libecap and Ronald N. Johnson, "Legislating Commons: The Navajo Tribal Council and the Navajo Range," *Economic Inquiry* 18 (Jan. 1980): 72; Graham D. Taylor, *The New Deal and American Indian Tribalism* (Lincoln: University of Nebraska Press, 1980), p. 127; Henry F. Dobyns, "Experiment in Conservation: Erosion Control and Forage Production on the Papago Indian Reservation in Arizona," in *Human Problems in Technological Change*, p. 214.

15. *Indians at Work*, 15 Nov. 1933, pp. 12–13; L. Schuyler Fonaroff, "Conservation and Stock Reduction on the Navajo Tribal Range," *Geographical Review* 53 (Apr. 1963): 206, 211–12; Taylor, *New Deal and American Indian Tribalism*, pp. 126–27; Philip, *John Collier's Crusade*, p. 188; Peter Iverson, *The Navajo Nation* (Westport, Conn.: Greenwood Press, 1981), p. 27; Spicer, "Sheepmen and Technicians," pp. 192–93.

16. Donald L. Parman, *The Navajos and the New Deal* (New Haven, Conn.: Yale University Press, 1976), p. 48; Libecap and Johnson, "Legislating Commons," p. 72; Richard White, *The Roots of Dependency* (Lincoln: University of Nebraska Press, 1983), pp. 262, 265–66; Fonaroff, "Conservation and Stock Reduction," p. 213; Taylor, *New Deal and American Indian Tribalism*, pp. 127–28; Philip, *John Collier's Crusade*, p. 188; *Indian Truth*, Oct. 1935, p. 3.

17. *Indians at Work*, 1 Aug. 1934, p. 38, and 1 Oct. 1934, pp. 13–14; Kelly, *Navajo Indians*, pp. 161–62; White, *Roots of Dependency*, pp. 263–65; Parman, *Navajos*, pp. 63–67; interview with E. R. Fryer, former general superintendent of the Navajo Reservation from 1935 to 1942, conducted by Donald L. Parman, 21 July 1970, pp. 2–3, 27; Fonaroff, "Conservation and Stock Reduction," pp. 210–11; *Annual Report of the Secretary of the Interior*, 1936, p. 188.

18. Kelly, *Navajo Indians*, p. 162; White, *Roots of Dependency*, pp. 291–314; Parman, *Navajos*, pp. 172–76; Libecap and Johnson, "Legislating Commons," p. 73; Hugh G. Calkins and D. S. Hubbell, "A Range Conservation Demonstration in the Land of the Navajos," *Soil Conservation* 6 (Sept. 1940): 67.

19. Gerald R. Ogden, "U.S. Department of Agriculture Activities in Navajoland: The 1930s," *Associates NAL Today* 3 (Jan.–June 1978): 22; Cecil T. Blunn, "Improvement of the Navajo Sheep," *Journal of Heredity* 31 (Mar. 1940): 105–6.

20. *Annual Report of the Secretary of the Interior*, 1935, p. 126; Blunn, "Improvement of the Navajo Sheep," pp. 99, 108; Parman, *Navajos*, pp. 127–30; Ogden, "U.S. Department of Agriculture Activities in Navajoland," pp. 22–23.

21. William O. Roberts, "Successful Agriculture within the Reservation Framework," *Applied Anthropology* 2, no. 3 (1943): 39; *Annual Report of the Secretary of the Interior*, 1934, p. 111; R. Douglas Hurt, *The Dust Bowl: An Agricultural and Social History* (Chicago: Nelson-Hall, 1981), pp. 106–8.

22. *Indians at Work*, 15 Sept. 1934, pp. 10–12; *Annual Report of the Secretary of the Interior*, 1935, pp. 124–25.

23. *Indians at Work*, 1 Jan. 1935, pp. 19–20, and 1 May 1936, pp. 28–29.

24. Ibid., 15 Jan. 1939, p. 39; James O. Fine, "An Analysis of Factors Affecting Agricultural Development on the Fort Totten Indian Reservation" (Master's thesis, University of North Dakota, 1951), pp. 43–47; *Annual Report of the Secretary of the Interior*, 1935, p. 125; *Indians at Work*, Jan. 1936, p. 3, 1 Apr. 1936, p. 17, 1 May 1936, p. 29, and May 1937, p. 11. By 30 June 1940, only twenty-one of the fifty-five Indians on the Fort Totten Reservation, who had received a total of 641 cattle, were delinquent in their payments. Eleven delinquents owed 122 head, but they were no longer raising cattle.

25. *Indians at Work,* 15 Jan. 1934, pp. 22–30, 1 June 1934, p. 36, 1 May 1935, pp. 37–41, 15 May 1936, pp. 42–46, 1 Apr. 1937, p. 21, and Jan. 1939, pp. 42–46.

26. Harry T. Getty, "The San Carlos Indian Cattle Industry," University of Arizona, Anthropological Papers no. 7 (1963), pp. 38, 43–44; Harry T. Getty, "San Carlos Apache Cattle Industry," *Human Organization* 20 (Winter 1961/62): 182–83; Veronica E. Tiller, *The Jicarilla Apache* (Lincoln: University of Nebraska Press, 1983), p. 177.

27. *Indians at Work,* 1 Feb. 1938, pp. 5–6, May 1938, p. 14, and July 1939, pp. 19–20.

28. Ibid., 15 Aug. 1935, p. 22; S. D. Aberle, "The Pueblo Indians of New Mexico: Their Land, Economy and Civil Organization," *Memoirs of the American Anthropological Association* 50 (Oct. 1948): 32; Rolf W. Bauer, "The Papago Cattle Economy," in *Food, Fiber and the Arid Lands,* ed. William G. McGinnies, Bram J. Goldman, and Patricia Paylore (Tucson: University of Arizona Press, 1971), pp. 89–90.

29. U.S. National Resources Board, *General Conditions and Tendencies Influencing the Nation's Land Requirements,* pt. 10: *Indian Land Tenure, Economic Status, and Population Trends* (Washington, D.C.: Government Printing Office, 1935), pp. 15–19, 58; *Indians at Work,* 1 Sept. 1935, pp. 23–24, 15 Oct. 1934, pp. 5–6, 1 Dec. 1936, p. 37, and 1 Apr. 1937, pp. 22–24; *Indian Truth,* Feb. 1936, p. 3.

30. *Indians at Work,* 1 Dec. 1936, p. 39; James A. Vlasich, "Transitions in Pueblo Agriculture, 1938–1948," *New Mexico Historical Review* 55 (Jan. 1980): 28–29.

31. *Indians at Work,* 15 Dec. 1935, pp. 37–41; James E. Officer, "Arid-Lands and the Indians of the American Southwest," in *Food, Fiber and the Arid Lands,* p. 65.

32. *Indians at Work,* 15 Nov. 1933, pp. 28–30, 15 Jan. 1934, pp. 32–33, 15 June 1935, pp. 39–42, 15 June 1936, pp. 46–48, 1 Aug. 1936, pp. 14–17, 1 Apr. 1937, pp. 26–29, and Nov. 1939, p. 1.

33. Ibid., 1 May 1934, pp. 19–22, 15 Nov. 1935, p. 42, and 1 May 1936, p. 32; *Indian Truth,* Oct. 1937, p. 4; Taylor, *New Deal and American Indian Tribalism,* p. 132.

34. *Indians at Work,* 1 Aug. 1935, p. 42, and 1 May 1936, pp. 31–32; U.S. National Resources Board, *Indian Land Tenure,* p. 2.

35. *Indians at Work,* 1 Dec. 1936, pp. 17–19; *Indian Truth,* Jan. 1940, p. 3; *Report,* 1944, p. 239; Taylor, *New Deal and American Indian Tribalism,* p. 132. Surplus lands were returned to tribal domain in Minnesota, Montana, Oklahoma, North Dakota, and South Dakota.

36. Prucha, *Great Father,* 2:1007; *Indian Truth,* May–June 1940, p. 2, and Dec. 1940, pp. 1–2; *Annual Report of the Secretary of the Interior, 1940,* p. 375; *Indians at Work,* Feb. 1941, p. 19.

37. *Indians at Work,* Aug. 1939, pp. 17–18, and Oct. 1939, pp. 25–27; Spicer, "Sheepmen and Technicians," pp. 191, 198; Fryer interview, pp. 8, 11, 45; Parman, *Navajos,* p. 244.

38. *Indians at Work,* Aug. 1939, pp. 15–16; Libecap and Johnson, "Legislating Commons," p. 72; Spicer, "Sheepmen and Technicians," p. 192; Dewey Dismuke, "Acoma and Laguna Indians Adjust Their Livestock to Their Range," *Soil Conservation* 6 (Nov. 1940): 130–31.

39. *Indians at Work,* Dec. 1940, pp. 7–10; *Annual Report of the Secretary of the Interior,* 1940, p. 376. One sheep or goat equals one unit, while one cow equals four units, one horse equals five units, and one burro equals three units.

40. Edward F. Castetter and Willis H. Bell, *Yuman Indian Agriculture* (Albuquerque: University of New Mexico Press, 1951), pp. 86–89, 130, 155; *Indian Truth,* May–June 1941, pp. 1–3; Robert John Stahl, "Farming among the Kiowa, Comanche, Kiowa-Apache and Wichita" (Ph.D. diss., University of Oklahoma, 1978), p. 260; Leslie Hewes, "Indian Land in the Cherokee Country of Oklahoma," *Economic Geography* 18 (Oct. 1942): 408–9.

41. *Report,* 1941, p. 426; *Report,* 1942, p. 241; Prucha, *Great Father,* 2:993–94, 1004–5; Philip, *John Collier's Crusade,* pp. 208, 210.

42. Prucha, *Great Father,* 2:1009–10; Parman, *Navajos,* p. 31; *Annual Report of the Secretary of the Interior,* 1933, p. 109.

43. Taylor, *New Deal and American Indian Tribalism,* pp. 126, 133; Kenneth R. Philip, "Termination: A Legacy of the Indian New Deal," *Western Historical Quarterly* 14 (Apr. 1983): 176; Edward Everett Dale, *The Indians of the Southwest* (Norman: University of Oklahoma Press, 1949), p. 22; Kelly, "Indian Reorganization Act," p. 312; Parman, *Navajos,* p. 30; Prucha, *Great Father,* 2:1011–12; Michael L. Lawson, "Indian Heirship Lands: The Lake Traverse Experience," *South Dakota History* 12 (Winter 1982): 225. By 1945, more than 6,000 cattle had been repaid to the original 40,000 head revolving cattle pool. The cattle pool enabled approximately 10,000 Indians to become stockmen. Collier estimated that beef production increased 2300 percent, and crop production jumped 400 percent during the first twenty years of the Indian Reorganization Act (see William T. Hagan, *American Indians* [Chicago: University of Chicago Press, 1979], p. 158).

44. Philip, "Termination," pp. 177–78; *Annual Report of the Secretary of the Interior,* 1940, pp. 360–69, 375–76; Philip, *John Collier's Crusade,* p. 186; Kelly, "Indian Reorganization Act," pp. 298–99. The Indians who benefited the most from the return of surplus lands were those on the Kiowa, Comanche, and Apache reservations in Oklahoma, the Pine Ridge Reservation in South Dakota, the Grand Portage Reservation in Minnesota, the Standing Rock Reservation in North and South Dakota, the Colorado River Reservation in Arizona, and the Flathead Reservation in Montana.

45. Kelly, "Indian Reorganization Act," pp. 306–8; Taylor, *New Deal and American Indian Tribalism,* p. 121.

46. Parman, *Navajos,* pp. 185–86; idem, "Indian and the Civilian Conservation Corps," pp. 51, 54–56; Bromert, "Sioux and the Indian—CCC," pp. 354–56; *Indian Truth,* Nov.–Dec. 1942, p. 3.

47. *Report,* 1941, pp. 414, 423, 450; *Report,* 1944, p. 242; *Report,* 1945, pp. 234, 236, 239; Tom Sasaki and John Adair, "New Land to Farm: Agricultural Practices among the Navaho Indians of New Mexico," in *Human Problems in Technological Change,* pp. 107–9. By 1945, only 7 million of 56 million acres of Indian lands were arable. Indian farmers used 32 million acres of their lands for grazing purposes; crop production reached $9,458,000; and sales from beef and dairy cattle totaled $15,039,000.

CHAPTER 12. THE TERMINATION ERA

1. *Report,* 1946, pp. 363–64.
2. Ibid., pp. 364–65.

3. Ibid., pp. 356, 363, 366–68, 371–72; *Report,* 1948, p. 377; *Report,* 1949, p. 346.

4. *Report,* 1946, p. 369; *Indian Truth,* Mar.–Aug. 1946, p. 3, Jan.–Feb. 1948, p. 3, and Mar.–Apr. 1948, p. 4; *Report,* 1947, p. 356. Some progress in stock reduction had been achieved: between 1933 and 1947, the Navaho reduced their goats from 173,000 to 56,000, their horses from 44,000 to 35,000, their cattle from 21,000 to 11,000, and their sheep from 570,000 to 358,600 (see Peter Iverson, *The Navajo Nation* [Westport, Conn.: Greenwood Press, 1981], p. 54).

5. Edward Everett Dale, *The Indians of the Southwest* (Norman: University of Oklahoma Press, 1949), pp. 237–38, 240–41.

6. Ibid., p. 243–44, 253–54.

7. *Report,* 1947, pp. 345–46, 354–55; *Report,* 1948, p. 371; *Report,* 1949, p. 347.

8. *Report,* 1947, p. 352; *Report,* 1948, p. 381; *Report,* 1949, p. 341.

9. Francis Paul Prucha, *The Great Father,* vol. 2 (Lincoln: University of Nebraska Press, 1984), pp. 1013–14; Larry W. Burt, *Tribalism in Crisis: Federal Indian Policy, 1953–1961* (Albuquerque: University of New Mexico Press, 1982), p. 20; Clayton R. Koppes, "From New Deal to Termination: Liberalism and Indian Policy, 1933–53," *Pacific Historical Review* 46 (Nov. 1977): 556.

10. Richard Shifter, "Indian Title to Land," *American Indian* 7 (Spring 1954): 38; Graham D. Taylor, "The Tribal Alternative to Bureaucracy: The Indian New Deal, 1933–1945," *Journal of the West* 13 (Jan. 1974): 134; Kenneth R. Philip, "Termination: A Legacy of the Indian New Deal," *Western Historical Quarterly* 14 (Apr. 1983): 166, 174–76; Frederick J. Stefon, "The Irony of Termination, 1943–1958," *Indian Historian* 11 (Summer 1978): 4; Prucha, *Great Father,* 2:1015–16; Burt, *Tribalism in Crisis,* pp. 4–5.

11. Prucha, *Great Father,* 2:1028–30, 1036–41; Philip, "Termination," p.165; Burt, *Tribalism in Crisis,* p. 45.

12. Prucha, *Great Father,* 2:1043–44; Koppes, "From New Deal to Termination," p. 556; Burt, *Tribalism in Crisis,* p. 19; Stefon, "Irony of Termination," p. 8.

13. Prucha, *Great Father,* 2:1044–47; Burt, *Tribalism in Crisis,* p. 26; Stefon, "Irony of Termination," p. 7. Fourteen days later, on 15 Aug., Congress passed Public Law 280, which transferred civil and criminal jurisdiction over tribal lands to five states. This law actually began the process of termination.

14. Prucha, *Great Father,* 2:1048–49; Burt, *Tribalism in Crisis,* pp. 30, 97.

15. Prucha, *Great Father,* 2:1047, 1056–58; Burt, *Tribalism in Crisis,* pp. 95, 107; Stefon, "Irony of Termination," p. 9.

16. Prucha, *Great Father,* 2:1048, 1058–59; Burt, *Tribalism in Crisis,* p. 126.

17. James O. Fine, "An Analysis of Factors Affecting Agricultural Development on the Fort Totten Indian Reservation" (Master's thesis, University of North Dakota, 1951), pp. 59–62.

18. Carl K. Eicher, "An Approach to Income Improvement on the Rosebud Sioux Indian Reservation," *Human Organization* 20 (Winter 1961/62): 192–93, 195.

19. Ruth Hill Useem and Carl K. Eicher, "Rosebud Reservation Economy," in *The Modern Sioux,* ed. Ethel Nurge (Lincoln: University of Nebraska Press, 1970), pp. 13, 16–21; Vernon D. Malan and Ernest L. Schusky, "The Dakota Indian Community: An Analysis of the Non-Ranching Population of the Pine Ridge Reservation," South Dakota Agricultural Experiment Station, Bulletin 505 (n.d.), p. 9; U.S., Department of Health, Education and Welfare, *Indians on Federal Reservations in the United States—A Digest—Aberdeen Area,* no. 615, pt. 3 (June 1959), pp. ix,

13, 15, 21, 23. On The Standing Rock Reservation in North and South Dakota, 75 percent of the land was leased, while approximately 50 percent of the land was leased on the Cheyenne River, Lower Brule, and Pine Ridge reservations in South Dakota.

20. U.S., Department of Health, Education and Welfare, *Indian Reservations in the United States—A Digest—Billings Area,* no. 615, pt. 2 (Sept. 1958), p. 3; Ralph E. Ward et al., "Indians in Agriculture: Cattle Ranching on the Crow Reservation," Montana Agricultural Experiment Station, Bulletin 522 (July 1956), pp. 6–8, 10–12, 14–15, 17, 25–26, 50.

21. Ward, "Indians in Agriculture," p. 32; Walter U. Fuhriman, "Economic Opportunities for Indians," in *The West in A Growing Economy: Proceedings of the Western Farm Economics Association,* 14–17 July 1959, Logan, Utah, p. 194.

22. Ward, "Indians in Agriculture," pp. 33–34, 46.

23. *Indian Reservations in the United States—A Digest—Billings Area,* p. 1, 5, 7, 9–10, 15; Sidney J. Tietema, Ralph E. Ward, and Chester B. Baker, "Indians in Agriculture II: Cattle Ranching on the Blackfeet Reservation," Montana Agricultural Experiment Station, Bulletin 532 (June 1957), pp. 8–9, 11, 19, 21, 31, 43. In the northern Great Plains, 66 percent of the Flathead and Fort Belknap Reservation, more than 50 percent of the Tongue River and Wind River reservations, and 80 percent of the Fort Peck Reservation lands were leased to whites.

24. Robert John Stahl, "Farming among the Kiowa, Comanche, Kiowa-Apache and Wichita" (Ph.D. diss., University of Oklahoma, 1978), pp. 276, 291; U.S., Department of Health, Education and Welfare, *Indians on Federal Reservations—A Digest—Oklahoma City Area,* no. 615, pt. 5 (June 1960), pp. 1, 3, 5, 7, 9, 13.

25. Philip Reno, *Mother Earth, Father Sky, and Economic Development* (Albuquerque: University of New Mexico Press, 1981), p. 20; *Indian Truth,* June–Aug. 1951, p. 1, and Jan.–Feb. 1952, p. 2; Gary D. Libecap and Ronald N. Johnson, "Legislating Commons: The Navajo Tribal Council and the Navajo Range," *Economic Inquiry* 18 (Jan. 1980): 69–70, 73–74, 83; *Navajo Yearbook,* 1958, p. 71.

26. *Navajo Yearbook,* 1958, pp. 71, 74.

27. Ibid., pp. 61, 82–83, 101, 105, 108; Ned O. Thompson, "Economic Opportunities for Indians," in *The West in a Growing Economy,* p. 201.

28. Charles Cooper, "New Hope for the Apache," *American Forests* 61 (Sept. 1955): 25; Harry T. Getty, "The San Carlos Indian Cattle Industry," Anthropological Papers of the University of Arizona no. 7 (1963), pp. 45–46.

29. Getty, "San Carlos Indian Cattle Industry," pp. 49–50.

30. Ibid., pp. 52–56, 62, 66–67.

31. Ibid., pp. 67–68, 73–76.

32. Veronica E. Tiller, *The Jicarilla Apache* (Lincoln: University of Nebraska Press, 1983), pp. 181, 183–86, 200.

33. *Indian Truth,* Mar.–May 1951, pp. 6–7; Henry F. Manuel, Juliann Ramon, and Bernard L. Fontana, "Dressing for the Window: Papago Indians and Economic Development," in *American Indian Economic Development,* ed. Sam Stanley (The Hague: Mouton Publishers, 1978), pp. 527–28, 531.

34. Manuel et al., "Dressing for the Window," pp. 530, 532; U.S., Department of Health, Education and Welfare, *Indians on Federal Reservations in the United States—A Digest—Phoenix Area,* no. 615, pt. 6 (Jan. 1961), pp. 13, 27.

35. U.S., Department of Health, Education and Welfare, *Indians on Federal Reservations in the United States—A Digest—Portland Area,* no. 615, pt. 1 (June 1958), pp. viii, 1, 3, 7, 13, 15; *Indian Truth,* Jan.–Feb. 1956, p. 4 and Winter

1958/59, p. 4. In 1956, 66 percent of the 56 million acres of reservation lands was under tribal control; 30 percent was allotted, and 4 percent was reserved by the government for special purposes, such as schools and demonstration areas. Of this acreage, 60 percent was useable only for grazing; 30 percent, for grazing and timber; 5 percent, for cultivation; and 5 percent was wasteland. Less than 1 percent of the acreage was irrigated (see John Province, "Tenure Problems of the American Indian," in *Land Tenure*, ed. Kenneth H. Parsons, Raymond J. Penn, and Philip M. Raup [Madison: University of Wisconsin Press, 1956], p. 422).

36. U.S., Department of Agriculture, *Yearbook of Agriculture*, 1958, p. 99; *Indian Truth*, Apr.–July 1955, pp. 1–2.

37. Alan L. Sorkin, "Trends in Employment and Earnings of American Indians," in *American Indians: Facts and Future* (New York: Arno, 1970), pp. 115–17.

38. Peter Dorner, "Needed: A New Policy for the American Indians," *Land Economics* 37 (May 1961): 163–64.

39. Ibid., p. 171.

40. Ibid., pp. 171–72; William G. Hayden, "Oklahomans for Indian Opportunity, Incorporated and Economic Development for Nonreservation Indian People," in *American Indians: Facts and Future*, pp. 434, 437–38; Joan Dorthy Travis, "Agrarian Problems and Prospects of the Colorado Indian Reservation" (Master's thesis, University of California at Los Angeles, 1968), pp. 60–65.

41. C. G. d'Easum, "Changing Hardships to Opportunity," *Extension Service Review* 37 (Mar. 1966): 11–12.

42. Ronald Trosper, "American Indian Relative Ranching Efficiency," *American Economic Review* 68 (Sept. 1978): 503, 508, 513.

43. L. Schuyler Farnoff, "Conservation and Stock Reduction on the Navajo Tribal Range," *Geographical Review* 53 (Apr. 1963): 217; *Navajo Yearbook*, 1961, p. 163.

44. Victor Chikezie Uchendu, "Seasonal Agricultural Labor among the Navaho Indians: A Study in Socio-Economic Transition" (Ph.D. diss., Northwestern University, 1966), pp. 2, 132, 214, 219, 281; *Navajo Yearbook*, 1961, pp. 165–66.

45. David F. Aberle, "A Plan for Navajo Economic Development," in *American Indians: Facts and Future*, pp. 260–61.

46. C. W. Loomer, "Land Tenure Problems in the Bad River Indian Reservation of Wisconsin," Wisconsin Agricultural Experiment Station, Bulletin 188 (Dec. 1955), p. 4; Alan L. Sorkin, *American Indians and Federal Aid* (Washington, D.C.: Brookings Institution, 1971), p. 71; William H. Gilbert and John L. Taylor, "Indian Land Questions," *Arizona Law Review* 8 (Fall 1966): 123.

47. Sorkin, *American Indians and Federal Aid*, pp. 67–70; Manuel, "Dressing for the Window," p. 529; Harry T. Getty, "San Carlos Apache Cattle Industry," *Human Organization* 20 (Winter 1961/62): 185.

48. Sorkin, *American Indians and Federal Aid*, pp. 68–69. For one extension-service success story see Verl B. Matthews, "Utes Welcome Extension Help," *Extension Service Review* 42 (Apr. 1971): 6–7.

49. Sorkin, *American Indians and Federal Aid*, pp. 70, 78–79, 182.

50. Helen W. Johnson, "American Indians in Transition," U.S. Department of Agriculture, Agricultural Economic Research Service, *Agricultural Economic Report 283* (Apr. 1975), pp. 1, 14; William H. Veeder, "Federal Encroachment of Indian Water Rights and the Impairment of Reservation Development," in *American Indians: Facts and Future*, p. 460.

CHAPTER 13. QUAGMIRES

1. Michael S. Laird, "Water Rights: The *Winters* Cloud over the Rockies: Indian Water Rights and the Development of Western Energy Resources," *American Indian Law Review* 7, no. 1 (1979): 158, 164, 169; Belinda K. Orem, "Paleface, Redskins and the Great White Chiefs in Washington: Drawing the Battle Lines over Western Water Rights," *San Diego Law Review* 17 (1980): 488; "Indian Reserved Water Rights: The *Winters* of our Discontent," *Yale Law Review* 88 (July 1979): 1711; Philip W. Dufford, "Water for Non-Indians on the Reservation: Checkerboard Ownership and Checkerboard Jurisdiction," *Gonzaga Law Review* 15, no. 1 (1979): 95.

2. Robert G. Dunbar, *Forging New Rights in Western Waters* (Lincoln: University of Nebraska Press, 1983), pp. 59–60; Richard Trudell, ed., *Indian Water Policy in a Changing Environment* (Oakland, Calif.: American Indian Lawyer Training Program, 1982), p. 25; Robert Peregay, "Jurisdictional Aspects of Indian Reserved Water Rights in Montana and on the Flathead Reservation after Adsit," *American Indian Culture and Research Journal* 7, no. 1 (1983): 42.

3. Trudell, *Indian Water Policy*, p. 26; Dunbar, *Forging New Rights*, pp. 60–63; Charles T. Dumars, Marilyn O'Leary, and Albert E. Utton, *Pueblo Indian Water Rights: Struggle for a Precious Resource* (Tucson: University of Arizona Press, 1984), pp. 3–4; Peregay, "Jurisdictional Aspects," p. 42.

4. Trudell, *Indian Water Policy*, p. 27.

5. Ibid., pp. 27–28; Laird, "Water Rights," p. 160; Dumars, *Pueblo Indian Water Rights*, pp. 14–15.

6. Norris Hundley, Jr., "The 'Winters' Decision and Indian Water Rights: A Mystery Reexamined," *Western Historical Quarterly* 13 (Jan. 1982): 29; Trudell, *Indian Water Policy*, pp. 28–29; James L. Merrill, "Aboriginal Water Rights," *Natural Resources Journal* 20 (Jan. 1980): 61.

7. *Winters* v. *United States*, 207 US 564 (1908); Trudell, *Indian Water Policy*, p. 29; Norris Hundley, Jr., "The Dark and Bloody Ground of Indian Water Rights: Confusion Elevated to Principle," *Western Historical Quarterly* 9 (Oct. 1978): 463–64; Hundley, "The 'Winters' Decision," p. 18; David H. Getches, "Water Rights on Indian Allotments," *South Dakota Law Review* 26 (Summer 1981): 406; Laird, "Water Rights," pp. 161, 163; Charles Dumars and Helen Ingram, "Congressional Quantification of Indian Reserved Rights: A Definitive Solution or a Mirage," *Natural Resources Journal* 20 (Jan. 1980): 17; Dufford, "Water for Non-Indians," pp. 98–99, 115; Dunbar, *Forging New Rights*, p. 196; *161 Fed 829* (1908).

8. Hundley, "The 'Winters' Decision," pp. 33, 35; Merrill, "Aboriginal Water Rights," p. 58; Francis Paul Prucha, *The Great Father*, vol. 2 (Lincoln: University of Nebraska Press, 1984), p. 1181; Hundley, "Dark and Bloody Ground of Indian Water Rights," pp. 465–67.

9. Hundley, "Dark and Bloody Ground of Indian Water Rights," pp. 467–69; idem, "The 'Winters' Decision," p. 35.

10. *Arizona* v. *California*, 373 US 546 (1963); Hundley, "Dark and Bloody Ground of Indian Water Rights," pp. 470–71.

11. *Arizona* v. *California*, 373 US 546 (1963); Laird, "Water Rights," pp. 166, 169.

12. *Arizona* v. *California*, 373 US 546 (1963); Hundley, "Dark and Bloody Ground of Indian Water Rights," pp. 475–77; Prucha, *Great Father*, 2:1181; Dumars et al., *Pueblo Indian Water Rights*," pp. 4–5; H. S. Burness et al., "United States

Reclamation Policy and Indian Water Rights," *Natural Resources Journal* 20 (Jan. 1980): 807.

13. Gwendolyn Griffith, "Indian Claims to Underground Water: Reserved Rights or Beneficial Interest?" *Stanford Law Review* 33 (Nov. 1980): 105–10.

14. Hundley, "Dark and Bloody Ground of Indian Water Rights," p. 476; Merrill, "Aboriginal Water Rights," pp. 66–67; *Cappaert v. United States,* 426 US 128 (1975); Griffith, "Indian Claims to Underground Water," pp. 116, 118–19.

15. Griffith, "Indian Claims to Underground Water," p. 120.

16. Merrill, "Aboriginal Water Rights," pp. 54–55, 59–64, 67–68; Laird, "Water Rights," p. 159; Dumars et al., *Pueblo Indian Water Rights,* pp. 11, 16–17, 55, 111–19.

17. Merrill, "Aboriginal Water Rights," pp. 68–69; Michael F. Lamb, "Adjudication of Indian Water Rights: Implementation of the 1979 Amendment to the Montana Water Use Act," *Montana Law Review* 41 (Winter 1980): 95; Trudell, *Indian Water Policy,* p. 144; Peregay, "Jurisdictional Aspects," p. 68; A. Patrick Maynez, "Pueblo Indian Water Rights: Who Will Get the Water?" *New Mexico v. Aamodt,*" *Natural Resources Journal* 18 (Jan. 1978): 639–58; Dumars and Ingram, "Congressional Quantification," pp. 19, 43; Rennard Strickland, "American Indian Water Law Symposium," *Tulsa Law Review* 15 (Summer 1980): 703; "Indian Reserved Water Rights," p. 1712. For an overview of the Navajo Indian Irrigation Project that shows the complex relationship between Indian water rights and federal policy see Michael L. Lawson, "The Navajo Indian Irrigation Project: Muddied Past, Clouded Future," *Indian Historian* 9 (Winter 1976): 19–28; and Philip Reno, *Mother Earth, Father Sky, and Economic Development* (Albuquerque: University of New Mexico Press, 1981), pp. 65–81.

18. *United States v. Powers,* 305 US 533 (1939); Hundley, "Dark and Bloody Ground," pp. 474–75; Trudell, *Indian Water Policy,* pp. 39–40; Jack D. Palma II, "Considerations and Conclusions Concerning the Transferability of Indian Water Rights," *Natural Resources Journal* 20 (Jan. 1980): 91–100; Dufford, "Water for Non-Indians," pp. 95–131; Dumars et al., *Pueblo Indian Water Rights,* pp. 97–103.

19. Dumars et al., *Pueblo Indian Water Rights,* p. 72; Hundley, "Dark and Bloody Ground of Indian Water Rights," pp. 478–81; Trudell, *Indian Water Policy,* p. 41; *Colorado River Conservation District v. United States,* 424 US 800 (1976).

20. Ethel J. Williams, "Too Little Land, Too Many Heirs: The Indian Heirship Land Problem," *Washington Law Review* 46, no. 4 (1971): 711–12.

21. Ibid., pp. 713, 715–16.

22. Reno, *Mother Earth,* pp. 2, 19; William T. Hagan, *American Indians* (Chicago: University of Chicago Press, 1979), p. 77; *The Navajo Nation: An American Colony,* Report of the United States Commission on Civil Rights, 1975, p. 21; Lorraine Turner Ruffing, "Navajo Economic Development: A Dual Perspective," in *American Indian Economic Development,* ed. Sam Stanley (The Hague: Mouton Publishers, 1978), p. 27.

23. *Navajo Nation,* p. 132; Lorraine Turner Ruffing, "Navajo Economic Development Subject to Cultural Constraints," *Economic Development and Culture Change* 24 (Apr. 1976): 614–16.

24. Ruffing, "Navajo Economic Development Subject to Cultural Constraints," pp. 616, 619.

25. Seth Kantor, "Indian Wool Makes the Grade," *Nation's Business* 60 (Jan. 1972): 65–66; Reno, *Mother Earth,* pp. 43–44; *Navajo Nation,* pp. 42–43.

26. James M. Goodman and Gary L. Thompson, "The Hopi-Navajo Land Dispute," *American Indian Law Review* 3 (1975): 413; Ruffing, "Navajo Economic

Development: A Dual Perspective," pp. 28, 66–68; Ronald N. Johnson and Gary D. Libecap, "Agency Costs and the Assignment of Property Rights: The Case of Southwestern Indian Reservations," *Southern Economic Journal* 47 (Oct. 1980): 333; Reno, *Mother Earth,* p. 43; Gary D. Libecap and Ronald N. Johnson, "Legislating Commons: The Navajo Tribal Council and the Navajo Range," *Economic Inquiry* 18 (Jan. 1980): 69–70, 80–81.

27. Libecap and Johnson, "Legislating Commons," pp. 76–78, 80, 83–84.

28. Roy W. Meyer, *The Village Indians of the Upper Missouri* (Lincoln: University of Nebraska Press, 1977), pp. 253–54.

29. James J. Harris, *The North Dakota Indian Reservation Economy: A Descriptive Study,* North Dakota Economic Studies no. 11, Bureau of Business and Economics Research, University of North Dakota (Aug. 1975), pp. 21, 88, 99.

30. Ibid., p. 88.

31. George Williamson Norton, "A Model for Indian Reservation Agricultural Development: The Case of the Sisseton-Wahpeton Sioux" (Ph.D. diss., University of Minnesota, 1979), pp. 1, 7–8, 12; George W. Norton, K. William Easter, and Terry L. Roe, "American Indian Farm Planning: An Analytical Approach to Tribal Decision Making," *American Journal of Agricultural Economics* 62 (Nov. 1980): 689–90.

32. Raymond J. DeMallie, "Pine Ridge Economy: Cultural and Historical Perspectives," in *American Indian Economic Development,* pp. 264–65.

33. Ibid., pp. 282–83.

34. Ibid., pp. 287–88.

35. Susan Work, "The 'Terminated' Five Tribes of Oklahoma: The Effect of Federal Legislation and Administrative Treatment on the Government of the Seminole Nation," *American Indian Law Review* 6 (Summer 1978): 109–10.

36. Theodore W. Taylor, *American Indian Policy* (Mt. Airy, Md.: Lomond Publications, Inc., 1983), pp. 6, 90–92.

37. Larry W. Burt, *Tribalism in Crisis: Federal Indian Policy, 1935–1961* (Albuquerque: University of New Mexico Press, 1982), p. 130. For a survey of the Sagebrush Rebellion, see Richard D. Lamm and Michael McCarthy, *The Angry West: A Vulnerable Land and Its Future* (Boston, Mass.: Houghton Mifflin Co., 1982).

38. *1982 Census of Agriculture,* pt. 51, p. 4.

39. Ibid., pp. 4–5, 387. The 2,268 Indian farmers who sold more than $10,000 annually operated a total of 2,647,299 acres in 1982.

40. Hagan, *American Indians,* pp. 168, 174; Michael L. Lawson, *Dammed Indians: The Pick-Sloan Plan and the Missouri River Sioux, 1944–1980* (Norman: University of Oklahoma Press, 1982); Michael L. Lawson, "Federal Water Projects and the Indians: The Pick-Sloan Plan, A Case Study," *American Indian Culture and Research Journal* 7, no. 1 (1983): 23–40.

CHAPTER 14. EPILOGUE

1. Paul Weatherwax, *Indian Corn in Old America* (New York: Macmillan Co., 1954), pp. 58–59.

2. Efraim Hernandez Xolocatzi, "Maize Granaries in Mexico," *Harvard University, Botanical Museum Leaflets* 13 (17 Jan. 1949): 154–55; Weatherwax, *Indian Corn in Old America,* p. 86.

3. Richard A. Yarnell, *Aboriginal Relationships between Culture and Plant*

Life in the Upper Great Lakes Region, Anthropological Papers no. 23, Museum of Anthropology, University of Michigan (Ann Arbor, 1964), p. 119.

4. Gregory Waselkov, "Prehistoric Agriculture in the Central Mississippi Valley," *Agricultural History* 51 (July 1977): 518; James E. Fitting, "Regional Cultural Development," in *Northeast,* ed. Bruce G. Trigger, vol. 15 of *Handbook of North American Indians,* ed. William C. Sturtevant (Washington, D.C.: Smithsonian Institution, 1978), p. 44.

5. *Report,* 1879, p. iv.

Bibliographical Note

Scholars and students who want to pursue further study of the agricultural history of the American Indians should consult the notes for the chapters that relate to their work. I have tried to be as extensive as possible with my documentation, and there is no need to list every source once again. In addition to those works, however, researchers will find these bibliographies indispensable: Cecil L. Harvey, *Agriculture of the American Indian: A Selected Bibliography* (n.p., 1979); Everett E. Edwards and Wayne D. Rasmussen, *A Bibliography of the Agriculture of the American Indians,* United States Department of Agriculture, Miscellaneous Publication no. 447 (Jan. 1942); and Francis Paul Prucha's *A Bibliographical Guide to the History of Indian-White Relations in the United States* (Chicago: University of Chicago Press, 1977) and *Indian-White Relations in the United States: A Bibliography of Works Published 1975–1980* (Lincoln: University of Nebraska Press, 1982). For sources at the National Archives see Edward E. Hill, *Guide to Records in the National Archives of the United States Relating to American Indians* (Washington, D.C.: National Archives and Records Service, 1981).

More general sources include Douglas S. Byers, ed., *Environment and Subsistence,* vol. 1 of *The Prehistory of the Tehuacan Valley,* ed. Douglas S. Byers (Austin: University of Texas Press, 1967). For two excellent introductions to Mesoamerican agriculture see Stuart Struever, ed., *Prehistoric Agriculture* (Garden City, N.Y.: Natural History Press, 1971), and Charles A. Reed, ed., *Origins of Agriculture* (The Hague: Mouton Publishers, 1977). Other useful sources on Mesoamerican agriculture include Peter D. Harrison and B. L. Turner II, eds., *Pre-Hispanic Agriculture* (Albuquerque: University of New Mexico Press, 1978); B. L. Turner and Peter D. Harrison, *Pulltrouser Swamp: Ancient Maya Habitat, Agriculture and Settlement in Northern Belize* (Austin: University of Texas Press, 1983); Robert C. West, ed., *Natural Environments and Early Cultures,* vol. 1 of *Handbook of Middle American Indians,* ed. Robert Wauchope (Austin: University of Texas Press, 1964); and Kent V. Flannery, ed., *Guilá Naquitz: Archaic Foraging and Early Agriculture in Oaxaca, Mexico* (Orlando, Fla.: Academic Press, 1986). The most recent studies of the domestication of corn can be found in Richard I. Ford, ed., *Prehistoric Food Production in North America,* Anthropological Papers no. 75, Museum of Anthropology, University of Michigan (Ann Arbor, 1985). Paul C. Mangelsdorf also provides important insights into this subject in "The Mystery of Corn: New Perspectives," *Proceedings of the American Philosophical Society* 127, no. 4 (1983). For differing interpretations see George W. Beadle, "The Ancestry of Corn," *Scientific American* 242 (Jan. 1980): 112–19; and Hugh H. Iltis, "From Teosinte to Maize: The Catastrophic Sexual Transmutation," *Science* 222 (25 Nov. 1983): 886–94. An earlier but still useful study of the domestication of corn can be found in Paul Weatherwax, *Indian Corn in Old America* (New York: Macmillian Co., 1954).

Richard I. Ford's *Prehistoric Food Production in North America* and Stuart Struever's *Prehistoric Agriculture* provide important essays on Indian agriculture in the eastern portion of the continental United States. Martha Potter's *Ohio's Prehistoric Peoples* (Columbus: Ohio Historical Society, 1968) provides an intensive survey of the prehistoric people in a portion of the Old Northwest. Also of importance is William S. Webb and Raymond S. Baby, eds., *The Adena People No. 2* (Columbus: Ohio Historical Society, 1957). For more recent studies see David S. Brose and N'omi Greber, eds., *Hopewell Archaeology* (Kent, Ohio: Kent State University Press, 1979), as well as Patty Jo Watson, ed., *Archaeology of the Mammoth Cave Area* (New York: Academic Press, 1970). Also useful are Richard A. Yarnell, *Aboriginal Relationships between Culture and Plant Life in the Upper Great Lakes Region,* Anthropological Papers no. 23, Museum of Anthropology, University of Michigan (Ann Arbor, 1964); Bruce G. Trigger, *The Huron Farmers of the North* (New York: Holt, Rinehart & Winston, 1969); Arthur D. Parker, *Iroqouis Uses of Maize and Other Plant Foods,* New York State Museum, Bulletin 114 (Albany: University of the State of New York, 1910); Albert Ernest Jenks, "The Wild Rice Gatherers of the Upper Great Lakes," *Nineteenth Annual Report of the Bureau of American Ethnology, 1897–98,* pt. 2 (Washington, D.C.: Government Printing Office, 1900).

For a study of the southeastern United States see Charles Hudson, *The Southeastern Indians* (Knoxville: University of Tennessee Press, 1976). Three studies by John R. Swanton are particularly useful: *The Indians of the Southeastern United States,* Smithsonian Institution, Bureau of American Ethnology, Bulletin 137 (Washington, D.C.: Government Printing Office, 1946); *Early History of the Creek Indians and Their Neighbors,* Smithsonian Institution, Bureau of American Ethnology, Bulletin 73 (Washington, D.C.: Government Printing Office, 1922); and *Indian Tribes of the Lower Mississippi Valley and Adjacent Coast of the Gulf of Mexico,* Smithsonian Institution, Bureau of American Ethnology, Bulletin 43 (Washington, D.C.: Government Printing Office, 1911). See also Carl Ortwin Sauer, *Sixteenth Century North America* (Berkeley: University of California Press, 1971), and William N. Fenton and John Gulick, *Symposium on Cherokee and Iroquois Culture,* Smithsonian Institution, Bureau of American Ethnology, Bulletin 180 (Washington, D.C.: Government Printing Office, 1961).

Alfonso Oritz, ed., provides a thorough introduction to the Indian agriculturists of the southwestern United States in *Southwest,* vol. 9 of *Handbook of North American Indians,* ed. William C. Sturtevant (Washington, D.C.: Smithsonian Institution, 1979). Also important are Emil W. Haury, *The Hohokam: Desert Farmers and Craftsmen* (Tucson: University of Arizona Press, 1976); George F. Carter, *Plant Geography and Culture History in the American Southwest,* Viking Fund Publications in Anthropology no. 5 (New York, 1945); Donald D. Brand, ed., *Symposium on Prehistoric Agriculture,* University of New Mexico Bulletin 296, no. 5 (1936); Edward F. Castetter and Willis H. Bell, *Pima and Papago Indian Agriculture* (Albuquerque: University of New Mexico Press, 1942), and *Yuman Indian Agriculture*

(Albuquerque: University of New Mexico Press, 1951); and W. W. Hill, *The Agricultural and Hunting Methods of the Navaho Indians,* Yale University Publications in Anthropology no. 18 (1938).

For an introduction to Indian agriculture in the Great Plains see Preston Holder, *The Hoe and the Horse on the Plains* (Lincoln: University of Nebraska Press, 1974); Waldo R. Wedel, *Environment and Native Subsistence Economies in the Central Great Plains,* Smithsonian Institution, Miscellaneous Collections, vol. 101, no. 3 (Washington, D.C.: Smithsonian Institution, 1941); George F. Will and George E. Hyde, *Corn among the Indians of the Upper Missouri* (Lincoln: University of Nebraska Press, 1964): Gilbert Livingstone Wilson, *Agriculture of the Hidatsa Indians: An Indian Interpretation,* University of Minnesota Studies in the Social Sciences, Bulletin no. 9 (1917); and Lynn Marie Allex, "Prehistoric and Early Historic Farming and Settlement Patterns," *South Dakota History* 13 (Spring/Summer 1983).

Sources relating to landholding by Indians that are particularly useful include C. Daryll Forde, "Hopi Agriculture and Land Ownership," *Royal Anthropological Institute of Great Britain and Ireland* 61 (1931); Ralph M. Linton, "Land Tenure in Aboriginal America," in *The Changing Indian,* ed. Oliver La Farge (Norman: University of Oklahoma Press, 1942); Chester E. Eisinger, "The Puritans' Justification for Taking the Land," *Essex Institute Historical Collection* 84 (Apr. 1948); William T. Hagan, "Justifying Dispossession of the Indian: The Land Utilization Argument," in *American Indian Environments: Ecological Issues in Native American History,* ed. Christopher Vecsey and Robert W. Venables (Syracuse, N.Y.: Syracuse University Press, 1980); Anthony F. C. Wallace, "Political Organization and Land Tenure among the Northeastern Indians, 1600–1830," *Southwestern Journal of Anthropology* 13 (Winter 1957); John M. Cooper, "Land Tenure among the Indians of Eastern and Northern North America," *Pennsylvania Archaeologist* 8 (July 1938); George S. Snyderman, "Concepts of Land Ownership among the Iroquois and Their Neighbors," in *Symposium on Local Diversity in Iroquois Culture,* ed. William N. Fenton, Smithsonian Institution, Bureau of American Ethnology, Bulletin 149 (Washington, D.C.: Government Printing Office, 1951); and W. W. Hill, "Notes on Pima Land Law and Tenure," *American Anthropologist* 38 (Oct.–Dec. 1936).

For the matter of land cessions see Charles C. Royce, "Indian Land Cessions in the United States," *Eighteenth Annual Report of the Bureau of American Ethnology to the Secretary of the Smithsonian Institution, 1896–'97,* pt. 2 (Washington, D.C.: Government Printing Office, 1899); J. P. Kinney, *A Continent Lost—A Civilization Won: Indian Land Tenure in America* (Baltimore, Md.: Johns Hopkins Press, 1937); Dwight L. Smith, "The Land Cession Treaty: A Valid Instrument of Transfer of Indian Title," in *This Land of Ours: The Acquisition of the Public Domain* (Indianapolis: Indiana Historical Society, 1978); Wilcomb E. Washburn, "The Moral and Legal Justification for Dispossession of the Indians," in *Seventeenth Century America: Essays in Colonial History,* ed. James Morton Smith (Chapel Hill: University of North Carolina Press, for the Institute of Early American History and Culture, 1959); and Francis Paul Prucha, *American Indian Policy in the Formative Years: The Indian Trade and Intercourse Acts, 1790–1834* (Cambridge, Mass.: Harvard University Press, 1962).

The best sources for Indian agriculture on the reservations are the *Annual Reports of the Commissioner of Indian Affairs.* For a guide that will provide easy access to these documents see J. A. Jones, "Key to the Annual Reports of the United States Commissioner of Indian Affairs," *Ethnohistory* 2 (Winter 1955). These reports must be used with caution. Invariably, federal agents and governmental farmers praised their own work while condemning that of the men who preceeded them, only to have their successors follow suit. Even so, these reports provide the best information about annual agricultural activities on the reservations. Researchers should also consult the *Annual Reports of the Board of Indian Commissioners* and the *Annual Reports of the Indian Rights Association.* Two periodicals—*Indian Truth* and *Indians at Work*—also provide useful information relating to agriculture.

General Indian policy is treated extensively in Francis Paul Prucha, *The Great Father,* 2 vols. (Lincoln: University of Nebraska Press, 1984). Other useful studies are Reginald Horsman, *Expansion and American Indian Policy, 1783–1812* (East Lansing: Michigan State University Press, 1967); Ronald N. Satz, *American Indian Policy in the Jacksonian Era* (Lincoln: University of Nebraska Press, 1975); Edmund Jefferson Danziger, Jr., *Indians and Bureaucrats: Administering Reservation Policy during the Civil War* (Urbana: University of Illinois Press, 1974); William T. Hagan, *The Indian Rights Association: The Herbert Welsh Years, 1882–1902* (Tucson: University of Arizona Press, 1985); Lawrence F. Schmeckebier, *The Office of Indian Affairs: Its History, Activities and Organization* (Baltimore, Md.: Johns Hopkins Press, 1927); Robert M. Utley, *The Indian Frontier and the American West, 1846–1890* (Albuquerque: University of New Mexico Press, 1984); and Donald J. Berthrong, *The Cheyenne and Arapaho Ordeal: Reservation and Agency Life in Indian Territory, 1875–1907* (Norman: University of Oklahoma Press, 1976).

Specific studies relating to Indian agriculture include Richard White, *The Roots of Dependency* (Lincoln: University of Nebraska Press, 1983); Leonard A. Carlson, *Indians, Bureaucrats, and Land: The Dawes Act and the Decline of Indian Farming* (Westport, Conn.: Greenwood Press, 1981); Tom L. Sasaki, *Fruitland, New Mexico: A Navaho Community in Transition* (Ithaca, N.Y.: Cornell University Press, 1960); Gilbert C. Fite, "Development of the Cotton Industry by the Five Civilized Tribes in Indian Territory," *Journal of Southern History* 15 (Aug. 1949); Daniel H. Usner, Jr., "American Indians on the Cotton Frontier: Changing Economic Relations with Citizens and Slaves in the Mississippi Territory," *Journal of American History* 72 (Sept. 1985); and Herbert T. Hoover, "Arikara, Sioux and Government Farmers: The American Indian Agricultural Legacies," *South Dakota History* 13 (Spring/Summer, 1983).

Important studies of Indian cattle raising are William T. Hagan, "Kiowas, Comanches, and Cattlemen, 1867–1906: A Case Study of the Failure of U.S. Reservation Policy," *Pacific Historical Review* 40 (1971); Martha Buntin, "Beginning of the Leasing of Surplus Grazing Lands on the Kiowa and Comanche Reservations," *Chronicles of Oklahoma* 10 (Sept. 1932); Edward Everett Dale, "Ranching on the Cheyenne-Arapaho Reservation, 1880–1885," *Chronicles of Oklahoma* 6 (Mar. 1928); Robert M. Burrill, "The Establishment of Ranching on the Osage Indian Reservation," *Geographical Review* 62 (Oct. 1972); Harry T. Getty, "The San Carlos Indian Cattle Industry," University of Arizona, Anthropological Papers no. 7 (1963); Rolf W. Bauer, "The Papago Cattle Economy," in *Food, Fiber and the Arid Lands,* ed. William G. McGinnies, Bram J. Goldman, and Patricia Paylore (Tucson: University of Arizona Press, 1971).

Readers who are interested in the allotment process should consult Paul W. Gates, "Indian Allotments Preceding the Dawes Act," in *The Frontier Challenge: Responses to the Trans-Mississippi West,* ed. John G. Clark (Lawrence: University Press of Kansas, 1971); D. S. Otis, *The Dawes Act and the Allotment of Indian*

Lands, ed. Francis Paul Prucha (Norman: University of Oklahoma Press, 1973); Frederick E. Hoxie, *A Final Promise: The Campaign to Assimilate the Indians, 1880–1920* (Lincoln: University of Nebraska Press, 1984); Michael L. Lawson, "Indian Heirship Lands: The Lake Traverse Experience," *South Dakota History* 12 (Winter 1982); Ethel J. Williams, "Too Little Land, Too Many Heirs: The Indian Heirship Land Problem," *Washington Law Review* 46, no. 4 (1971); Donald J. Berthrong, "Legacies of the Dawes Act: Bureaucrats and Land Thieves at the Cheyenne-Arapaho Agencies of Oklahoma," *Arizona and the West* 21 (Winter 1979); and Janet A. McDonnell's "Land Policy on the Omaha Reservation: Competency Commissions and Forced Fee Patents," *Nebraska History* 63 (Fall 1982), and "Competency Commissions and Indian Land Policy, 1913–1920," *South Dakota History* 11 (Winter 1980).

For an excellent analysis of the failure of Indian policy by the eve of the New Deal see Lewis Meriam, *The Problem of Indian Administration* (Baltimore, Md.: Johns Hopkins Press, 1928). Additional insights into the twentieth-century reform movement can be gained from Kenneth R. Philip, *John Collier's Crusade for Indian Reform, 1920–1954* (Tucson: University of Arizona Press, 1977). Lawrence C. Kelly has written three important works: *John Collier and the Origins of Indian Policy Reform* (Albuquerque: University of New Mexico Press, 1983); *The Navajo Indians and Federal Indian Policy, 1900–1935* (Tucson: University of Arizona Press, 1968); and "The Indian Reorganization Act: The Dream and the Reality," *Pacific Historical Review* 44 (Aug. 1975). See also Graham D. Taylor, *The New Deal and American Indian Tribalism* (Lincoln: University of Nebraska Press, 1980); Laurence M. Hauptman, *The Iroquois and the New Deal* (Syracuse, N.Y.: Syracuse University Press, 1981); Peter Iverson, *The Navajo Nation* (Westport, Conn.: Greenwood Press, 1981); Donald L. Parman, *The Navajos and the New Deal* (New Haven, Conn.: Yale University Press, 1976); Larry W. Burt, *Tribalism in Crisis: Federal Indian Policy, 1953–1961* (Albuquerque: University of New Mexico Press, 1982); and Kenneth R. Philip, "Termination: A Legacy of the Indian New Deal," *Western Historical Quarterly* 14 (Apr. 1983).

The best surveys of the water problems of the American Indians are Norris Hundley, Jr., "The Dark and Bloody Ground of Indian Water Rights: Confusion Elevated to Principle," *Western Historical Quarterly* 9 (Oct. 1978), and "The 'Winters' Decision and Indian Water Rights: A Mystery Reexamined," *Western Historical Quarterly* 13 (Jan. 1982). In addition see Richard Trudell, ed., *Indian Water Policy in a Changing Environment* (Oakland, Calif.: American Indian Lawyer Training Program, 1982), and Charles T. DuMars, Marilyn O'Leary, and Albert E. Utton, *Pueblo Indian Water Rights: Struggle for a Precious Resource* (Tucson: University of Arizona Press, 1984).

Index

283

—leeks, 32
—maygrass, 12
—orache, 29
—peaches, 32, 41
—peanuts, 3
—peas, 32
—popcorn, 32
—potatoes, 32
—pumpkins: at contact, 27, 29, 31, 32, 35; prehistoric, 2, 4, 9
—squash: at contact, 12, 18, 24, 31, 32, 33, 37, 40, 41, 238 n. 1; prehistoric, 4, 6, 7, 9, 11, 12, 16, 17
—sumpweed, 11, 12, 13
—sunflowers, 11, 12, 13, 29, 33, 34
—tobacco, 4, 18, 27, 31, 32, 33, 41, 47, 48, 246
—tomatoes, 3, 9
—watermelon, 33, 41
—wild rice, 37–38
—*Zamia,* 27
Crow Creek Agency, 133, 147
Crow Creek Reservation, 148, 166, 172
Crownpoint Subagency, 210
Crow Reservation, 202, 220
Crow Tribal Council, 202
Cumberland Plateau, 3
Curtis Act, 261 n. 21

Dakota Territory, 116, 117, 124, 125, 131, 132, 133
Dawes, Henry L., 137
Dawes Act, 142, 260 n. 5; problems with, 151, 152, 172, 231; provisions of, 137–38
Dead Indian Land Act, 262 n. 33
Dearborn, Henry, 87, 99
Delaware Indians, 93, 99
De Soto, Hernando, 27, 28, 29
Devil's Lake Agency, 125, 133
Devil's Lake Reservation, 147, 166. *See also* Fort Totten Reservation
Dole, William P., 115, 116
Duck Valley Reservation, 176

Early Woodland Period, 12
Education, 96–112, 151, 155–57, 168; boarding schools, 127, 134, 258 n. 37; demonstration farms, 109, 262 n. 4; manual-labor schools, 101–2, 109, 110, 111; missionary schools, 100–102, 106, 109, 111, 127
El Riego Cave, 7
Emergency Feed Grain Program, 204
Emmons, Glenn L., 207
European influences, 31, 32–41

Farm Credit Administration, 186
Farmers Home Administration, 202, 212, 223, 224, 226
Farm Security Administration, 188
Federal Farm Board, 172
Federal Surplus Relief Corporation, 181, 183, 188
Fertilizer myth, 38–39
Figs, 41
First Mesa Tewa Pueblo Indians, 71
Five Civilized Tribes: and Civil War, 108, 113, 118; lands of, 92, 123, 133, 140, 198; problems of, 145, 151, 226, 260 n. 5, 261 n. 21, 264 n. 25, 266 n. 12
Flacco cultural phase, 4
Flathead Agency, 139, 147, 154, 271 n. 23
Florence Mound site, 12
Florida (state), 27, 28
Flournay Livestock and Real Estate Co., 143
Fort Apache, 146
Fort Apache Agency, 197
Fort Apache Reservation, 190
Fort Belknap Agency, 146
Fort Belknap Reservation, 184, 215, 271 n. 23
Fort Berthold Agency, 139
Fort Berthold Reservation, 148, 155, 158, 166, 223, 224
Fort Gibson, 102, 103, 114
Fort Hall Reservation, 184, 207, 209, 210, 216
Fort Leavenworth Agency, 101, 106
Fort McDowell, 197
Fort Peck Reservation, 208, 271 n. 23
Fort Riley, 107
Fort Smith, 104, 114
Fort Totten Reservation, 184, 201, 224, 267 n. 24. *See also* Devil's Lake Reservation
Fort Towson, 103, 104, 106
Fort Van Buren, 104
Fort Walton Mound site, 15
Fort Washita, 104
Fort Wingate, 182
Fox Indians, 107, 130, 148, 260 n. 5

Garlic, 32
Georgia, 15, 90, 93, 98
Gila River Reservation, 167, 172, 197
Goosefoot, 11
Gourds, 2, 3, 7, 12, 32
Granaries. *See* Storage facilities
Grand Portage Reservation, 269 n. 44
Grand Ronde Agency, 117
Grant, Ulysses S., 199–21, 126
Great Nemaha Subagency, 104, 105, 147
Great Plains, 57–63, 68, 113, 183; en-